AF789141

NEUROIMMUNE CIRCUITS, DRUGS OF ABUSE, AND INFECTIOUS DISEASES

ADVANCES IN EXPERIMENTAL MEDICINE AND BIOLOGY

Editorial Board:

NATHAN BACK, *State University of New York at Buffalo*

IRUN R. COHEN, *The Weizmann Institute of Science*

DAVID KRITCHEVSKY, *Wistar Institute*

ABEL LAJTHA, *N. S. Kline Institute for Psychiatric Research*

RODOLFO PAOLETTI, *University of Milan*

Recent Volumes in this Series

Volume 484
PHYLOGENETIC PERSPECTIVES ON THE VERTEBRATE IMMUNE SYSTEM
Edited by Gregory Beck, Manickam Sugumaran, and Edwin L. Cooper

Volume 485
GENES AND PROTEINS UNDERLYING MICROBIAL URINARY TRACT VIRULENCE: Basic Aspects and Applications
Edited by Levente Emődy, Tibor Pál, Jörg Hacker, and Gabriele Blum-Oehler

Volume 486
PURINE AND PYRIMIDINE METABOLISM IN MAN X
Edited by Esther Zoref-Shani and Oded Sperling

Volume 487
NEUROPATHOLOGY AND GENETICS OF DEMENTIA
Edited by Markus Tolnay and Alphonse Probst

Volume 488
HEADSPACE ANALYSIS OF FOODS AND FLAVORS: Theory and Practice
Edited by Russell L. Rouseff and Keith R. Cadwallader

Volume 489
HEMOPHILIA CARE IN THE NEW MILLENNIUM
Edited by Dougald M. Monroe, Ulla Hedner, Maureane R. Hoffman, Claude Negrier, Geoffrey F. Savidge, and Gilbert C. White II

Volume 490
MECHANISMS OF LYMPHOCYTE ACTIVATION AND IMMUNE REGULATION VIII
Edited by Sudhir Gupta

Volume 491
THE MOLECULAR IMMUNOLOGY OF COMPLEX CARBOHYDRATES—2
Edited by Albert M. Wu

Volume 492
NUTRITION AND CANCER PREVENTION: New Insights into the Role of Phytochemicals
Edited under the auspices of the American Institute for Cancer Research

Volume 493
NEUROIMMUNE CIRCUITS, DRUGS OF ABUSE, AND INFECTIOUS DISEASES
Edited by Herman Friedman, Thomas W. Klein, and John J. Madden

A Continuation Order Plan is available for this series. A continuation order will bring delivery of each new volume immediately upon publication. Volumes are billed only upon actual shipment. For further information please contact the publisher.

NEUROIMMUNE CIRCUITS, DRUGS OF ABUSE, AND INFECTIOUS DISEASES

Edited by

Herman Friedman
Thomas W. Klein
University of South Florida College of Medicine
Tampa, Florida

and

John J. Madden
Georgia Mental Health Institute
Emory University
Atlanta, Georgia

Springer Science+Business Media, LLC

Library of Congress Cataloging-in-Publication Data

Neuroimmune circuits, drugs of abuse, and infectious diseases/edited by Herman Friedman, Thomas W. Klein, and John J. Madden.
p. cm.—(Advances in experimental medicine and biology; v. 493)
Includes bibliographical references and index.
ISBN 978-1-4757-8686-6 ISBN 978-0-306-47611-2 (eBook)
DOI 10.1007/978-0-306-47611-2
1. Drugs of abuse—Immunology—Congresses. 2. Immunosuppression—Congresses. 3. Infection—Congresses. 4. HIV infections—Congresses. 5. Neuroimmunology—Congresses. I. Friedman, Herman, 1931- II. Klein, Thomas W. III. Madden, John J. IV. Symposium on Neuroimmune Circuits, Infectious Diseases, and Drugs of Abuse (7th: 1999: Bethesda, Md.) V. Series.
RM316 .N485 2001
616.8'0479—dc21

00-052729

ISBN 978-1-4757-8686-6

©2001 Springer Science+Business Media New York
Originally published by Kluwer Academic/Plenum Publishers, New York in 2001
Softcover reprint of the hardcover 1st edition 2001

http://www.wkap.nl/

10 9 8 7 6 5 4 3 2 1

A C.I.P. record for this book is available from the Library of Congress

All rights reserved

No part of this book may be reproduced, stored in a retrieval system, or transmitted in any form or by any means, electronic, mechanical, photocopying, microfilming, recording, or otherwise, without written permission from the Publisher

PREFACE

Introduction and Perspectives

This volume is based on the proceedings of the 7th annual symposium on the topic Neuroimmune Circuits, Infectious Diseases and Drugs of Abuse, Bethesda, Maryland, October 7–9, 1999. This symposium, as in the past, focused on newer knowledge concerning the relationship between the immune and nervous systems with regards to the effects of drugs of abuse and infections, including AIDS, caused by the immunodeficiency virus. Presentations discussed the brain-immune axis from the viewpoint of drugs of abuse rather than from the subject of the brain or immunity alone. The major aim of this series of conferences has been to clarify the consequences of immunomodulation induced by drugs of abuse in regards to susceptibility and pathogenesis of infectious diseases, both in man and in various animal model systems.

The recreational use of drugs of abuse such as morphine, cocaine, and marijuana by large numbers of individuals in this country and around the world has continued to arouse serious concerns about the consequences of use of such drugs, especially on the normal physiological responses of an individual, including immune responses. Much of the recent data accumulated by investigators show that drugs of abuse, especially opioids and cannabinoids, markedly alter immune responses in human populations as well as in experimental animals, both *in vivo* and *in vitro*. It is now widely recognized that many drugs of abuse are associated with increased susceptibility to infectious diseases, especially opportunistic intracellular microbial infections, including infection by the AIDS virus. A concerted effort is now being made to determine the mechanisms whereby drugs of abuse increase susceptibility of infected individuals, be they humans or experimental animals, to opportunistic infectious agents. In particular, it is now known that drugs such as morphine, marijuana and cocaine enhance the susceptibility of humans as well as animals to infections by bacteria, viruses, protozoa, or fungi. Various infectious agents have shown that drugs of abuse increase susceptibility. Recently, the relationship between immunomodulation induced by drugs of abuse and effects on specific receptors, especially for opioids and cannabinoids, as well as on signal transduction mechanisms, have been reported.

Many investigators continue to believe that it is important to focus attention on the accumulating new information concerning the effects of recreational drugs of abuse and modulation of the immune response, especially that pertaining to resistance mechanisms important in infections. Studies concerning how illicit drugs of abuse affect immunity are urgent because of the worldwide epidemic of AIDS caused by the human immunodeficiency virus, resulting in a collapse of the immune system so as to make in individual susceptible to opportunistic infections that would otherwise not cause disease in health individuals. Among the possible mechanisms involved are direct and indirect effects of these drugs of abuse on the brain-immune axis. This axis is known to be a highly complex and dynamic interactive system. The effects of various drugs of abuse on this interactive system between the brain and the immune mechanism is known to have multiple consequences, depending upon the specific sites of action of a drug, duration of exposure and underlying neurological and behavioral status of the exposed individual to the drug. It is widely accepted that an understanding of the effects of a drug of abuse on the immune response, especially the neuroimmune axis and neuroendocrine system in general, involves many interdisciplinary areas, including behavioral, pharmaceutical, neurological, anatomical and cellular and molecular subjects. A multidisciplinary approach is needed to elucidate the interactions of drugs of abuse with the brain-immune axis, especially as these interactions impact on susceptibility or resistance to infections.

The symposium highlighted discussion of effects of illicit drugs on the brain immune system, especially effects on central nervous system infections, including effects on susceptibility to the AIDS virus. There were many reviews concerning the interaction of opiates with immune mechanisms, including the effects of opiates on specific receptors as well as receptors for cytokines and chemokines. A presentation of the "cross-talk" between chemokine and opioid receptors affecting cell migration and macrophage function was also given. Since there is now much interest in programmed cell death or apoptosis, it was pointed out that opioids promote T cell apoptosis through the JMK and signal transduction pathways. The involvement of cannabinoid receptors, especially CB1 and CB2 receptors, was discussed in detail in several presentations, as well as discussion of the effects of cocaine on immune cells and a discussion of the effects of cocaine abuse on HIV replication. The role of substance P receptor mediating macrophage responses was also discussed.

It is hoped by the organizers of this series of symposia, including the symposium in Bethesda on which this volume is based, that the publications of these proceedings will further inspire interest in this rapidly developing field of basic biomedical science in terms of immune modulation, drugs of abuse and the neuroimmune circuit. It is hoped that the publication of this volume will further the understanding of newer knowledge concerning the impact of drugs of abuse on the brain-immune axis and its relationship to immunomodulation and infection, especially that caused by the AIDS virus. The organizing committee of this symposium is thanked for invaluable assistance contributing to the success of the meeting. The editors of this volume also especially thank Ms. Ilona Friedman for continued invaluable assistance for once again serving as editorial coordinator and managing editor of this symposium proceedings. We also thank the National Institute on Drug Abuse for financial support and especially Dr. Charles Sharpe, Project Director of NIDA for Immunology, for continuing interest and invaluable assistance.

H. Friedman
T. W. Klein
and
J. J. Madden

February 2000

CONTENTS

TARGETING THE BRAIN'S IMMUNE SYSTEM: A PSYCHOPHARMACOLOGICAL APPROACH TO CENTRAL NERVOUS SYSTEM INFECTIONS

Phillip K. Peterson,[1,2] Genya Gekker,[1,2] Shuxian Hu,[1,2] Philip S. Portoghese,[3] Wen S. Sheng,[1] and James R. Lokensgard[1,2]

[1]Institute on Brain and Immune Disorders, Minneapolis Medical Research Foundation
[2]University of Minnesota Medical School
[3]Department of Medicinal Chemistry
University of Minnesota
Minneapolis, MN 55455

INTRODUCTION

Historically, the brain has been considered an "immunologically privileged site." An absence of lymphocytes within the parenchyma of the brain and a highly specialized blood-brain barrier that restricts entry of immunocytes and immune mediators are key determinants of this privileged status. In recent years, however, this concept of the brain has been evolving. It is now known that activated lymphocytes gain ready access to the brain parenchyma[1] and that the central nervous system (CNS) itself is populated by resident macrophages, i.e., microglia, which function as the brain's immune system.[2] Nevertheless, when one considers the extraordinary number of pathogens that have neurotropic properties (Table 1), the brain might be more properly considered an "immunologically underprivileged site."

From a clinical perspective, infections of the CNS are of major importance, not only because of the large number of microorganisms that have a predilection for this organ system but also based upon the seriousness of many of these infections. Rabies virus infection, for example, remains uniformly fatal, a distinction held by few if any other pathogens. Even for CNS infections for which antimicrobial therapy is available, mortality remains high. For example, the case fatality of antibiotic-treated pneumococcal meningitis is 20-30%,[3] and herpes simplex virus (HSV) encephalitis is fatal in over 30% of adult patients, despite treatment with acyclovir.[4]

Certain pathogens that cause CNS infection target microglial cells, most notably HIV.[5] In addition to providing harbor for such organisms, microglia appear to contribute to brain damage.[6] Because of these properties, our laboratory has been interested in the role of microglia in the pathogenesis of CNS infections. Upon discovery that certain psychoactive drugs alter the function of microglia, our attention has recently turned to studies of the potential benefit of psychoactive drugs in treatment of CNS infections.

Neuroimmune Circuits, Drugs of Abuse, and Infectious Diseases
Edited by Herman Friedman *et al.*, Kluwer Academic/Plenum Publishers, 2001

Table 1. Neurotropic infectious agents

BACTERIA	**PARASITES**	
Streptococcus pneumoniae	*Cysticercus*	
Haemophilus influenzae	*Toxoplasma gondii*	
Neisseria meningitidis	*Trypanosoma*	
Listeria monocytogenes	*Entamoeba histolytica*	
Mycobacterium tuberculosis	*Echinococcus*	
Borrelia burgdorferi	*Schistosoma*	
Treponema pallidum	*Strongyloides sterocorales*	
Leptospira	*Angiostrongylus cantonensis*	
Brucella	**VIRUSES**	
Rickettsia	Arboviruses	Rabies virus
Mycoplasma	Herpes simplex virus	Measles virus
FUNGI	Cytomegalovirus	Rubella virus
Crytococcus neoformans	Varicella zoster virus	JC virus
Coccidioides immitis	Epstein-Barr virus	Hendra virus
Histoplasma capsulatum	Herpes B virus	
Blastomyces dermatitidis	Retroviruses (HTLV-1, HIV)	
Candida	Influenza virus	
Zygomycetes	Mumps virus	
Sporothrix schenckii	Lymphocytic choriomeningitis virus	
	PRIONS	

MICROGLIA: A DOUBLE-EDGED SWORD

Many of the biological features of microglia were elucidated in the early decades of the twentieth century. In a treatise on microglia published in 1932,[7] Pio del Rio-Hortega argued persuasively for the mesodermal origin of these cells, i.e., arising from bone marrow-derived blood monocytes. He also described the remarkable capacity of microglia to differentiate from a ramified to an amoeboid (motile) form in response to brain injury and infection.

After an eclipse which lasted almost half a century, the recognition in the 1980's that microglia are the principal brain cell type which supports productive infection of HIV reignited scientific interest in brain macrophages.[5] Additionally, the observations that reactive glial cells and cytokine expression are histopathologic features of Alzheimer's disease[8] spurred interest in the potential involvement of activated microglia in this neurodegenerative disorder. Presently, a large body of evidence supports the notion that microglia participate not only in defense and repair of the CNS but also in brain damage.[2] Work in our laboratory, as well as in many others, has focused on the key mediators of these beneficial and deleterious processes, i.e., cytokines and free radicals.

OPIOIDS AND MICROGLIA

It has been known for over a century that exogenous opioids have immunomodulatory properties.[9] Research in the past several decades has shown that drugs of this class, i.e., μ-opioid receptor agonists, can manifest their immunomodulatory activities directly on immune cells or indirectly via CNS-endocrine pathways.[10] Using primary human microglial cell cultures, we have demonstrated that treatment with morphine suppresses microglial cell migration toward the activated complement component C5a.[11] This "anti-inflammatory" property of morphine involves activation of μ-opioid receptors expressed by microglia cells.

Recently, Grimm et al[12] have shown that morphine inhibits chemotaxis of blood monocytes toward the β-chemokine macrophage inflammatory protein-1α. In studies of human microglia, we recently have found that morphine treatment inhibits chemotaxis toward RANTES.[13] Additionally, treatment of microglia with morphine suppressed the production of this β-chemokine.

HIV is known to gain entry into monocytes/macrophages via CD4 and β-chemokine coreceptors on these cells.[14] Thus, we have been interested in the effects of several classes of opioid receptor ligands on HIV infection of human microglia. Although the exogenous μ-opioid receptor agonist morphine appears to have little, if any, influence on HIV-1 expression in microglial cell cultures, the recently described endogenous μ-opioid receptor ligand endomorphin-1 was found to potentiate replication of HIV-1 in microglia.[15] This proviral effect of endomorphin-1 was blocked by naloxone as well as by the relatively μ-opioid receptor selective antagonist β-funaltrexamine, suggesting that activation of μ-opioid receptors is involved in this effect of endomorphin-1 on microglia.

In contrast to our findings with μ-opioid receptor agonists, treatment of human microglia[16] and blood monocyte-derived macrophages[17] with κ-opioid receptor ligands suppresses expression of HIV-1 in vitro. This antiviral property is blocked by nor-BNI, a κ-opioid receptor antagonist, suggesting that in addition to μ-opioid receptors, macrophages also express κ-opioid receptors.

Taken together, the results of our in vitro studies of opioids on microglia imply that the function of these brain macrophages could be altered by the administration of opioids of the μ- and κ-opioid receptor classes and that some of these effects could have positive consequences in patients with certain types of viral infection of the CNS, such as, patients with HIV-1-related dementia. In in vivo studies, treatment of swine[18] and mice[19] with morphine was found by other investigators to have a beneficial effect on the neuropathogenesis of suid herpes virus-1 and HSV infections, respectively. While the effect of morphine on herpes virus infections of the CNS may or may not involve an interaction with microglia, nevertheless these animal studies support the potential therapeutic role of opioids in viral infections of the brain.

OTHER PSYCHOACTIVE DRUGS

In studies paralleling those demonstrating immunodulatory effects of opioids, other groups of investigators have demonstrated similar influences of cannabinoids,[20] cocaine,[21] and benzodiazepines.[22] Little work, however, has been focused on the effects of these psychopharmacologic agents on human microglia. Because these cells are the principal brain cell type supporting HIV replication within the brain parenchyma, our laboratory has investigated the effects of several such drugs on HIV-1 expression in microglial cell cultures. In the case of benzodiazipines, we demonstrated that agents which act through peripheral benzodiazepine receptors potently suppress HIV-1 expression in microglial cell cultures.[23] The mechanism underlying this antiviral property of benzodiazepines involves, at least in part, the inhibition of the regulatory protein nuclear factor-κB.

Recently, we have extended our studies of the effects of psychoactive agents on HIV-1 expression in human microglia to other classes of drugs. As shown in figure 1, both thalidomide (a CNS sedative) and St. John's wort (a natural product with antidepressant properties) inhibit HIV-1 expression in microglial cell cultures. While the mechanisms underlying the antiviral effects of these drugs are presently unknown, nonetheless these in vitro findings support the concept that psychoactive drugs can be exploited to treat CNS infections.

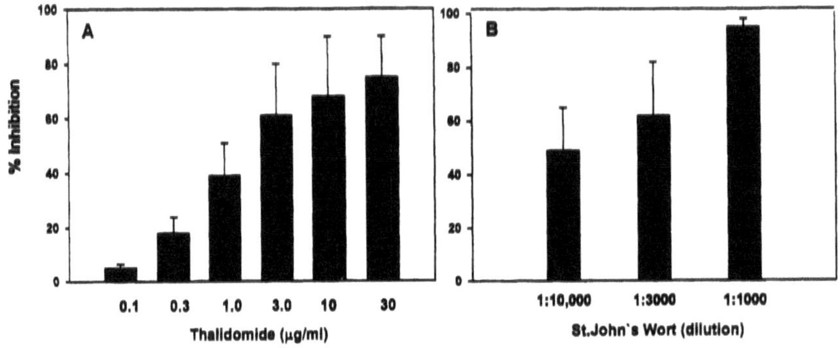

Figure 1. Effects of (A) thalidomide and (B) St. John's wort on HIV-1 expression in human microglial cell cultures. HIV-1 SF-162 was added to primary human microglial cells that were treated simultaneously with thalidomide or St. John's wort at the indicated concentrations, and HIV-1 expression was assessed by assaying for HIV-1 p24 antigen (Ag), using previously described methods (16). Data are expressed as percentage inhibition of viral expression relative to control (drug-free) cultures.

CONCLUSION

Antimicrobial therapy is urgently needed for many CNS infections. The search for new therapeutic agents will be guided by pharmacologic considerations, such as, penetrability of the blood-brain barrier. By their nature, psychoactive drugs fulfill this important pharmacologic feature. Our preliminary studies of the effects of opioids, benzodiazepines, thalidomide, and St. John's wort on microglia suggest that drugs which target the brain's immune system may be useful in treatment of certain types of viral infection of the brain. The work of other investigators[24] suggests that this concept can be extended to cannabinoid derivatives and treatment of bacterial infection of the CNS, as well.

Whether the brain is regarded as an immunologically "privileged" or "underprivileged" site, it is clear that the CNS and immune system are "connected," i.e., these systems possess elegant but complex pathways of communication involving neuropeptide and cytokine mediators. The observation that neuropsychiatric symptoms are a common manifestation of many CNS infections and the proposed association of psychiatric illnesses, such as, schizophrenia and bipolar disorder, with viral agents[25] provide evidence of the linkage between infection and the neuroimmune axis. Within this conceptual framework, the search for new treatments of CNS infections properly includes testing of derivatives of psychoactive drugs that target the brain's immune system.

ACKNOWLEDGEMENTS

This work was supported in part by U.S. Public Health Service grants DA-04381, DA-09924, and MH-57617.

REFERENCES

1. W.F. Hickey, Leukocyte migration into the central nervous system, in: *In Defense of the Brain: Current Concepts in the Immunopathogenesis and Clinical Aspects of CNS Infections*, P.K. Peterson and J.S. Remington, eds., Blackwell Science, Malden, MA (1997).
2. P. K. Peterson, G. Gekker, S. Hu, and C. C. Chao, Microglia: "a double-edged sword," in: *In Defense of the brain: Current Concepts in the Immunopathogenesis and Clinical Aspects of CNS Infections*, P.K. Peterson and J. S. Remington, eds., Blackwell Science, Malden, MA (1997).
3. A. Schuchat, K. Robinson, J.D. Wenger, et al, Bacterial meningitis in the United States in 1995, *N. Engl. J. Med.* 337:970 (1997).
4. R. J. Whitley and A. Arvin, Herpes simplex infections of the central nervous system, in: *In Defense of the Brain: Current Concepts in the Immunopathogenesis and Clinical Aspects of CNS Infections*, P. K. Peterson and J. S. Remington, eds., Blackwell Science, Malden, MA (1997).
5. S. Koenig, H.E. Gendelman, J. M. Orenstein, et al. Detection of AIDS virus in macrophages in brain tissue from AIDS patients with encephalopathy, *Science* 233:1089 (1986).
6. C. C. Chao, S. Hu, and P. K. Peterson, Glia-mediated neurotoxicity, in: *In Defense of the Brain: Current Concepts in the Immunopathogenesis and Clinical Aspects of CNS Infections*, P. K. Peterson and J. S. Remington, eds., Blackwell Science, Malden, MA (1997).
7. P. del Rio-Hortega, Microglia, in: *Cytology and Cellular Pathology of the Nervous System*, W. Penfield, ed., P. B. Hober, New York (1932).
8. P. L. McGeer and E. G. McGeer, The inflammatory response system of brain: implications for therapy of Alzheimer and other neurodegenerative diseases. *Brain Res. Rev.* 21:195 (1995).
9. J. Cantacuzene, Nouvelles researches sur le mode de destruction des virions dans l'organisme, *Ann. Inst. Pasteur* 12:273 (1898).
10. T. K. Eisenstein and M. E. Hilburger, Opioid modulation of immune responses: effects on phagocyte and lymphoid cell populations, *J. euroimmunol.* 83:36 (1998).
11. C. C. Chao, S. Hu, K. B. Shark, et al, Activation of mu opioid receptors inhibits microglial cell chemotaxis, *J. Pharmacol. Exp. Ther.* 281:998 (1997).
12. M. C. Grimm, B. Ben-Baruch, D. D. Taub, et al, Opiate inhibition of chemokine-induced chemotaxis, *Ann. N. Y. Acad. Sci.* 840:9 (1998).
13. S. Hu, C. C. Chao, C. C. Hegg, S. Thayer, and P. K. Peterson, Morphine inhibits human microglial cell production of and migration toward RANTES, *J. Psychopharmacol.*, in press.
14. G. Alkhatib, C. Combadiese, C. C. Broder, et al, CC CKR5: a RANTES, MIP-1α, MIP-1β receptor as a fusion cofactor for macrophage-topic HIV-1, *Science* 272:1955 (1996).
15. P. K. Peterson, G. Gekker, H. Hu, et al, Endomorphin-1 potentiates HIV-1 expression in human brain cell cultures: implication of an atypical μ-opioid receptor, *Neuropharmacol.* 38:273 (1999).
16. C. C. Chao, G. Gekker, S. Hu, et al, κ opioid receptors in human microglia downregulate human immunodeficiency virus expression, *Proc. Natl. Acad. Sci. (USA)* 93:8051 (1996).
17. C.C. Chao, G. Gekker, W. S. Sheng, H. Hu, and P. K. Peterson, U50, 488 inhibits HIV-1 expression in acutely infected monocyte-derived macrophages, *Drug Alc. Dependence*, in press.

18. J. M. Risdahl, P. K. Peterson, C.C. Chao, C. Pijoan, and T. W. Molitor, Effects of morphine dependence on the pathogenesis of swine herpes virus infection, *J. Infect. Dis.* 167:1281 (1993).
19. N. C. Alonzo and D. J. J. Carr, Morphine reduces mortality in mice following ocular infection with HSV-1, *Immunopharmacol.* 41:187 (1999).
20. T. W. Klein, C. Newton, and H. Friedman, Cannabinoid receptors and immunity, *Immunol. Today* 19:373 (1998).
21. G. C. Baldwin, M.D. Roth, and D. P. Tashkin, Acute and chronic effects of cocaine on the immune system and the possible link to AIDS, *J. Neuroimmunol.* 83:133 (1998).
22. H. R. Bessler, M. Weizman, M. Gavish, I. Notti, and M. Djaldetti, Immunomodulatory effect of peripheral benzodiazepine receptor ligands on human mononuclear cells, *J. Neuroimmunol.* 38:19 (1992).
23. J. R. Lokensgard, G. Gekker, S. Hu, et al, Diazepam-mediated inhibition of human immunodeficiency virus type 1 expression in human brain cells, *Antimicrob. Ag. Chemother.* 41:2566 (1997).
24. R. Bass, D. Engelhard, V. Trembovler, and E. Shohami, A novel nonpsychotropic cannabinoid, HU-211, in the treatment of experimental pneumococcal meningitis, *J. Infect. Dis.* 173:735 (1996).
25. R. H. Yolken and E. F. Torrey, Viruses, schizophrenia, and bipolar disorder, *Clin. Microbiol. Rev.* 8:131 (1995).

MODEL SYSTEMS FOR ASSESSING COGNITIVE FUNCTION: IMPLICATIONS FOR HIV-1 INFECTION AND DRUGS OF ABUSE

Walter E. Zink, Jeffrey Boyle, Yuri Persidsky, Huangui Xiong and Howard E. Gendelman

The Center for Neurovirology and Neurodegenerative Disorders
University of Nebraska Medical Center
Omaha, NE 68198-5215, USA

ABSTRACT

Memory deficits are common among drug abusers and in those with chronic neurodegenerative disorders. Currently, the mechanisms through which diverse neurophysiologic processes alter memory are not known. This review describes the current systems and rationale for studying memory formation, consolidation, and recall. Special attention is given to physiologic (hippocampal long-term potentiation) and behavioral animal models. The principles and methods described can be applied to studies of diverse clinical disorders.

INTRODUCTION

Upon entry to the central nervous system (CNS) a noxious agent may initiate an inflammatory response in the brain or directly alter neural function(s) [fig.1]. In the first model, an inciting event (molecule, microbe, or immunogen) is transported to the brain. The insult activates resident brain mononuclear phagocytes (MP) (microglia, perivascular, and brain macrophages) to initiate an inflammatory response. Virus (for example, the human immunodeficiency virus, HIV) microbes, and/or cellular proteins (for example beta-amyloid) may prime the MP for activation. Proinflammatory molecules and/or chemokines manufactured from activated macrophages trigger immune responses including intracellular killing, antigen presentation, leukocyte recruitment and/or secretory processes. Secreted products from brain MP then compromise neuronal function.[1,2] Indeed, a number of neurodegenerative processes such as Alzheimer's disease (AD) and HIV-1 associated dementia (HAD) can be affected by MP neurotoxic activities.

In an alternative, but not mutually exclusive model for neurodegeneration, peripheral toxins and/or drugs outside the brain influence neuronal function. Such toxins cross the blood-brain-barrier, bind receptors in the neuronal membrane or cytosol, and in so doing, alter neuronal function. Neuronal poisons of this nature are extensively characterized, including organophaophates, lead, toluene, among others. Corticosteroids can also affect neuronal function and have been shown to alter mood and behavior in higher doses.[3] In cell cultures and in animals, steroids induce neurotoxicity[4,5] and neuronal dropout.[6,7] Chronic neurologic impairments associated with alcohol, cannabinoids, nicotine, cocaine, opiates, amphetamines, LSD, psilocybin, PCP, and others also cause neural toxicity by interacting with specific neuronal receptors. Abused substances influence neural viability, as evidenced by drug-dependent sensory alterations and/or addictive potential of drug(s). Abused drugs may exhibit regional and circuit-specific damage within the brain, and cause long-term alterations in neuronal function.

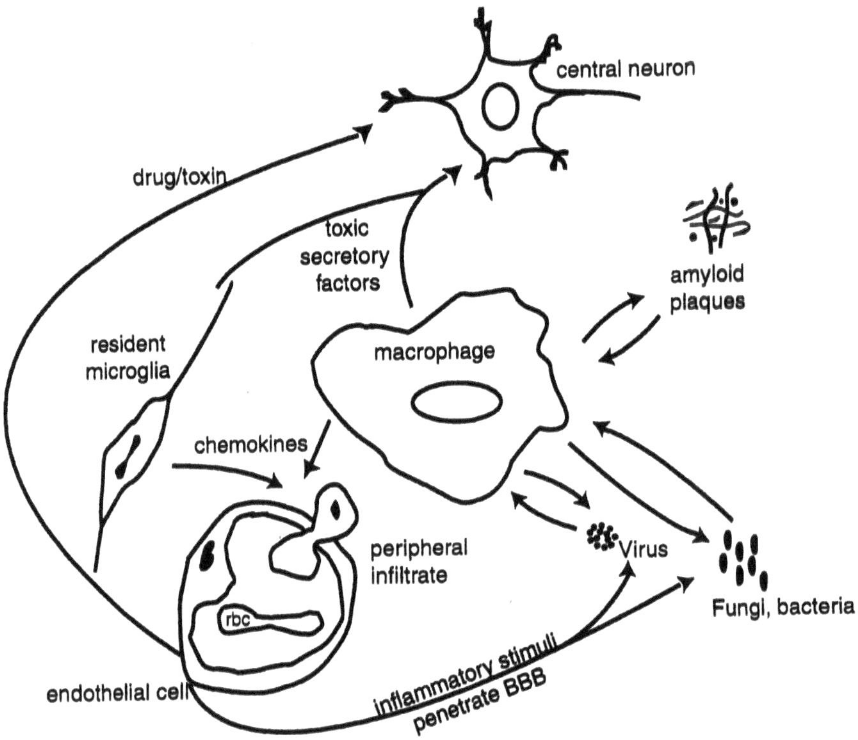

Figure 1 shows an inflammatory cascade that can effect neural function in HAD and AD. Brain MP secrete toxic factors, chemokines, and proinflammatory cytokines. Pathways for direct neurotoxicity are illustrated.

Research within the Center for Neurovirology and Neurodegenerative Disorders (CNND) is aimed at uncovering the mechanism(s) through which the innate immune system impairs central nervous sytem (CNS) function. Drugs of abuse may damage neurons alone or in synergy with underlying immune processes. At the CNND, molecular, cellular, and animal models are being applied to investigate potential interactions between opiate administration and progressive HIV-1 infection in brain.

Many potential sites of interaction between opiates and underlying HIV-1 disease exist. First, opiates may alter transmigration of peripheral monocytes into brain. Second, opiates may alter spread of viral infection within brain or other tissue reservoirs. Third, opiates may alter MP function such as antigen presentation, intracellular killing, phagocytosis, or effector function. Fourth, opiates may act in synergy with MP-secreted neurotoxic products. Finally, opiate-stimulated signal transduction may stimulate MP-independent signal transduction compromising function of intraneuronal compartments such as the neuronal membrane, synaptic vesicles, pre-or post-synaptic cytosol, or nucleus. The CNND has constructed *in vitro* and *in vivo* models to address these hypotheses, including neuronal and glial morphology, electrophysiology, and signal transduction. Findings from murine model systems are substantiated in primary human culture systems.

Animal Model Systems

Multiple experiments systems attempt to quantify cognitive function in rodents. These include tests of learning and memory, anxiety, aggression, helplessness, emotion, nocioception, addiction, tolerance, withdrawal, gait, and hyperactivity. Such rodent testing systems can map regional brain function, test effects of CNS-acting drugs, and study the aging process. Behavior systems used with histological and molecular analyses can describe phenotypes of transgenic and knockout mice.[11,12,13] Previous works have demonstrated the utility of behavior systems to analyze memory function(s) in animal models of disease, some of which are shown in Table 1. In general, systems for animal cognitive function use either positive or negative reinforcement for motivation. Latency often is used as an output, either time to reward or time to escape from an aversive stimulus (AS). Short-term memory, acquisition, consolidation, recall, and attention paradigms are shown in Table 2. In most systems, if training occurs before disease (initiated by a microbe or other noxious stimuli), output will describe consolidation and/or memory recall. If training occurs after treatment, attention and/or memory acquisition can be described. Importantly, impairments detected in one system may not correlate with those of others.

The earliest systems used for investigating rodent memory were developed soon after publication of Pavlov's landmark description of classical conditioning in dogs. Stone's early work with rodents focused on positive-reinforcement-driven recall of complex light patterns[14] and negative reinforcement-driven platform escape tasks.[15] Shortly thereafter, positive reinforcement-driven guillotine-door mazes were created,[16] serving as forerunners to more modern T-, Y-, plus-, and radial arm mazes. Passive avoidance (PA) and the Morris water maze (MWM) are the most commonly used contemporary memory systems. Rats generally make better subjects for behavior analysis than mice because of their greater size, strength, and endurance. However, mice have a natural tendency toward exploratory behavior that can prove useful in positive-reinforcement tests. Analysis of transgenic and knockout mice also provides important information. However, mouse strains are highly variable. For example, the C57BL/6 strain was reported to perform better than other strains in MWM tests.[13] Consideration of strain-specific differences in learning may be crucial to investigators intending do develop transgenic or knockout models for disease processes. Mice have been used in all the experimental systems described in Table 1.

Table 1. Rodent memory analysis systems

System	Aversive Stimulus	Positive Stimulus	Creator
Light Discrimination (LD)	---		Stone, 1929
Active Avoidance (AA)	Electric shock, noise	---	Stone, 1929
Passive Avoidance (PA)	Electric shock, noise	---	Miller, 1948
Guillotine Mazes (T-maze, Y-maze)	---	Food	Leeper, 1932 Jarvik, Kopp 1967
Olfactory Discrimination (OD)	---	Food	Bowens & Alexander, 1967
Radial Arm Maze (RAM)	---	Food	Olton, 1976
Barnes Platform Maze (BPM)	---	Bedding	Barnes, 1979
Morris Water Maze (MWM)	Cold water	---	Morris, 1981
Elevated Mazes	Height	---	Handley & Mithani, 1984

Table 2. Prototypical paradigms for analysis of memory subprocesses in mice and rats

Component of memory	Behavior system
Working memory (short-term)	RAM
Memory Acquisition	MWM, daily repeated trials
Memory Consolidation	PA, single trial 24 hrs. after training
Memory Recall (long-term)	MWM, 1-3 probe trials during peak treatment effect

RAM = radial arm maze, MWM = Morris Water Maze, PA = Passive Avoidance

The Radial Arm Maze (RAM)

The radial arm-maze (RAM) assesses working memory and the ability to interpret and retain spatial information for the purpose of completing an immediate task. This test is based on remembering lists of spatial locations. The prototypical RAM apparatus consists of a central octagonal platform with eight arms radiating from it like spokes of a wheel. Each arm is a corridor separated from the central platform by a guillotine door. The walls of the central platform have cues on the inside so that the rodent can recognize previously entered doors. Each corridor is separated from neighboring corridors by a barrier, preventing the animal from jumping from one corridor to the next. The rodent must pass through the central platform to gain access to different corridors in order to receive a reward.

When studying working memory, no pre-treatment acclimation is necessary for RAM. Following treatment, rodents are deprived of food for 24 hours prior to each session. At the start of testing, a small, rapidly consumable food morsel (or pellet) is placed at the end of each arm distal to the center platform. It is necessary that the food be consumed in seconds, does not satiate a food-deprived rodent, and has no more than a faint odor. Empirical trials using untreated controls are necessary prior to experimentation to determine the optimal food type and size. A hungry rodent is introduced to the center platform with the guillotine doors closed. After an arbitrary interval, all eight guillotine doors are opened. The rodent runs to the end of the corridor and eats the food pellet. While the rodent eats, the experimenter closes doors to the seven unchosen corridors. Eventually, the rodent returns to the center platform and the single remaining open guillotine door is closed behind the fed, but still hungry animal. After an arbitrary interval the process is repeated. The process continues until all the food is gone or until a pre-determined number of choices are made. It has been suggested that the final guillotine door should not be closed until the rodent turns its back, to reduce the potential for consecutively choosing the same or adjacent doors.[17] Analysis of RAM may be tricky because even random motion will result in a high percentage of success in early choices. In some variations, only four arms are baited for this reason. Output data is frequently presented as "numbers of repeated entries" (errors) at a given timepoint or "cumulative errors" over multiple timepoints. Because RAM does not quantify rodent speed, the test is particularly useful when describing rodents with superimposed motor impairments.

RAM sessions are conducted *after* treatment and are not incorporated into long-term memory stores. However spatial cues must be 'held' in the brain during trials and used to formulate choices. RAM forces the rodent to remember which arms have been depleted of food based only on spatial cues. Application to task is the basis of working memory, a process that may involve hippocampal circuits.

Morris Water Maze (MWM)

The best described animal model for memory acquisition and storage is the MWM, first published in 1981[18] and recapitulated in detail in 1984[19]. Like the RAM, the MWM system relies on a rodent's ability to apply spatial cues to complete a task. A typical MWM apparatus consists of a circular pool 100cm in diameter and 110 cm in height with a video camera suspended 170cm above the liquid surface. The pool is filled with white or black liquid in contrast to the tested rodent. Input gathered from the camera travels to a computer where tracking software records the path of the rodent by light/dark contrast detection. The camera is mounted so that the detection field perfectly represents the pool. The detection apparatus divides the field into four quadrants. A square platform (13cm x 13cm) is placed in one of the quadrants and the detection system is programmed to recognize the corresponding region in the detection field. The so-called "escape platform" must be the same color as the liquid. The escape platform is either visible 5cm above liquid level or is submerged immediately beneath the surface depending on the type of analysis desired. Motivation for finding the platform is generated by maintaining the liquid at 18-19°C. The decrease in escape latency shown in figure 2 demonstrates the motivational effect of cold water.

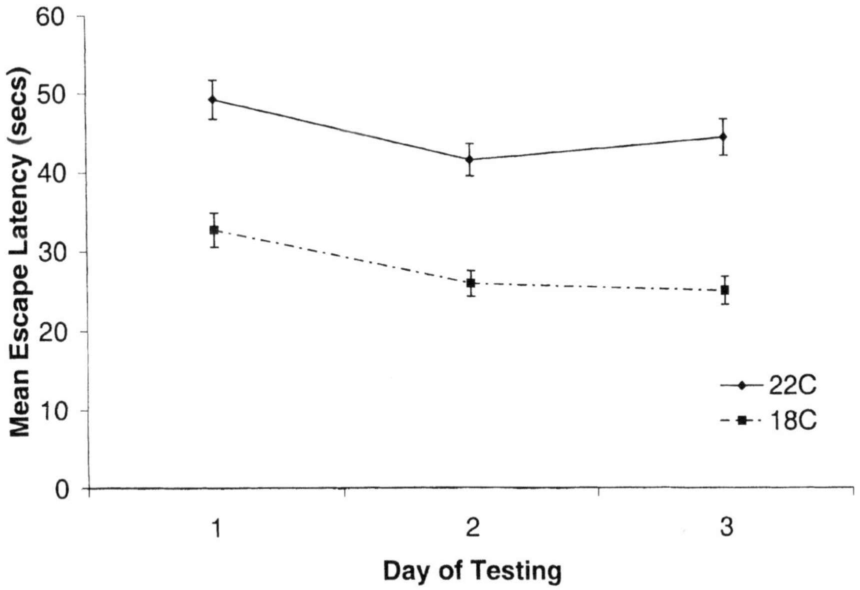

Figure 2 demonstrates the effect of water temperature on mean latency time. Two groups of trained mice (male C.B-17/IcrCrl-SCID-bgBR, 6 weeks old) were tested for escape latency on three consecutive days. The testing was identical for both groups with the exclusion of water temperature. Error bars represent standard error of the mean.

The rodent is placed in one of the four quadrants, determined arbitrarily. The animal must use extra-mazal cues to orient itself and ultimately find the escape platform. Trials are terminated if the rodent is unable to find the platform in 90 seconds [fig.3a] or if the detection system perceives that the animal has reached the escape.[fig.3b] Final output includes the actual path traveled, latency time, total distance traveled, and time spent in each quadrant. For analysis purposes, a latency time of 90 seconds is assigned to every rodent unable to find the platform during a given trial. Average speed can be calculated simply by dividing the distance traveled by latency time. Probe trials describe frequency of time spent in a given quadrant, necessitating Chi-squared analysis.

Experimental design depends on the component of memory to be studied. For study of memory acquisition, output can be accomplished in a single day. *After* treatment, rodents are placed in the MWM apparatus for 5-16 separate trials in the same day. Latency time to hidden platform is measured, generating a learning curve [fig.3c]. After many trials, the rodent will cease to reduce mean latency time to escape platform. When asymptote is reached, subsequent trials provide no new information. The number of trials to asymptote differs from experiment to experiment depending on rodent species, strain, litter, and gender.[20,21] Treatment groups with

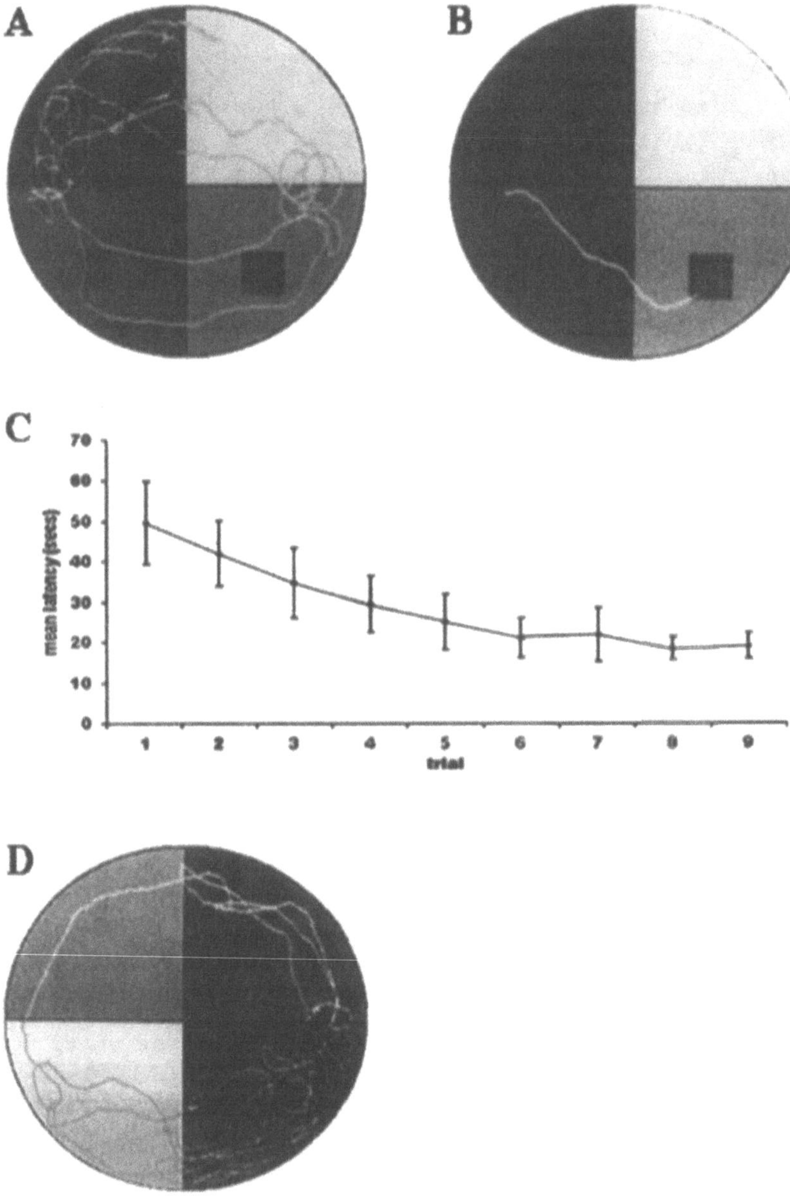

Figure 3 A shows the path of an untrained mouse in the MWM. Figure 3B shows the path of a trained mouse. Figure 3C shows a typical learning curve. All eight timepoints were conducted on the same day with the same group of mice. Figure 3D shows the path of trained, untreated mouse during a probe trial. Path similarities between the mice shown in 3A and 3D underscore the need for extensive statistical analyses to delineate differences between treatment groups.

impaired memory acquisition generate learning curves with reduced slopes compared to untreated controls. When testing memory acquisition at multiple post-treatment points, the platform is moved to different quadrants from day to day in a pseudo random fashion (and detection system adjusted accordingly) to eliminate residual memory from confounding analysis. Still, learning curves generated at later points tend to be steeper due to task and place familiarity.

Multiple daily precautions must be included in experimental design to ensure differences in MWM output truly represent memory alterations and not exhaustion, hypothermia, or alterations in motor, motivational, or sensory function. Each rodent must be given adequate rest periods between trials, usually under a heat lamp to help recovery from the cold liquid. To control for hypothermia, rodents are placed in the apparatus only after rectal temperature returns to 37-38°C (the pre-testing temperature). When treatments alter core body temperature, different tactics must be employed. Average speed during hidden platform trials can be used to control for exhaustion and motor function. Sensory function, motivation, and motor function can be measured using visible platform trials. The platform is placed in a different quadrant than in the hidden platform trials and protrudes 2-5cm above liquid level. Sometimes an odd-shaped, easily noticed object must be placed on the center of the platform to protrude well above fluid level to attract attention. Frequently, having found a hidden platform in a given quadrant during 5-16 previous trials, rodents become unwilling to investigate an object in a different location, regardless of perception, motivation, or motor function. In these situations, separate mice are treated and used to generate control data.

Alternatively, "probe" trials (also called "transfer tests") can be used to study memory recall. In memory recall studies, rodents are trained in the hidden platform system *before* treatment. At each of three days prior to treatment, rodents are placed in the MWM apparatus for 5-16 trials/day, such that asymptote is reached at each day. The platform is left in the same quadrant for every training session. After treatment, rodents are again placed in the apparatus, however the platform is removed and rodents are allowed to swim freely with no potential for escape, unbeknownst to them, for a single trial lasting 90 seconds [fig.3d]. The percentage of time spent in the quadrant formerly containing the submerged platform is quantified. Visible platform controls for motor, motivation, and sensory function may be employed as described above. Motor function also can be inferred from swim speed in the hidden platform and probe trials. In the memory recall system, hypothermia and fatigue are not factors.

Passive Avoidance Tests (PAT)

PAT is based on the idea that a period of memory consolidation following learning, deficits of which can be explored independent of the preceding process of memory acquisition or the subsequent process of memory recall. This idea is based on early studies in cycloheximide and acetoxycycloheximide treated-mice. Subcutaneous acetocycloheximide administration inhibits brain protein synthesis up to 95.%[22,23] Cycloheximide and its congeners are preferred to puromycin and other well-known protein synthesis inhibitors due to greater CNS penetrance. Mice were trained in passive avoidance or visual discrimination (Deutsch carousel)[24,25] systems prior to or after treatment. Cycloheximide does not affect acquisition of information in mice treated prior to training. Cycloheximide-treated mice completed working memory tasks as well as untreated controls for up to 3 hours but suffered subsequent amnesia. As brain protein synthesis returned to normal, memory returned.[26] Interestingly, studies conducted using intracerebral injection of the transcription inhibitor actinomycin D showed similar temporal performance patterns.[27] These studies delineated the functional processes of working memory, consolidation, and memory recall.

Two styles of passive avoidance exist, the "step-through" and "step down" methods. The more prevalent step-through system consists of a bread basket-sized box separated by a guillotine door into two chambers.[fig.4] Walls of the shock chamber are shaded black. The start chamber is shaded white and illuminated by a dim bulb (e.g.12V bulb often found under car hoods). The chambers are separated from the shock box by a wall containing a translucent guillotine door. The floor in the shock chamber consists of a grid of exposed metal rods (usu. brass or stainless steel) through which current can be administered (e.g. 0.8mA). The floor of the dark start box is not electrified. To acclimate the rodent to the apparatus prior to treatment and testing, the animal is placed in the start box with the guillotine door open and the light in the shock chamber illuminated. Each rodent is left in the apparatus for a short duration (e.g. 90 seconds) whether or not it enters the shock chamber. The mouse or rat usually wanders into the shock chamber presumably due to a preference for darkness. Latency time is measured and used to determine exploratory tendency of the rodent strain. Percentage of time spent in each chamber during acclimation establishes place preference. Usually, the amount of time spent in the shock chamber increases with successive acclimation trials. Acclimation should be conducted at least once, 24 hours prior to testing and may be repeated multiple times as determined by empirical standardization of untreated controls.

After treatment, the rodent is placed in the start chamber for 20-60sec. The guillotine-door in then lifted, allowing the mouse/rat access into the dark chamber. The rodent is administered a scrambled foot shock through the floor shortly after entering the darkened chamber. In some systems, repeated or constant shock is applied until the animal escapes back through the guillotine door into the start chamber. Escape latency can be measured as an internal control for gross sensory and motor function. Entrance latency is the primary output variable. The step-down method is similar in principle but uses a single chamber design with an insulated escape platform suspended slightly above the surface of a metal-grid floor in lieu of a start chamber. In both systems, single trial recordings applied at least 24 hours after treatment appear to reflect memory consolidation.

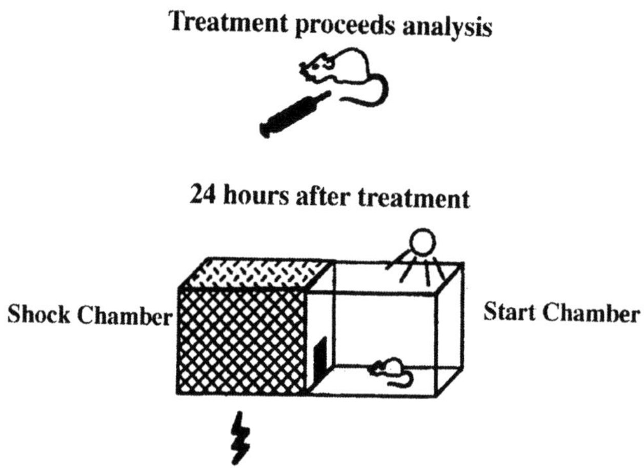

Figure 4 shows a schematic diagram of a passive avoidance apparatus. The mouse is placed in the apparatus at least 24 hours following treatment to test memory consolidation.

Brain Imaging and Cognitive Testing

MR techniques are useful in assessing in vivo changes in tissue architecture, energy consumption, and catabolite production, including magnetic resonance spectroscopy (MRS, also called magnetic resonance spectroscopic imaging, MRSI) and functional magnetic resonance imaging (fMRI). These descriptive techniques hold great utility for analysis of disease progression owing to greater sensitivity to subtle metabolic and physiologic processes. MR modalities and gross regional morphologic and blood flow studies are poised to provide descriptive longitudinal information describing deficits in higher brain function such as learning and memory. These technologies can be applied to studies of chronic drug abuse, Alzheimer's type dementia, dementia of vascular origin (stroke), and other neurodegenerative processes.

Vasogenic edema is the process of extracellular water accumulation in brain parenchyma secondary to vascular compromise. Cytotoxic edema is the accumulation of intracellular water in resident brain cells and can be used as a non-specific marker of cellular damage. Conventional T_2-

weighted MRI detects vasogenic edema but is relatively insensitive to cytotoxic edema.[28] The technique of diffusion weighted imaging (DWI) measures the rate constant for proton diffusion through total brain water (apparent diffusion coefficient of water, ADC_W) and has shown utility as a prognostic indicator following stroke.[29,30] During cytotoxic edema, neurons and glia swell and ADC_W values decline compared to normal. By noting the relative shifts of ADC_W and T_2-weighted scans, a "tissue signature" may be created for various brain disorders.

Other MR modalities can be used to describe the longitudinal metabolic and physiologic changes in the brain during memory formation. Proton magnetic resonance spectroscopy (^{1}H SI) has been used to quantify small molecular markers of neuronal damage within parenchyma, including choline (Cho), total creatine (tCr, includes all phosphorylation states), and myoinositol (MI). Because Cho is converted to acetylcholine (ACh) by the neuronal enzyme acetylcholine acetyltransferase (ChAT), in vivo quantification of Cho dynamics may serve as a useful indicator of region specific activity at cholinergic synapses. ^{1}H SI also can be used to quantify levels of N-acetyl aspartate (NAA), a marker for mature neurons.

Additional neuroimaging techniques may provide useful correlation information describing CNS glucose uptake and blood flow during memory acquisition and recall. Regional glucose uptake serves as a marker for tissue metabolic activity and can be quantified using ^{13}C-labeled spectroscopy in conjunction with a glucose metabolic tracer, such as [18F]-2-fluoro-2-deoxy-D-glucose. Glucose uptake studies are currently being used to describe progression of memory impairment during AD.[31] Correlation between cerebral blood flow (CBF) and cognitive deficit also has been studied in humans using positron emission tomography (PET), single photon emission computed tomography (SPECT), and magnetic resonance imaging (MRI).

BRAIN PHYSIOLOGY AND MEMORY: THE HIPPOCAMPUS

Descriptive studies have correlated memory consolidation with alterations in hippocampal morphology and neurobiology, particularly in the dentate gyrus. Increased glycoprotein[32] and ribosome synthesis,[33] increased density of mossy fiber projections into the dentate gyrus,[34] and a doubling in the number of dendritic spines per dentate gyrus granule cell[35,36] correlate with demonstrable learning. The hippocampus is suspected as the docking site for memory, a neural Ellis Island, where input waits in the form of transiently potentiated synapses.[37] Memory consolidation is believed to be the process of mapping docked memories to other brain regions for long-term storage.[38]

The clinical case of a 1957 human neurosurgical patient who underwent bilateral excision of the medial temporal lobes, including the rostral hippocampus, amygdala, afferent and efferent limbic circuitry, and parahippocampal gyrus, has provided valuable information for subsequent research into memory acquisition and storage.[39] After resection, the patient was cognizant and functional, was able to acquire very-short term memories, but suffered an inability to transfer information from short to long term memory. Later studies of homologous structures in rodents have shown that the hippocampus may act as a docking site for memories eventually transferred elsewhere, perhaps during sleep. When considering animal data, it must be noted that rodent and human hippocampal structures differ somewhat. Rodent brains can not be divided into four lobes per se. The rodent hippocampus is tucked beneath the cortex, superficial to the rostral regions of the basal ganglia. Functionally, however the afferent and internal circuitry of the rodent hippocampus is identical in design to that of the human.

Significant intra-species variation in rodent hippocampal cellular morphology and architecture exists, making the hippocampus an intuitive candidate structure for processes involving a highly variable process such as memory. Two lines of experimental evidence strengthen the notion that the hippocampus is involved in learning and memory. First, studies creating lesions in the rodent hippocampus or adjacent regions (septum, fimbria/fornix, or parahippocampus) usually lead to impaired memory in spatial and working memory.[38] Second, implantation of recording and stimulating electrodes into living rats has revealed electrophysiologic changes associated with memory formation detected by behavior analysis. Post-synaptic electrophysiologic responses measured in the living rat change as a function of new memory formation.[40] Also, high-frequency stimulation (HFS) applied to dentate gyrus neurons via implanted electrodes can cause alterations in memory acquisition and recall.[41]

Neuroelectrophysiology

Respirating, firing hippocampal slices can be excised from mouse or rat brains immediately after euthanasia. Although alterations in memory are resolved functionally by behavior analysis, there exists only one widely accepted tissue model for memory formation. Hippocampal slices, bathed in isotonic solution for experimentation, provide substrate for these bench-top analyses. Long-term potentiation (LTP) is the 'strengthening' of a single synapse or group of synapses as a function of high frequency stimulation (HFS, also called "tetanic stimulation" or "tetany") of the presynaptic (afferent) neuron. LTP was initially described within the hippocampus by Bliss and Lomo,[42,43] but has been demonstrated in other tissues. The cerebral cortex appears to participate in long-term memory storage, however cortical recordings are difficult to achieve because of complex cytoarchitecture. The cortex has six layers. In contrast, the hippocampus has three layers with defined internal fiber connections, making hippocampal LTP technically feasible and reproducible. Intuitively, any mechanism for memory formation must allow for a lasting change in the brain that occurs as a function of transient stimuli. LTP offers such a mechanism and within the hippocampus, is highly reproducible. Well-sliced respirating hippocampal slices contain a circuitry that includes three sites for LTP recording. The synapse initially studied by Bliss and Lomo lies between afferent, presynaptic fibers of the perforant pathway and post-synaptic pyramidal neurons within the dentate gyrus. LTP can be generated at two other hippocampal synapses: between afferent fibers of the mossy pathway and pyramidal cells of the CA3 region, and between Schaffer collateral (afferent) branches of CA3 and pyramidal cells of the CA1 region. Prior to high frequency stimulation (HFS, 100mHz x 1sec), single-pulse stimulation of the perforant pathway fiber(s) using a stimulating electrode results in excitatory post-synaptic potential (EPSP). The EPSP represents the voltage change across the post-synaptic membrane as the permeability to cations, particularly Na^+, increases [fig.5]. EPSP magnitude depends on many factors including resting membrane potential. When HFS is applied to the perforant pathway fibers, subsequent low-frequency presynaptic stimulation generates an EPSP with a significantly steeper slope. Increased Ca^{++} flux into the dendritic spine[44] elicited by HFS is believed to initiate signal transduction that effects molecular changes manifest functionally as LTP. The principle channels involved in post-synaptic Ca^{++} influx are NMDA-sensitive glutamate-receptors[45] and L-type voltage dependent Ca^{++} channels (VDCC).[46] LTP is synapse specific, dependent on depolarization of the post-synaptic neuron, dependent on increases in post-synaptic Ca^{++} levels, saturable, and reversible.[47,48,49,50] Peculiarly, the mode of LTP influx, whether through glutamate-dependent channels or VDCCs, may precipitate different downstream signaling events. That is, Ca^{++} influx though the NMDA channel may have different downstream effects than influx through VDCCs.[51,52]

Using electrophysiologic recordings from hippocampal slices to model memory formation has limitations. The first is that *ex vivo* electrophysiology relies on high frequency stimulation as a technique for induction of EPSP potentiation. Presumably high-frequency afferent stimulation of the perforant pathway, mossy fibers, or Schaffer collateral fibers does not occur *in vivo*. Second, despite evidence presented, the hippocampus may not be the most appropriate tissue for study of memory formation. Memories are stored for the lifetime of an organism but hippocampal LTP lasts hours to days in vivo.[53] Third, there is no direct evidence showing that LTP is involved in memory formation.

Studies monitoring LTP in trained animals have provided the best experimental evidence linking synaptic plasticity to memory formation. Dentate field potentials measured by electrodes permanently implanted into classically conditioned rabbits showed differences in dentate gyrus LTP in trained versus untrained groups.[54] A similar correlation was found in rats between LTP recorded at CA3 and RAM learning.[55] Conversely, animals given perforant pathway HFS before learning required significantly fewer trials than controls to reach the behavior conditioning criteria.[56] Stimulation of rabbit mossy fiber pathways projecting from pontine nuclei, cerebellar peduncle, and lateral reticular nucleus can be used as conditioned stimuli (CS) during classical conditioning.[57,58] Despite the limitations of *ex vivo* studies of synaptic plasticity, recording electrophysiologic recordings from in tact hippocampal slices from rodents is highly reproducible, avoids anesthesia, and describes functional neuronal changes in living tissue.

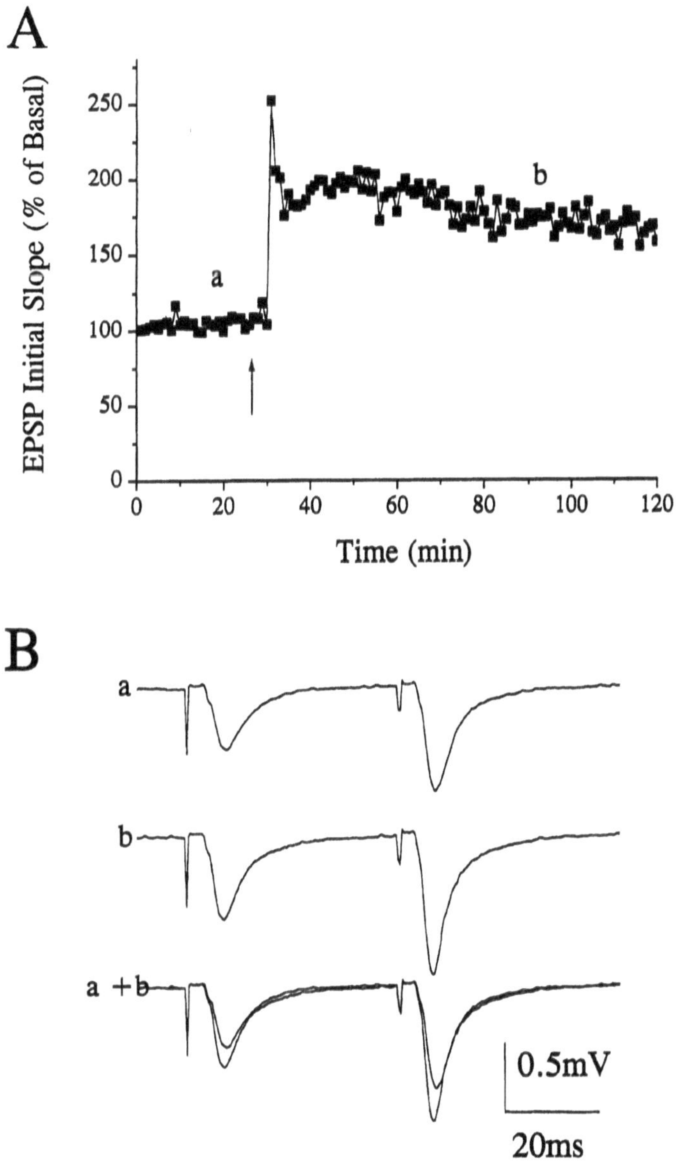

Figure 5 An example of long-term potentiation (LTP) recorded in the CA1 region of a rat hippocampal slice. LTP was induced by high frequency stimulation (HFS, 100 Hz,1000ms ¥ 2) of Schaffer-collateral pathway. Panel A illustrates the time course and magnitude of LTP. Each point in this graph is an average of three consecutive sweeps and represents normalized (as percentage of basal) field excitatory postsynaptic potential (EPSP). The graph plots the initial slope of falling phase of the evoked EPSPs recorded from CA1 dendrite field (stratum radium) in response to constant test stimuli, for 30 min before and 90 min following HFS. The HFS was delivered at the time indicated by the arrow. Panel B shows representative traces before and after the induction of LTP, taken at the times indicated by letters a and b respectively. Note the increase in slope of evoked EPSPs following HFS.

Long Term Potentiation: A Model for Memory and Recall

LTP can be observed for hours after application of HFS to a single afferent pathway *ex vivo* and can last for days *in vivo*. *In vivo* recordings are generated using implanted electrodes. The *ex vivo* response includes an early phase, lasting no longer than three hours, and a later phase.

Inhibitors of transcription[59,60] and translation[61,62] do not effect LTP up to three hours after HFS, but block persistence of LTP beyond three hours. Distinct cellular processes likely underlie each phase. The early phase may represent Ca^{++} dependent vesicle or receptor modification, while the late phase may represents expression of new gene products. Both phases likely depend on common upstream signaling events. Calcium-calmodulin dependent kinase II (CaMKII), the major protein in postsynaptic densities,[63] is a candidate signal transduction branching point.

The temporal sensitivity of LTP to inhibitors of protein and RNA synthesis shares tantalizing congruity with early studies of memory consolidation in cycloheximide and actinomycin D treated mice. It is important to note that a hearty late phase LTP response can be induced by application of three widely spaced HFS trains. This technique holds critical value in the study of phase-specific LTP.

Evidence supporting a pre-synaptic basis for LTP focuses on activation of a presynaptic Ca++/calmodulin dependent adenylyl cyclase (AC).[64,65,66] Presynaptic effector mechanisms for LTP may involve altered expression or regulation of proteins critical for vesicle-membrane fusion.[67,68,69,70] Differential activity of vesicle-fusion proteins may lead to an increase in the numbers of quantal neurotransmitter vesicles released into the synapse.

Separate though not contradictory evidence suggests that LTP represents changes at the post-synaptic terminal. High frequency stimulation may alter the post-synaptic membrane and/or intra-neuronal milieu, rendering neurotransmitter receptors more sensitive to ligand. Ca^{++} flux through NMDA receptor/channels expressed in the dendritic spine activates CaMKII.[71,72,73] CaMKII then phosphorylates the C-terminal domain of GluR1, a subunit of the AMPA-type glutamate receptor [fig.6].[74] Ca^{++}-dependent covalent modification of post-synaptic receptors may exemplify a fundamental mechanism underlying short-term neuronal plasticity. Covalent modification of previously expressed proteins explains how the post-synaptic response may be potentiated for the lifespan of the modified receptor. It is not well understood why large increases in post-synaptic calcium levels cause LTP while smaller increases, such as those observed following low frequency stimulation (LFS), show no effect or depression of the post-synaptic response. It may be that the magnitude of Ca^{++}-dependent signal transduction generated in the post-synaptic neuron following LFS is not great enough to stimulate widespread kinase activity. However, when exposed to input from multiple pre-synaptic neurons or from a single pre-synaptic neuron artificially stimulated with HFS, threshold is reached and LTP results. It has also been suggested that presynaptic stimulation creates protein-trafficking modifications in the post-synaptic terminal. [75]

Antibodies are available for delineating phosphorylation states of AMPA and NMDA subunits. Immunoprecipitation and immunoblot assays allow mechnanistic comparisons to electrophysiologic studies of early phase LTP. However, correlation between antibody-dependent identification of phosphorylation state and electrophysiology is difficult to achieve because LTP is synapse-specific. Immunoblot or immunoprecipitation studies of specific synaptic regions (e.g. dentate gyrus, CA1, or CA3) can not differentiate between pre-synaptic or post-synaptic protein expression. For this reason, pharmacologic induction of action potential must be employed. Application of NMDA to whole tissue slices in the presence of high extracellular $[Ca^{++}]$[76] allows for study of NMDA dependent LTP in a synapse-independent fashion. Similar activation of L-type Ca^{++} channels using high [KCl] also has been achieved.[77]

cAMP/PKA/CREB/CBP Pathways, Neuronal Signaling and LTP

CaMKII also may be important in late phase LTP by stimulating the cAMP-PKA-CREB-CBP transcription enhancement[64] [fig.6]. PKA is constitutively active in CA1 neurons,[78,79] but is inactive due to interaction between the catalytic (PKA_{CAT}) and regulatory (PKA_{REG}) subunits. In the presence of high levels of cytosolic cAMP, PKA_{CAT} dissociates from PKA_{REG} and becomes able to modify downstream signaling element. Unbound PKA_{CAT} can phosphorylate transcription machinery, including the cAMP response element binding protein (CREB), leading to CREB

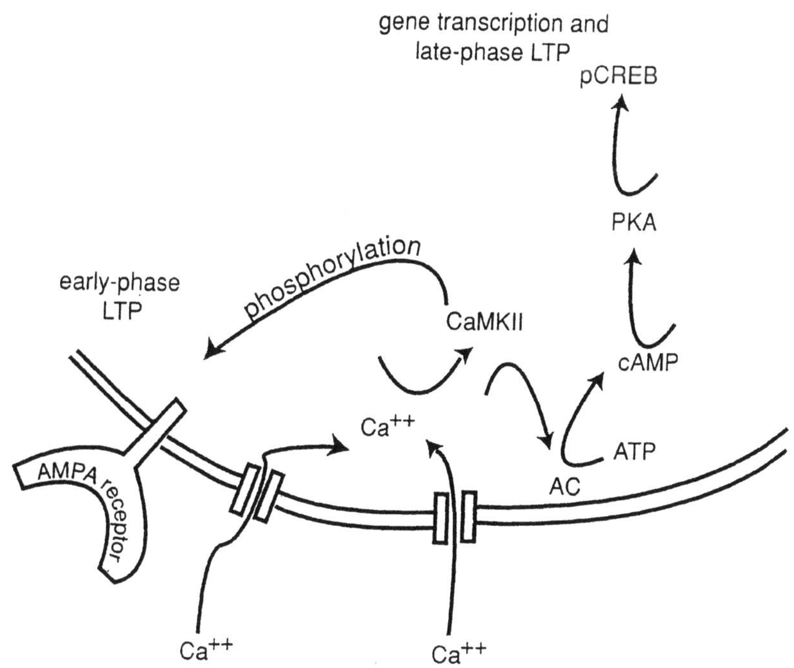

Figure 6 demonstrates mechanisms for signal transduction dependent early- and late-phase LTP.

dimerization, nuclear translocation, and enhanced transcription of CRE-containing genes. Inhibition of cAMP-dependent protein kinase (PKA) blocks late-phase LTP in a temporal pattern similar to that observed by application of transcription and translation inhibitors.[80] Based on evidence suggesting that increased intracellular Ca^{++} activates CREB/CBP dependent transcription,[81,82] the cAMP/PKA/CREB/CBP pathway represents a potential pathway of information flow from synapse to nucleus. CREB-dependent gene expression may represent the molecular basis for long-term LTP. Based on this rationale, it is imperative to assess post-synaptic cAMP, PKA, and CREB/CBP activities in experimental systems.

Afterhyperpolarization (AHP) is a Ca^{++} dependent membrane hyperpolarization observed immediately following a brief, single depolarizing pulse. "AHP suppression" is an electrophysiologic technique used as an indirect marker of post-synaptic cAMP activity. AHP has a slow decay constant (>1sec) and therefore can be studied more easily than more rapid phenomena. AHP pertains to cAMP signaling in that activation of the cAMP pathway abrogates AHP. Because of this relationship, magnitude of AHP suppression can be used as an electrophysiologic assessment of cAMP levels in the post-synaptic neuron. Though highly sensitive, AHP suppression is not a cAMP-specific response. AHP is also suppressed by muscarinic agonists via a cAMP-independent mechanism. To show cAMP-specificity, factor-dependent AHP suppression must be reversed using pharmacologically inactive cAMP analogues like RP-cAMPS.

PKA contributions to post-synaptic events can be quantified using PKA inhibitors and stimulators. PKA blocking studies relying on competition between RP-cAMPS and cAMP at the PKA regulatory site should control for the weak competitive antagonism of RP-cAMPS at other enzymes regulated by cyclic phosphonucleotides, such as cGMP dependent protein kinase (PKG). Another method for studying PKA activity takes advantage of interaction between the PKA catalytic (PKA_{CAT}) and regulatory subunits (PKA_{REG}). Because PKA_{CAT} is active only when dissociated from KA_{REG}, intracellular electrophysiologic recordings can be taken using pipettes containing excessively high concentrations of the PKA_{REG}.[83] Excess PKA_{REG} subunits shift the equilibrium of PKA_{CAT} to favor the bound, catalytically inactive form even in the presence of high cAMP concentrations. Lastly, immunoprecipitation and immunoblot studies also can help identify specific components involved in induction of LTP with the same caveats discussed above.

cAMP response element binding protein (CREB) is a transactivator that binds so-called cAMP responsive elements (CREs) and enhances downstream transcription. CREB was discovered by its interaction with the somatostatin gene.[84] Activation of CREB by cAMP/PKA was discovered shortly thereafter.[85] CREs share in common stretches of DNA containing a specific eight-nucleotide sequence (TGACGTCA) that interacts with the CREB to enhance downstream transcription. CREB/CBP dependent expression systems exist for studying neuronal transcription *in vivo* and *in vitro*. A recently published transgenic mouse model incorporates multiple copies of CREB downstream of a neuron-specific enolase promoter and a tetracycline regulator. CREB expression can be induced in the hippocampus and other brain regions making this model a crucial tool in uncovering the molecular basis of late phase LTP and learning.[86] *In vitro* reporter gene systems have been described in which primary rat cortical neurons are transfected with a CRE-CAT (chloramphenicol acetyltransferase) using a modified calcium phosphate transfection procedure.[87] Early passage PC-12 cells have also been used for similar analysis.[88] The somatostatin promoter can serve as a functional CRE in functional clones. Microinjection provides an alternative to transfection. CRE reporter systems in primary neuronal and PC12 cells (immortal, dividing pheochromocytoma cells) can be used in conjunction with fura-2 or fluo-3 Ca^{++} imaging techniques and confocal microscopy to study cellular localization of signaling components.

Long-term Depression and Memory Function

The phenomenon of long-term depression (LTD) has also been described as a mechanism for memory storage in the hippocampus[89] and elsewhere. LTD is similar in principle to LTP except that the EPSP slope generated at the post-synaptic terminal is decreased following HFS. The events of LTP and LTD can be described collectively and generally as "functional neuronal plasticity." Two types of LTD exist. Heterosynapitc LTD describes activity-dependent depression of inactive synapses. Homosynaptic LTD describes similar depression of active synapses. Heterosynaptic LTD is the more widely characterized form. Ca^{++} influx through the NMDA receptor/channel/channel appears to be crucial for formation of LTD. However it is not understood whether Ca^{++} acts as by activating voltage dependent channels or signal transduction machinery.[90] Although LTD exists as a second potential mechanism for memory storage or as a modulator of stored information, the contribution of each type of LTD to hippocampal function in vivo is unclear.

Cooperativity

Cooperativity is the mechanism through which glutamatergic axons A and B, both of which synapse with the same post-synaptic site C, induce a potentiated response. Neither induce a potentiated reponse when firing alone. The NMDA subtype glutamate receptor is necessary for hippocampal LTP formation at dentate gyrus and CA1 synapses[88,89] and may be involved in a molecular mechanism for cooperativity. The NMDA receptor/channel is a Na^{+}/Ca^{++} channel dependent on simultaneous activation by both voltage and ligand.[90,91] At hyperpolarized voltages (-70mV), Mg^{++} occupies the cation pore of the NMDA receptor/channel.[92] At resting voltage, NMDA is insensitive to glutamate. When depolarization commences (potential rises to -40mV), Mg^{++} dissociates from the NMDA receptor/channel. Subsequent glutamate binding results in opening of the NMDA receptor/channel, inward flow of Ca^{++}. If cell C expresses AMPA- and NMDA-subtype glutamate receptors, glutamate secreted only from cell A can activate AMPA channels, but does not activate NMDA receptors owing to presence of Mg^{++} in the NMDA-dependent pore [fig.7a]. In contrast, simultaneous release of neurotransmitter from neurons A and B triggers a more vigorous AMPA response, raising the membrane voltage high enough for Mg^{++} dissociation from the NMDA receptor/channel complex [fig.7b]. When dissociated from Mg^{++}, binding of glutamate to NMDA receptor/channels allows influx of cations including Ca^{++}. In this way, cooperative afferent input can cause a massive increase in intracellular Ca^{++} in the post-synaptic terminal. The dual dependence of the NMDA receptor/channel on voltage and ligand, and the dependence on the NMDA receptor/channel as a source of inward Ca^{++} flux, make the NMDA receptor/channel appealing for explaining synergy between converging signals from two spatially related afferent axons. Cooperative afferent input may represent the physiologic correlate of HFS. NMDA antagonists can impair memory *in vivo*.[93] The synapse between afferent mossy fibers and post-synaptic neurons of the CA3 region are insensitive to NMDA antagonists. NMDA-independent LTP has also been reported at extra-hippocampal sites, suggesting that glutamatergic transmission through NMDA is sufficient for LTP formation at some synapses. NMDA is not a necessary component of the LTP pathway.

OPIATES AND HIV-1: EFFECTS ON COGNITIVE FUNCTION

Drugs of abuse may interact with the pathogenesis of HIV-1 associated neurologic dysfunction at many sites. First, drugs of abuse may interact with receptors expressed by brain MPs to alter transmigration across the blood-brain barrier, spread of infection, antigen presentation, activation threshold, or secretory function. Second, drugs of abuse may interact with MP neurotoxic products to accelerate retraction of processes, metabolic decline, and non-specific neuronal deterioration. Third, drugs of abuse may alter the synaptic milieu. Fourth, drugs of abuse may interact with injured neurons to alter signal transduction in intraneuronal compartments such as the neuronal membrane, synaptic vesicle, cytosol, or nucleus. Currently, the CNND employs the MWM, electrophysiology on murine hippocampal explants, and immunohistochemistry to investigate cognitive, neurobiological, and histopathological deficits in a murine model for HIV-1 encephalitis (HIVE). Briefly, HIVE is induced by injecting HIV-1 infected human monocyte-derived macrophages (MDM) into the basal ganglia of 2-6 week-old SCID mice. Ongoing studies suggest that injection of HIV-1 infected MDM results in transient dementia, diminished LTP, and little histologic change in mice (unpublished data). Effect of opiates on progression of HIVE also is under investigation in the murine model. Morphine pellets are implanted in the dorsal subcutaneous space immediately after anesthetic administration and preceding intracranial inoculation. Whether the same analysis systems can be used to quantitatively assess HIV-1 morphine interactions is not yet clear.

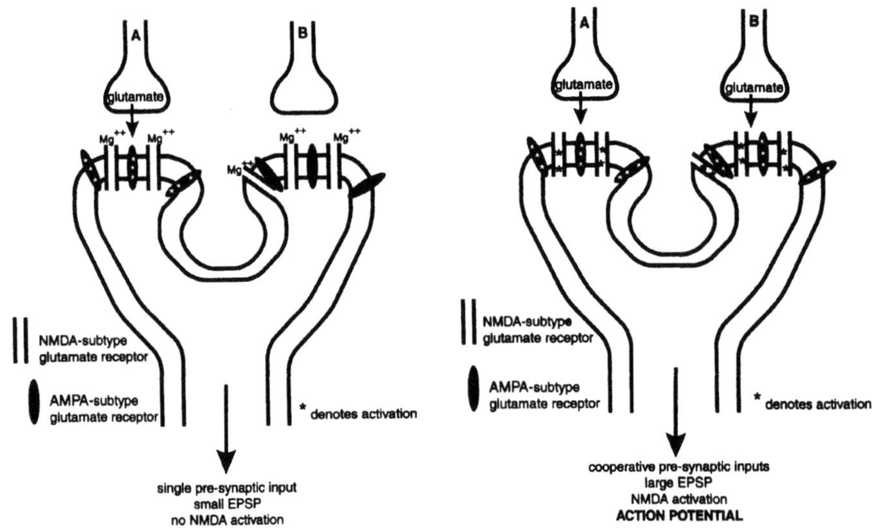

Figure 7a shows two pre-synaptic glutamatergic neurons in contact with a single dendrite. When stimulated, neuron A secretes glutamate. Figure 7b shows the same two presynaptic neurons firing simultaneously. The sum EPSP generated is greater in amplitude, causes Mg^{++} to release from NMDA-receptor ion channels, and allows NMDA receptors to become sensitive to glutamate.

CONCLUSIONS

The study of memory is a rapidly evolving field. Technical proficiency is required, from laboratory bench to patient's bedside, to dissect the mechanisms through which disease effects learning and memory. Molecular, cellular and animal model systems are all needed to determine how drugs of abuse, microbial infections, and neurodegenerative disorders effect cognitive function in man.

ACKNOWLEDGEMENTS

The authors appreciate the administrative, technical and intellectual support made to this work by Robin Taylor, Clancy Williams, and Jialin Zheng.

REFERENCES

[1] Xiong H, Zheng J, Thylin M, Gendelman HE. 1999. Unraveling the mechanims of neurotoxocity in HIV type 1-associated dementia: inhibition of neuronal synaptic transmission by macrophage secretory products. *AIDS Res Hum Retrov.* 15(1):57-63

[2] Zheng J, Gendelman HE. 1997. The HIV-1 associated dementia complex: a metabolic encephalopathy fueled by viral replication in mononuclear phagocytes. *Curr Opin Neurol.* 10(4):319-25

[3] Mellon SH. 1994. Neurosteroids: biochemistry, modes of action, and clinical relevance. *J Clin Endocrinol Metab.* 78:1003-8

[4] McIntosh LJ, Sapolsky RM. 1996. Glucocorticoids may enhance oxygen radical-mediated neurotoxicity.

Neurotoxiology. 17(3-4):873-882
[5] Chan, et. al. 1996. Endocrine modulators of neuronal death. Brain Pathol. 6(4):481-91 Mellon SH. 1994. Neurosteroids: biochemistry, modes of action, and clinical relevance. *J Clin Endocrinol Metab*. 78:1003-8
[6] Limoges J, Persidsky Y, Bock P, Gendelman HE. 1997. Dexamethasone worsens the neuropathology of human immunodeficiency virus type 1 encephalitis in SCID mice. *J Infec Dis*. 175(6):1368-81
[7] McIntosh LJ, Cortopassi KM, Sapolsky RM. 1998. Glucocorticoids may alter antioxidant enzyme capacity in the brain: kainic acid studies. *Brain Res*. 791(1-2):215-22
[8] Ghorpade A, et. al. Human immunodeficiency virus neurotropism: an analysis of viral replication and cytopathicity for divergent strains in monocytes and microglia. *J Virol*. 72(4):3340-50
[9] Persidsky Y, Gendelman HE. 1997. Development of laboratory and animal model systems for HIV-1 encephalitis and its associated dementia. *J Leuk Biol*. 62(1):100-6
[10] Persidsky Y, et. al. 1996. Human immunodeficiency virus encephalitis in SCID mice. *Am J Pathol*. 149(3):1027-53
[11] Crawley JN. 1999. Behavioral phenotyping of transgenic and knockout mice: experimental design and evaluation of general health, sensory functions, motor abilities, and specific behavioral tests. *Brain Res*. 835(1):18-26
[12] Steele PM, Medina JF, Nores WL, Mauk MD. 1998. Using genetic mutations to study the neural basis of behavior. *Cell*. 95(7):879-82
[13] Picciotto MR, Wickman K. 1998. Using knockout mice to study neurophysiology and behavior. *Physiol Rev*. 78(4):1131-63
[14] Stone CP. 1929. The age factor in animal learning: II. Rats on a multiple light discrimination box and a difficult maze. *Genet Psychol Monog*. 6:125-202
[15] Stone CP. 1929. The age factor in learning: I. Rats in the problem box and the maze. *Genet Psychol Monog*. 5:1-130
[16] Leeper RL. 1932. The reliability and validity of maze experiments with white rats. *Genet Psychol Monog*. 7:137-245
[17] Olton DS. 1987. The radial arm maze as a tool in behavioral pharmacology. *Physiol and Behav*. 40:793-7
[18] Morris RGM. 1981. Spatial localization does not require the presence of local cues. *Learn Motiv*. 12:239-60
[19] Morris RGM. 1984. Developments of a water maze procedure for studying spatial learning in the rat. *J Neurosci Meth*. 11:47-60
[20] Crawley JN. 1999. Behavioral phenotyping of transgenic mice and knockout mice: experimental design and evaluation of general health, sensory functions, motor abilities, and specific behavioral tests. *Brain Res*. 835(1):18-26
[21] Rodgers RJ, Cole JC. 1993. Influence of social isolation, gender, strain, and prior novelty on plus-maze behaviour in mice. *Physiol and Behav*. 54(4):729-36
[22] Barondes SH, Cohen HD. 1968. Delayed and sustained effect of acetoxycycloheximide in mice. *Proc Nat Acad Sci USA*. 58:157-164
[23] Barondes SH, Cohen HD. 1968. Memory impairment after subcutaneous injection of acetoxycycloheximide. *Science*. 160:556-7
[24] Cohen HD, Barondes. 1968. Effect of acetoxycycloheximide on learning and memory of a light-dark discrimination. *Nature*. 218:271-3
[25] Barondes SH, Cohen HD. 1970. Cerebral protein synthesis inhibitors block long-term memory. *Int. Rev. Neurobiol*. 13:177-205
[26] Squire LR, Barondes SH. 1972. Variable decay of memory and its recovery in cycloheximide-treate mice. *Proc Nat Acad Sci USA* 69(6):1416-20
[27] Squire LH, Barondes SH. 1970. Actinomycin-D effects on memory at different times after training. *Nature* 225:649-50
[28] D'Olhaberriague L, et al. 1998. Preliminary clinical-radiological assessment of MR tissue signature model in human sroke. *J Neurol Sci* 156:158-66
[29] van Everdingen KJ, et al. 1998. Diffusion-weighted magnetic resonance imaging in acute stroke. *Stroke*. 29(9):1783-90
[30] Nagesh V, et al. 1998. Time course of ADC_W changes in ischemic stroke: beyond the human eye. *Stroke*. 29(9):1778-82
[31] Berent S, et al. 1999. Neuropsychological function and cerebral glucose utilization in isolated memory impairment and Alzheimer's disease. *J Psychiatr Res*. 33(1):7-16
[32] Doyle E, Nolan PM, Bell R, Regan CM. 1992. Intraventricular infusions of anti-neural cell adhesion molecules in a discrete posttraining period impair consolidation of a passive avoidance response in the rat. *J Neurochem* 59:1570-3
[33] O'Connell C, O'Malley A, and Regan CM. 1997. Transient, learning-induced ultrastructural changes in

spatially clustered dentate granule cells of the adult rat hippocampus. *Neuroscience*. 76:55-62
[34] Schwegler H, Crusio WE. 1995. Correlations between radial-maze learning and structural variations of septum and hippcampus in rodents. *Behav Brain Res*. 67:29-41
[35] Gomez RA, Pozzo-Miller LD, Aoki A, Ramirez OA. 1990. Long-term potentiation-induced synaptic changes in hippocampal dentate gyrus of rats with an inborn low or high learning capacity. *Brain Res*. 537(1-2):293-7
[36] O'Malley, O'Connell C, Regan CM. 1998. Ultrastructural analysis reveals avoidance conditioning to induce a transient increase in hippocampal dentate spine density in the 6 hour post-training period of consolidation. *Neuroscience*. 87(3):607-13
[37] Shapiro ML, Eichenbaum H. 1999. Hippocampus as a memory map: synaptic plasticiy and memory encoding by hippocampal neurons. *Hippocampus*. 9(4):365-85
[38] O'Keefe J, Nadel L. 1978. The hippocampus as a cognitive map. Clarendon Press. Oxford.
[39] Scoville WB, Milner B. 1957. Loss of recent memory after bilateral hippocampal lesions. *J Neurol Neurosurg Psychiat* 20:11-21
[40] van Hulzen ZJ, van der Staay. 1991. Spatial memory processing during hippocampal long-term potentiation in rats. *Physiol Behav*. 50(1):121-7
[41] Castro CA, Silbert LH, McNaughton BL, Barnes CA. 1989. Recovery of spatial learning deficits after decay of electrically induced synaptic enhancement in the hippocampus. *Nature*. 342:545-8
[42] Andersen P, Blackstad TW, Lomo T. 1966. Location and identification of excitatory synapses on hippocampal pyramidal cells. *Exp Brain Res*. 1:236-48
[43] Bliss TVP, Lomo T. Long-lasting potentiation of synaptic transmission in the dentate area of the anaesthetized rabit following stimulation of the perforant path. 1973. *J Physiol Lond*. 232:331-56
[44] Muller W, Connor JA. 1991. Dendritic spines as individual neuronal compartments for synaptic Ca++ responses. *Nature*. 354:73-6
[45] Jahr CE, Stevens CF. 1987. Glutamate activates multiple single channel conductances in hippocampal neurons. *Nature*. 325:522-5
[46] Grover LM, Teyler TJ. 1990. Two components of long-term potentiation induced by different patterns of afferent activation. *Nature*. 347:477-9
[47] Lynch G, Larson J, Kelso S, Barrioneuvo G, Schottler F. 1983. Intracellular injections of EGTA block induction of hippocampal long-term potentiation. *Nature*. 305:719-21
[48] Malinow R, Miller JP. 1986. Post-synaptic hyperpolarization during conditioning reversibly blocks induction of long-term potentiation. *Nature*. 320:529-30
[49] Kelso SR, Ganong AH, Brown TH. 1986. Hebbian synapses in hippocampus. *Proc Natl Acad Sci USA*. 83:5326-5330
[50] Malenka RC, Kauer JA, Zucker RC, Nicoll RA. 1988. Postsynaptic calcium is sufficient for potentiation of hippocampal synaptic transmission. *Science*. 242:81-4
[51] Bading H, Ginty DD, Greenberg ME. 1993. Regulation of gene expression in hippocampal neurons through distinct calcium signaling pathways. *Science*. 260:181-6
[52] Ghosh A, Greenberg ME. 1995. Calcium signaling in neurons: molecular mechanisms and cellular consequences. *Science*. 268:239-247
[53] Staubli U, Lynch G. 1987. Stable hippocampal long-term potentiation elicited by 'theta' pattern stimulation. *Brain Res*. 435(1-2):227-34
[54] Weisz DJ, Clark GA, Thompson RF. 1984. Increased responsivity of dentate granule cells during nictitating membrane response conditioning in rabbit. *Behav Brain Res*. 12:145-54
[55] Ishihara K, Mitsuno K, Ishikawa M, Sasa M. 1997. Behavioal LTP during learning in rat hippcampal CA3. *Behav Brain Res*. 83:235-8
[56] Berger TW. 1984. Long-term potentiation of hippocampal synaptic transmission affects rate of behavioral learning. *Science*. 224:627-30
[57] Steinmetz JE, Rosen DJ, Chapman PF, Lavond DG, Thompson RF. 1986. Classical conditioning of the rabbit eyelid response with mossy-fiber stimulation CS: I. Pontine nuclei and middle cerebral peduncle stimulation. *Behav Neurosci*. 100(6):878-87
[58] Lavond DG, Knowlton BJ, Steinmetz JE, Thompson RF. 1987. . Classical conditioning of the rabbit eyelid response with mossy-fiber stimulation CS: II. Lateral reticular nucleus stimulation. *Behav Neurosci*. 101(5):676-82
[59] Nguyen PV, Abel T, Kandel ER. 1994. Requirement of a critical period of transcription for induction of a late phase of LTP. *Science*. 265:1104-7
[60] Frey U, Frey S, Schollmeier F, Krug M. 1996. Influence of actinomycin D, a RNA synthesis inhibitor, on long-term potentiation in hippocampal neurons in vivo and in vitro. *J Physiol Lond*. 490(3):703-11
[61] Grecksch G, Matthies H. 1980. Two sensitive periods for the amnesic effect of anisomycin. *Pharmacol Biochem. Behav*. 12:663
[62] Frey U, Krug M, Reymann KG, Matthies H. 1988. Anisomycin, an inhibitor of protein synthesis,

blocks late phases of LTP phenomena in the hippocampal CA1 region in vitro. *Brain Res.* 452:57-65
[63] Kennedy MB, Bennett MK, Erondu NG. 1983. Biochemical and immunohistochemical evidence that the 'major postsynaptic density protein' is a subunit of a calmodulin-dependent protein kinase. *Proc Natl Acad Sci, USA.* 80:7357-7461
[64] Huang Y-Y, Li X-C, Kandel E. 1994. cAMP contributes to mossy fiber LTP by initiating both a covalently mediated early phase and a macromolecule dependent late phase. *Cell.* 79:69-79
[65] Weisskopf MG, Castillo PE, Zalutsky RA, Nicoll RA 1994 Mediation of hippocampal mossy-fiber long term potentiation by cAMP. *Science.* 265: 1878-1882
[66] Nicoll RA, Malenka RC. 1995. Contrasting properties of two forms of long-term potentiation in the hippocampus. *Nature.* 377, 115-8
[67] Applegate MD, Kerr DS, Landfield PW. 1987. Redistribution of synaptic vesicles during long-term potentiation n the hippocampus. *Brain Res.* 401(2):401-6
[68] Pozzo-Miller LD, et al. 1999. Impairments in high-frequency transmission, synaptic vesicle docking, and synaptic protein distribution in the hippocampus of BDNF knockout mice. *J Neurosci.* 19:4972-83
[69] Takahashi S, et al. 1999. Reduced hippocampal LTP in mice lacking a presynaptic protein: complexin II. *Eur J Neurosci.* 11(7):2359-66
[70] Janz R, et al. 1999. Essential roles in synaptic plasticity for synaptogyrin I and synaptophysin I. *Neuron.* 24(3):687-700
[71] Malenka R, et. al. 1989. An essential role for postsynaptic calmodulin and protein kinase activity in long-term potentiation. *Nature.* 340:554-7
[72] Malinow R, Schulman H, and Tsien RW. 1989. Inhibition of post-synaptic PKCor CAMKII blocks induction but not expression of LTP. *Nature* 345:862-5
[73] Petit D, Perlman S, Malinow R. 1994. Potentiated transmission and prevention of further LTP by increased CaMKII activity in postsynaptic hippocampal slice neurons. *Science.* 266:1881-5
[74] McGlade-McCulloh E, Yamamoto H, Tan SE, Brickey DA, Soderling TR. 1993. Phosphorylstion and regulation of glutamate receptors by calcium/calmodulin dependent protein kinase II. *Nature.* 362:640-642
[75] Frey U, Morris RGM. 1997. Synaptic tagging and long-term potentiaition. *Nature.* 385:533-6
[76] Lee HK, Kameyama K, Huganir RL, and Bear MF. NMDA induces long-term synaptic depression and dephosphorylation of GluR1 subunit of AMPA receptor in hippocampus. *Neuron.* 21:1151-62
[77] Hu S-C, Chrivia J, Ghosh A. 1999. Regulation of CBP-mediated transcription by neuronal calcium signaling. *Neuron.* 22:799-808
[78] Wang L-Y, Salter MW, MacDonald JF. 1991. Regulation of kainate receptors by cAMP-dependent protein kinase and phosphatases. *Science.* 253:1132
[79] Greengard P, Jen J, Nairn AC, Stevens CF. 1991. Enhancement of the glutamate response by cAMP dependent protein kinase in hippocampal neurons. *Science.* 253:1135
[80] Frey U, Huang Y-Y, Kandel ER. 1993. Effects of cAMP simulate a late stage of LTP in hippocampal neurons. *Science.* 260:1661-4
[81] Chawla S, Hardingham GE, Quinn DR, and Bading H. 1998. CBP: a signal-regulated transcriptional coactivator controlled by nuclear calcium and CaM kinase IV. *Science.* 281:1505-9
[82] Hardindham GE, Chawla S, Cruzalegui FH, and Bading H. 1999. Control of recruitment and transcription-activating function of CBP determines gene regulation by NMDA receptors and L type calcium channels. *Neuron.* 22:789-98
[83] Blitzer RD, Wong T, Nouranifar R, Iyengar R, Landau M. 1995. Postsynaptic cAMP pathway gates early LTP in hippocampal CA1 region. *Neuron.* 15:1403-14
[84] Montiminy MR, Bilezikjian LM. 1987. Binding of a nuclear protein to the cyclic-AMP response element of the somatstatin gene. *Nature.* 328:175-8
[85] Gonzales GA and Montiminy MR. 1989. Cyclic AMP stimulates somatostatin gene transcription by phosphorylation of CREB at serine 133. *Cell.* 59:675-80
[86] Chen J, et al. 1998. Transgenic animals with inducible, targeted gene expression in brain. *Mol Pharmacol.* 54(3):495-503
[87] Shieh PB, Hu S, Bobb, Timmusk T, Ghosh A. 1998. Identification of a signaling pathway involved in calcium regulation of BDNF expression. *Neuron.* 20:727-40
[88] Mark MD, Liu Y, Wong ST, Hinds TR, Storm DR. 1995. Stimulation of neurite outgrowth in PC12 cells by EGF and and KCl depolarization: a Ca++ independent phenomenon. *J Cell Biol* 130:701 710
[89] Levy WB, Steward O. 1983. Synapses as associative memory elements in the hippocampal formation. *Brain Res.* 175:233-45
[90] Bear MF, Abraham WC. 1996. Long-term depression in hippocampus. *Annu Rev Neurosci.* 19:437-62.
[91] Collingridge GL, Kehl SJ, McLennan H. 1983. The antagonism of amino acid-induced excitations of rat

hippocampal CA1 neurones. *J Physiol Lond.* 334:19-31.
[92] Jahr CE, Stevens CF. 1987. Glutamate activates multiple single channel conductances in hippocampal neurons. *Nature*. 325:522-5
[93] Nowak L, Bregestovski P, Ascher P, Herbert A, Prochiantz A. 1984.Magnesium gates glutamate activated channels in mouse central neurones. *Nature* 307:462-465
[94] Mayer ML, Westbrook GL, Guthrie PB. 1984. Voltage-dependent block by Mg^{2+} of NMDA receptors in spinal cord neurons. *Nature*. 309:263
[95] Ault B, Evans RH, Francis AA, Oakes DJ, Watkins JC. 1980. Selective depression of excitatory amino acid induced depolarizations by magnesium ions in isolated spinal cord preparations. *J Physiol Lond.* 307:413-28
[96] Morris RGM, Anderson E, Lynch GS, Baudry M. 1986. Selective impairment of learning and blockade of long-term potentiation by an N-methyl-D-aspartate receptor antagonist. *Nature*. 319:774-6

DIRECT AND INDIRECT MECHANISMS OF HIV-1 NEUROPATHOGENESIS IN THE HUMAN CENTRAL NERVOUS SYSTEM

Jean Hou and Eugene O. Major
Laboratory of Molecular Medicine and Neuroscience
36 Convent Dr., MSC 4164
Building 36, Room 5W21
Bethesda, MD 20892-4164

INTRODUCTION

HIV-1 infection of the central nervous system (CNS) can manifest itself in a wide range of clinical symptoms and pathological signs. The neuropathology associated with CNS invasion of HIV-1 includes macrophage infiltration, inflammation, astrocytosis, and ultimately, neuronal damage and cell loss. HIV-1 has the ability to rapidly penetrate the brain of most infected individuals, but results in a severe neurological disease in only a subset. It is estimated that approximately 20% to 30% of adults with advanced HIV infection will develop neurological impairments, and a significantly higher percentage of pediatric AIDS cases will demonstrate developmental abnormalities (1). AIDS related neurological complications have been documented as AIDS dementia complex (ADC), HIV-1 associated dementia (HAD), and HIV-1 related cognitive-motor complex as well as CNS lymphoma and opportunistic infections. In a small percentage of cases, the emergence of neurologic symptoms may actually presuppose a diagnosis of AIDS. As such, ADC, first recognized in 1986, has since been added to the list of the CDC's AIDS-defining conditions. The clinical manifestations of cortical and subcortical neuron loss in adult ADC include impaired memory function, motor deficits, and eventual dementia in the absence of any other opportunistic infection (1). CNS complications are particularly evident in children, as affected infants and children are slower to develop, and will exhibit motor abnormalities, followed by a rapidly progressing encephalopathy. A differential diagnosis of ADC can be difficult because the pathology and clinical symptoms mimic those of opportunistic CNS viral infections such as progressive multifocal leukoencephalopathy (PML) and cytomegalovirus (CMV) infection.

Although the symptoms are well recognized, the biological and molecular mechanisms responsible for the pathology and the occurrence of mild or severe dementia are unclear. HIV-1 is capable of productively infecting macrophages and monocytes, but only rarely infects neurons or astrocytes in situ (2). Furthermore, the percentage of infected cells is low in comparison to the total number of cells in the brain. The paradox, then, is how such a limited number of productively infected cells can lead to the severe neurologic deficits observed in AIDS encephalopathy. Also related to this is the structural and functional sequestration of the CNS from the body's immune system. Since there is no humoral immune response mounted against HIV-1 infection in the brain, it has been theorized that infected astrocytes may serve as a potential reservoir for virus in its latent state. Reactivation of HIV-1 and mechanisms of cellular amplification during the later stages of disease may result in the pathology and symptomology of ADC. Immune activation of glial cells is believed to play an important role in the emergence of neurological disorders seen in patients with ADC. The purpose of this review is to examine the interactions of HIV-1 with different cell types in the brain, in an attempt to understand the mechanisms responsible for the neurologic complications being seen in a significant number of AIDS patients.

Structural Proteins and Direct Neurotoxicity

Neuronal damage or cell loss due to apoptosis or neurotoxicity may be predominantly responsible for ADC. However, the decreasing number of neurons appears to work in conjunction with neuronal dysfunction and disruption of homeostasis due to viral and cellular products from infected brain macrophages and microglia. Neurologic deficits could be caused by synaptic destruction and degeneration of neurons in the brain; however, this HIV associated neuronal damage has been shown to occur in the absence of a productive viral infection of the neurons (3). Since the general consensus, then, is that HIV does not replicate in neurons in the brain of an infected individual, the mechanisms by which the virus causes neurodegeneration are unclear. In an attempt to address this discrepancy, the direct effects of HIV-1 structural and non-structural proteins on neurons have been considered. It has been postulated that binding of HIV-1 viral coat proteins, the envelope glycoprotein gp 120 in particular, to receptors on neurons may affect neural signaling, and therefore underlie viral neurotropism. Subsequent studies have shown that gp120 causes injury and apoptosis in human and rat neurons, both *in vivo* and *in vitro* (3).

Evidence for a direct mechanism of neurodegeneration also comes from studies involving Tat, a non-structural viral protein essential for viral replication, in conjunction with gp120. Both proteins have been shown to be neurotoxic individually, but it has been recently reported that the proteins may be synergistically neurotoxic by causing increases in intracellular calcium levels (4). Continuous exposure of neurons to tat and gp120 is not necessary to produce neurotoxicity. Even a transient exposure lasting for seconds will initiate neuronal cell death. Although the cellular mechanisms responsible for gp120-induced neurotoxicity are not fully understood, it has been shown that the interaction of gp120 with immune cells requires the participation of chemokine receptors. In addition to CD4, different gp120 isolates bind to the α and β chemokine receptors CXCR4 and CCR5, respectively, which are found on brain macrophages/microglia, some astrocytes, and neurons. Therefore, gp120 mediated neuronal cell death could not only occur via direct interaction with neurons but indirectly as well by binding to glial cells and inducing the release of neurotoxic factors.

Other HIV associated gene products have been postulated to have direct neurotoxic effects as well. The small HIV-1 accessory protein Vpr can be detected in the cerebrospinal fluid (CSF) of AIDS patients. The protein has the ability to form cation-specific channels across lipid bilayers, which can then potentially disrupt the ion gradient. In vitro experiments introducing purified Vpr and recombinant Vpr extracellularly, demonstrated that both forms could insert into the plasma membrane of rat hippocampal neurons. The increased permeability of the membrane to cations caused rapid depolarization and neuronal death (5).

Intracellular Signaling and Indirect Neurotoxicity

Despite some evidence for direct toxicity, the current opinion and majority of studies support the hypothesis that HIV-1 associated neuronal cell death occurs via an indirect mechanism. Taking into account the delicate chemical and physical interaction between neuronal and non- neuronal cell types in the brain, it becomes possible that the virus is mediating neurotoxicity by interacting with neighboring glial cells. The proposed mechanisms of cell death, to a large extent, have been attributed to the disruption of the neuron's homeostatic environment or through dysregulation of intracellular signaling pathways. For example, the HIV viral regulatory protein, Tat, appears to play an important part in the dysregulation of cytokines. Secreted and taken up by infected glial and microglial cells, Tat serves as a nuclear transactivator of viral transcription by increasing levels of transcription initiation and elongation. Furthermore, HIV-1 Tat protein in macrophages and astrocytes can activate the production of the cytokine TGF-β, which in turn activates the synthesis of TNF-α. This inflammatory cytokine is toxic to neurons as well as oligodendrocytes, and most likely constitutes the major pathway to HIV-1 induced neuronal destruction (6). Apoptosis of neuronal and non-neuronal cells has been demonstrated in the brains of ADC patients. Soluble factors such as Tat and TNF-α, alone or in combination can induce neuronal apoptosis by a pathway that involves oxidative stress and increased production of oxygen free radicals. Therefore, the increased microglial production of inflammatory cytokines, induced by HIV-1 Tat protein is consistent with the neurodegeneration and clinical dementia observed in ADC (6).

The downstream effects of Tat in the CNS are amplified by inducing the production of certain neurotoxic chemokines which can result in monocyte infiltration as well as a cascade of other intracellular signaling pathways. This provides an attractive explanation for the dichotomy between actual numbers of infected cells and severity of clinical signs. Rather than entry of the virus into the cells, the interaction of HIV-1 or gp 120 with cell surface receptors may be sufficient for cellular dysfunction (7). In this case, binding of the virus to the CD4 receptors on macrophages and microglia causes the production of diffusible factors, such as SDF-1α. Expression of the receptor for SDF-1α, CXCR4, is found in neurons and astrocytes, and studies have shown that SDF-1α induces neuronal apoptosis, possibly via a GTP-binding protein linked signaling pathway (8). SDF-1α can also act directly on astrocytes by inducing the production of other neurotoxic factors. Thus, infected macrophages can cause neurodegeneration by multiple pathways.

Microglial cells normally secrete neurotrophins that support the survival of neighboring neurons. Production of these factors in the normal CNS is abrogated by HIV-1 infection, which takes over protein synthesis machinery in target cells. Infected microglial cells, then, not only cease production of essential neurotrophic factors, but also begin secretion of neurotoxic factors (2). Disruption of intercellular interactions by macrophage secretory products can cause functional damage through disrupted signal transduction (9). In addition to neurotoxic viral factors such as gp 120, gp 41, and Tat, cellular macrophage secretory products such as TNF-α, nitric oxide (NO), and MCP-1 have also been shown to affect neuronal function. It has been suggested that astrocytes are able to downregulate the production of these macrophage neurotoxins. Therefore, HIV-1 infection of astrocytes may indirectly compromise the survival of neurons.

The most abundant cell type in the brain, astrocytes are essential for maintaining the homeostatic microenvironment surrounding neurons as well as providing structural support. As such, even restricted virus production can have major consequences. Early HIV markers such as nef can be detected in astrocytes, however structural proteins such as Gag and Env are very rarely observed, evidence of a limited or silent infection (10). Once infected, the astrocytes may disseminate the virus through their intercellular contact network or produce factors that are directly toxic to neighboring cells or that promote the chemoattraction of other susceptible cell types within the brain. Nef has been implicated in the alteration of neuronal function by increasing potassium ion currents, as well as by inducing apoptosis when overexpressed (11). These observations suggest that nef protein expression by infected astrocytes can directly or indirectly kill neurons (12).

Extracellularly secreted tat induces astrocytes to produce MCP-1, which, in turn, may facilitate the transmigration of leukocytes and monocytes across the blood brain barrier. This results in the upregulation of CCR5 expression, a cellular receptor for HIV-1. Thus, Tat secretion by astrocytes facilitates the entry of monocytes into the CNS and renders them more susceptible to HIV-1 infection (13). Because the number of infiltrating macrophages in the brain is directly correlated to the development of clinical dementia, this pathway is particularly important for studying HIV-1 related neurodegeneration. Studies have shown that levels of MCP-1 found in the CSF of patients diagnosed with ADC were markedly elevated as compared to non-demented controls. Furthermore, it appears that increasing degrees of dementia corresponded to higher levels of MCP-1 detected in the CSF (14). Subsequently, there has been a direct correlation made between the number of infiltrating and infected macrophages and microglia in the brain and the severity of neurologic impairments.

Whether the route of neuropathogenesis is direct or indirect, the final common pathway for neuronal damage appears to be glutamate-mediated neurotoxicity (15). Fluctuations in intracellular calcium levels have been largely implicated in HIV-1 induced neurotoxicity. Excessive stimulation of voltage dependent calcium channels or overactivation of NMDA receptors are involved in neuronal death not just in AIDS, but also in strokes or other neurodegenerative diseases. The NMDA receptors are a subtype of receptors for glutamate, the main excitatory neurotransmitter in the brain. Macrophage secreted toxins overstimulate the NMDA receptors, rendering cells more susceptible to calcium ion influx and the release of glutamate. Glutamate, in turn, will overexcite neighboring neurons in a cascade of overexcitation and injury. Astrocytes are critical in glutamate uptake and metabolism, a function that appears to be controlled by cytokines. Therefore, HIV-1 infection of astrocytes or cytokine dysregulation caused by macrophage secretory products may lead to increases in extracellular glutamate concentrations, and neuronal cell death. Furthermore, TNF-α inhibits glutamate uptake, and TGF-β interferes with the enzyme responsible for its metabolism in astrocytes. Since Tat activates the synthesis of both these chemokines, it also appears to play a crucial role in the final pathway leading to HIV-1 induced neurotoxicity.

CONCLUSION

Strategies for treating HIV-1 infected individuals have focused thus far on clearing the virus from systemic circulation. Treatment regimens such as highly active antiretroviral therapy, or HAART, have successfully brought viral copy numbers down to nearly undetectable levels. However, cessation of drug therapy is often followed by a rapidly rebounding viral load, indicating that the virus has not been completely cleared from the body (16). HIV-1 can remain latent in any number of anatomical locations, including the brain, where it has been demonstrated by immunocytochemical staining. Latently infected, resting CD4+ T cells will also give rise to infectious HIV upon stimulation, revealing their role as a potential reservoir of HIV-1. If silently infected astrocytes do indeed harbor latent HIV, then total eradication of the virus from the infected individual must include eradicating the virus from the CNS as well as from the rest of the body.

In the late stages of HIV-1 infection, ADC will become a common complication. Both the pathological consequences and virus levels in the brain are highly variable, and do not necessarily correlate with clinical severity. Rather, the best correlate of the severity of dementia seems to be the number of immune activated macrophages and microglia in the brain. The pathways by which HIV-1 causes neurodegeneration are numerous and still under debate. The complex structural and chemical interactions between neurons and the glial cells in the brain suggest that infection in any one of the pathways will ultimately lead to neuronal damage or loss. Furthermore, it is becoming increasingly clear that although HIV-1 is the causative agent for ADC, the dementia stems from inflammatory cell infiltration and cytokine dysregulation, rather than the virus levels in the CNS.

REFERENCES

1. Brew BJ (1999). AIDS dementia complex. *Neurologic Clinics* 17: 861-81.
2. Zink WE, Zheng J, Persidsky Y, Poluektova L, and Gendelman HE (1999). The neuropathogenesis of HIV-1 infection. *FEMS Immunology and Medical Microbiology* 26: 233-241.
3. Scorziello A, Florio T, Bajetto A, and Schettini G (1998). Intracellular signaling mediating HIV-1 gp120 neurotoxicity. *Cellular Signaling* 10: 75-84.
4. Nath A, Haughey NJ, Jones M, Anderson C, Bell JE, and Geiger JD (2000). Synergistic neurotoxicity by human immunodeficiency virus proteins tat and gp120: protection by memantine. *Annals of Neurology* 47: 186-194.
5. Piller SC, Jans P, Gage PW, and Jans DA (1998). Extracellular HIV-1 virus protein R causes a large inward current and cell death in cultured hippocampal neurons: Implications for AIDS pathology. *Proceedings of the National Academy of Science, USA* 95: 4595-4600.
6. Shi B, Raina J, Lorenzo A, Busciglio J, and Gabuzda D (1998). Neuronal apoptosis induced by HIV-1 Tat protein and TNF-α: potentiation of neurotoxicity mediated by oxidative stress and implications for HIV-1 dementia. *Journal of NeuroVirology* 4: 281-290.
7. Kaul M and Lipton SA (1999). Chemokines and activated macrophages in HIV gp120-induced neuronal apoptosis. *Proceedings of the National Academy of Sciences, USA* 96: 8212-8216.
8. Zheng J, Thylin MR, Ghorpade A, Xiong H, Persidsky Y, Cotter R, Niemann D, Che M, Zeng Y, Gelbard HA, Shepard RB, Swartz JM, and Gendelman HE (1999). Intracellular CXCR4 signaling, neuronal apoptosis and neuropathogenic mechanisms of HIV-1 associated dementia. *Journal of Neuroimmunology* 98: 185-200.
9. Zheng J, Ghorpade A, Niemann D, Cotter RL, Thylin MR, Epstein L, Swartz JM, Shepard RB, Liu X, Nukuna A, and Gendelman HE (1999). Lymphotropic virions affect chemokine receptor-mediated neural signaling and apoptosis: Implications for human immunodeficiency virus type 1-associated dementia. *Journal of Virology* 73: 8256-8267.
10. Gorry P, Purcell D, Howard J, and McPhee D (1998). Restricted HIV-1 infection of human astrocytes: potential role of nef in the regulation of virus replication. *Journal of NeuroVirology* 4: 377-386.
11. Ranki A, Nyberg M, Ovod V, Haltia M, Elovaara I, Raininko R, Haapaasalo H, Krohn K (1995). Abundant expression of HIV Nef and Rev proteins in brain astrocytes in vivo is associated with dementia. *AIDS* 9: 1001-1008.
12. Brack-Werner R (1999). Astrocytes: HIV cellular reservoirs and important participants in neuropathogenesis. *AIDS* 13: 1-22.
13. Weiss JM, Nath A, Major EO, and Berman JW (1999). HIV01 Tat induces monocyte chemoattractant protein-1-mediated monocyte transmigration across a model of the human blood-brain barrier and up-regulates CCR5 expression on human monocytes. *Journal of Immunology* 163: 2953-2959.
14. Conant K, Garzino-Demo A, Nath A, McArthur JC, Halliday W, Power C, Gallo RC, and Major EO (1998). Induction of monocyte chemoattractant protein-1 in HIV-1 Tat-stimulated astrocytes and elevation in AIDS dementia. *Proceedings of the National Academy of Sciences, USA* 95: 3117-3121.

15. Lipton SA, Sucher NJ, Kaiser PK, and Dreyer EB (1991). Synergistic effects of HIV coat protein and NMDA receptor-mediated neurotoxicity. *Neuron* 7: 111-18.
16. Hatano H, Vogel S, Yoder C, Metcalf JA, Dewar R, Davey RT, and Polis MA (2000). Pre-HAART HIV burden approximates post-HAART viral levels following interruption of therapy in patients with sustained viral suppression. *AIDS* 14: 1357-1363.

CHEMOKINE RECEPTORS ON BRAIN ENDOTHELIA – KEYS TO HIV-1 NEUROINVASION?

Milan Fiala,[1] Chandrasekhar Gujuluva,[1] Omri Berger,[1] Michael Bukrinsky,[2] Kwang Sik Kim,[3] Michael C. Graves[1]

[1]Departments of Medicine and Neurology, UCLA School of Medicine, Los Angeles, CA 90095
[2]The Picower Institute, Manhasset, NY 11030
[3]Division of Infectious Diseases, Children's Hospital of Los Angeles, Los Angeles, CA 90027

INTRODUCTION

One of the many blood-brain barrier (BBB)'s functions is to stop invasion by microorganisms from the blood into the brain, which is effective with most viruses but fails in case of HIV-1. It is possible that the critical step in HIV-1 neuroinvasion is the presence of HIV-1 receptors on brain endothelia. The role of host factors is also critical in the invasion of the meninges by pathogenic bacteria, such as *E. coli*. For example, we have shown that several microbial determinants contribute to invasion of the central nervous system (CNS) and identified specific cellular receptors for these determinants, which are present only on brain microvascular endothelial cells (BMVEC), suggesting that *E.coli* invasion into the CNS may occur via ligand-receptor interactions (1, 2).

HIV-1 is circulating in the blood as cell-free and cell-associated virus. Since HIV-1 penetrates into the cerebrospinal fluid early after the primary infection during the asymptomatic phase, when the plasma contains a high concentration of cell-free virus, the cell-free virus may be an important source of the CNS infection. The critical steps in the neuroinvasion by cell-free virus may involve HIV-1 binding to the receptors on brain endothelia, intra-cellular signaling, cytokine induction and endothelial remodeling. We speculate that mitigating the consequences of HIV-1 signaling in endothelia, in particular the induction of inflammatory cytokines, might stop the early neuroinvasion and ameliorate degradation of the BBB. The BBB could perhaps better resist the transmigration of mononuclear blood cells into the CNS in the late stage of AIDS.

RESULTS

Transcellular and paracellular routes of HIV-1 invasion across brain endothelia

We have been investigating the mechanisms of viral neuroinvasion using BMVEC and an *in vitro* BBB model constructed with these cells. In this BBB model we determined the kinetics of viral neuroinvasion and cytokine induction. Initially we studied the effect of tumor necrosis factor–α (TNF-α) on viral penetration. We showed that TNF-α increased both inulin permeability (inulin is a marker for paracellular transport) and viral penetration, thus implying a paracellular route for viral invasion (3). Later we observed that cocaine (cocaine was tested for its possible role as a co-factor in neuroAIDS) mimicked the effects of TNF-α on paracellular transport and viral penetration (4).

We also demonstrated in this BBB model that the barrier against HIV-1 is well preserved for approximately 1 - 4 h post-exposure but then breaks down. At 48 - 96 hr post-exposure the degree of invasion in an untreated BBB model approaches that in a model treated with TNF-α (Table 1). The increase in viral penetration across the untreated model 24-48 hr post-infection was not accompanied by any change in the paracellular (inulin) permeability of the model (unpublished data), suggesting that HIV-1 penetration may occur by the transcellular pathway (5).

Table 1. HIV-1 invasion across the blood-brain barrier (BBB) model

Pre-treatment of the BBB model	Virus invasion into the lower chamber (HIV-1 RNA copies) 1h	24 h	48 h
None	316± 17	8,164 ± 4,627	33,996 ± 6,822
TNF-α (10 ng/ml)	3,708	28,136	53,824
TNF-α (100 ng/ml)	5,346	50,052	99,848

BBB models were constructed in T.C. 24-well inserts with human brain microvascular endothelial cells (BMVEC) that reached confluence and low inulin permeability. Each model was treated as shown for 3 hrs. One million RNA copies of HIV-$1_{JR\text{-}FL}$ were then placed in the upper chamber. After the indicated time intervals, samples were obtained from the lower chamber and the concentration of HIV-1 RNA copies was determined by the Amplicor test (Roche Co.).

HIV-1 chemokine co-receptors, CCR3 and CXCR4, are present and functional on brain endothelia

Next we used flow cytometry and confocal microscopy to investigate the presence of CCR3, CCR5, CXCR4, and other chemokine receptors on BMVEC. All three main HIV-1 co-receptors were found by flow cytometry, of which CCR3 and CXCR4 were at a high density and CCR5 at a low density (Fig.1). Confocal microscopy revealed that a large proportion of CCR5 and CXCR4 was found in the endothelial cytoplasm and only a small fraction was on the plasma membrane (Fig.2). That CCR3 and CCR5 (both are the receptors for RANTES) on brain endothelia could play a role in angiogenesis was suggested when a chemotactic gradient of RANTES, macrophage inflammatory protein-

1α (MIP-1α) or macrophage inflammatory protein-1β (MIP-1β) increased BMVEC migration (6). In coronary endothelia, SDF-1 (the ligand for CXCR4) activated ERK and JNK (unpublished data), suggesting a role of CXCR4 in endothelial signaling.

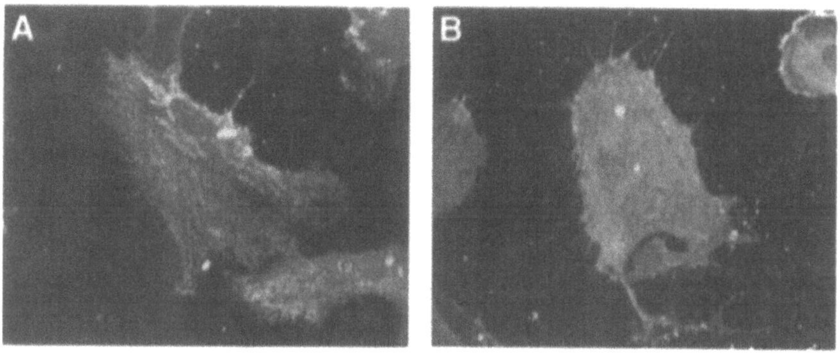

Fig.1 Chemokine receptor expression in BMVEC evaluated by flow cytometry.
Brain endothelial cells were lifted from the flask in a "cell-dissociation buffer" (GIBCO), washed, and stained using the monoclonal antibody to CCR3, CCR5 or CXCR4, and analyzed by flow cytometry. From Berger et al (1999) in Molecular Medicine 1999, with kind permission of The Picower Institute and the Plenum Press.

Fig.2 Confocal microscopy of subcellular localization of CXCR4 and CCR5 in brain endothelia.
Brain endothelia were grown on coverslips coated with glutaraldehyde-fixed gelatin. The cells were double-stained using, as the primary antibodies, monoclonal antibody to CXCR4 or CCR5 followed by the rabbit antibody to von Willebrand factor and, as the secondary antibodies, mouse and rabbit IgG labeled with FITC or Texas Red, respectively. From Berger et al (1999) in Molecular Medicine 5:795-805,1999, with kind permission of The Picower Institute and the Plenum Press.

Abortive HIV-1 infection of brain endothelia

Susceptibility of different cells to HIV-1 infection can range from complete resistance, to low-level and to high-level productive infection. The replication of the virus may be limited at the level of entry in those cells that lack CD4 or a chemokine co-receptor, but other blocks are well known to exist in subsequent steps of viral replication. Blocks to replication were shown at the step of reverse transcription, as demonstrated for quiescent $CD4^+$ T lymphocytes (7), nuclear translocation of the viral pre-integration complex, as shown for macrophages infected with certain X4 strains (8), and expression of the integrated provirus, as observed in U1 cells (9). We found BMVEC to be resistant to productive infection with the laboratory X4, R5, and X4R5 (dual-tropic) strains. The block to replication appears to occur at an early post-entry step, as early reverse transcription products could be detected in those cells, while no full-length DNA copies of the viral genome or intranuclear 2-LTR circle forms were found [Gujuluva, submitted for publication]. In particular, no evidence was found for virus transmission across the BMVEC barrier by budding on the abluminal side of brain endothelia, since no infectious virus was made in these cells.

HIV-1 signaling by mitogen-activated protein (MAP) kinase pathway in BMVEC

Five min after exposure, HIV-1 activates in BMVEC the MAP kinase pathway by inducing phosphorylation of mitogen-activated protein (MAP) kinase kinase (MEK), extracellular signal-regulated kinases 1 and 2 (ERK1/2), c-*Jun* N-terminal kinase, and p38 MAP kinase. Signaling via ERK1/2 phosphorylation is inhibited by pertussis toxin, demonstrating that HIV-1 signaling in BMVEC is dependent upon a G-protein coupled receptor (GPCR), such as a chemokine receptor (Fig. 3).

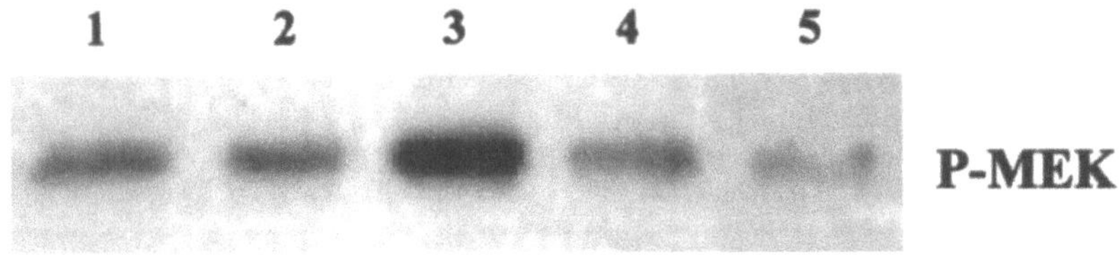

Fig.3 . Activation of MEK in BMVEC by HIV-1 and inhibition of MEK activation by pertussis toxin. 500,000 BMVEC in replicate 60 mm dish cultures were starved overnight and either not pre-treated (Lanes 1,2,3,and 4) or pre-treated with pertussis toxin (100 ng/ml for 90 min) (Lane 5). The cells were then either exposed to control medium for 5 or 15 min, respectively (Lanes 1 and 2) or HIV-1_{NL4-3} (15,000 copies per cell) for 5 min (lanes 3 and 5) or 15 min (Lane 4).

HIV-1 induces IL-6 in BMVEC

Beginning about 16 hr after HIV-1 exposure of BMVEC, HIV-1 exposure stimulates release of a high concentration of IL-6, as determined by ELISA (unpublished data). IL-6 was induced by HIV-1 in a prolonged fashion (up to 96 hr post-infection) despite absence of virus replication. Hofman et al (10) previously demonstrated that Tat may induce release of biologically active IL-6 in brain endothelia. Cytokine, such as IL-6, and other mediators induced by the virus or viral peptides could lead to structural remodeling of endothelia, which may underlie the decreased barrier function of endothelia for HIV-1 and possibly for monocyte/macrophages. Consequently, we are studying the effects of IL-6 on HIV-1 invasion across the BBB model.

Cocaine's role in HIV-1 neuroinvasion

Cocaine is an abused drug, which could contribute to the degradation of the BBB based on its ability to increase permeability of the BBB model (4). Our recent work suggests that cocaine may impair the BBB integrity by its effects on the actin cytoskeleton. Furthermore, cocaine may increase apoptosis of brain endothelia (4).

Our previous work also suggested that cocaine could influence HIV-1 neuroinvasion by modulation of mononuclear/endothelial cell interactions (11) as well as Thl/Th2 cytokine balance (12), the latter involving effects on the hypothalamic-pituitary-adrenal axis (13). *In vitro* cocaine was shown to enhance endothelial adhesion molecules, intercellular adhesion molecule-1 (ICAM-1), vascular cell adhesion molecule-1 (VCAM-1), and endothelial leukocyte adhesion molecule-1 (ELAM-1), on BMVEC. Furthermore, cocaine also increased secretion of TNF-α and interleukin-6 (IL-6) in BMVEC/monocyte cocultures, and of CC- {macrophage inflammatory protein-1α (MIP-1α) and monocyte chemoattractant protein-1 (MCP-1)} and CXC- chemokines {interleukin-8 (IL-8) and interferon-inducible protein-10 (IP-10)} in monocyte cultures (4). Finally, cocaine treatment enhanced monocyte migration across the BBB model (11).

In subjects addicted to cocaine, cocaine infusion increased interferon-γ (IFN-γ) and decreased interleukin-10 (IL-10) secretion by peripheral blood mononuclear cells (12). Accordingly, the modulation of the immune system and the BBB by cocaine could increase virus entry into the CNS by several mechanisms. First, the increase in IFN-γ could modulate chemokine receptor expression on endothelia, as was shown in case of CCR1 and CCR3 in neutrophils (14). Second, the increase in IL-6 secretion produced by cocaine could act synergistically with the virus-induced IL-6 to modulate the BBB function. Third, the induction of chemokines and pro-inflammatory cytokines could facilitate mononuclear cell transmigration into the CNS. Fourth, the damage to the barrier function would allow direct entry of the virus into the brain.

CONCLUSIONS

HIV-1 binding to receptors on brain endothelia, possibly CCR3 and CXCR4, may initiate a cascade of effects that culminate in increased transcellular penetration of the BBB by the virus and invasion of the CNS. Cocaine may open the paracellular route across the BBB and modulate the immune system to increase HIV-1 neuroinvasion. The BBB is, indeed, a barrier that is worth preserving against virus neuroinvasion.

ACKNOWLEDGMENTS

The preparation of this review was supported by the NIH grants DA10442, HL63065, HL63639, and the NS 26126 subcontract to MF, and RO1 NS 26310 and HLK61951 to KSK.

REFERENCES

1. Huang, S. H., C. Wass, Q. Fu, N. V. Prasadarao, M. Stins, and K. S. Kim. 1995. Escherichia coli invasion of brain microvascular endothelial cells in vitro and in vivo: molecular cloning and characterization of invasion gene ibe10. *Infect Immun 63:4470.*

2. Prasadarao, N. V., C. A. Wass, S. H. Huang, and K. S. Kim. 1999. Identification and characterization of a novel Ibe10 binding protein that contributes to Escherichia coli invasion of brain microvascular endothelial cells. *Infect Immun 67:1131.*
3. Fiala, M., D. J. Looney, M. Stins, D. D. Way, L. Zhang, X. Gan, F. Chiappelli, E. S. Schweitzer, P. Shapshak, M. Weinand, M. C. Graves, M. Witte, and K. S. Kim. 1997. TNF-alpha opens a paracellular route for HIV-1 invasion across the blood-brain barrier. *Molecular Medicine 3:553.*
4. Zhang, L., D. Looney, D. Taub, S. L. Chang, D. Way, M. H. Witte, M. C. Graves, and M. Fiala. 1998. Cocaine opens the blood-brain barrier to HIV-1 invasion. *J Neuro Virol 4:619.*
5. Bishop, N. E. 1997. An Update on Non-clathrin-coated Endocytosis. *Rev Med Virol 7:199.*
6. Berger, O. 1999. Chemokine receptors on coronary and brain endothelia. *submitted.*
7. Zack, J. A., S. J. Arrigo, S. R. Weitsman, A. S. Go, A. Haislip, and I. S. Chen. 1990. HIV-1 entry into quiescent primary lymphocytes: molecular analysis reveals a labile, latent viral structure. *Cell 61:213.*
8. Schmidtmayerova, H., M. Alfano, G. Nuovo, and M. Bukrinsky. 1998. HIV-1 T-lymphotropic strains enter macrophages via a CD4+-and CXCR-4-mediated pathway: Replication is restricted at a postentry level. *J Virol 72:4633.*
9. Pomerantz, R. J., D. Trono, M. B. Feinberg, and D. Baltimore. 1990. Cells nonproductively infected with HIV-1 exhibit an aberrant pattern of viral RNA expression: a molecular model for latency. *Cell 61:1271.*
10. Hofman, F., M. Dohadwala, A. Wright, D. Hinton, and S. Walker. 1994. Exogenous tat protein activates central nervous system-derived endothelial cells. *Journal of Neuroimmunology 54:19.*
11. Gan, X., L. Zhang, D. Taub, S. Chang, M. Stins, K. Kim, D. Way, M. Weinand, M. Witte, M. Graves, and M. Fiala. 1999. Cocaine enhances endothelial adhesion molecules and leukocyte migration. *Clin Immunol 91:68.*
12. Gan, X., L. Zhang, T. Newton, S. Chang, W. Ling, V. Kermani, O. Berger, M. Graves, and M. Fiala. 1998. Cocaine infusion increases interferon-gamma and decreases interleukin-10 in cocaine-dependent subjects. *Clin Immunol Immunopathol 89:181.*
13. Gan, X., L. Zhang, C. Metz, R. Bucala, S. Chang , V. Kermani, T. Newton, and M. Fiala. 1998. Modulation of cortisol, DHEA and MIF by IV cocaine. *FASEB J.*
14. Bonecchi, R., N. Polentarutti, W. Luini, A. Borsatti, S. Bernasconi, M. Locati, C. Power, A. Proudfoot, T. N. Wells, C. Mackay, A. Mantovani, and S. Sozzani. 1999. Up-regulation of CCR1 and CCR3 and induction of chemotaxis to CC chemokines by IFN-gamma in human neutrophils. *J Immunol 162:474.*

NEUROTROPHIC FACTOR REGULATION OF HUMAN IMMUNODEFICIENCY VIRUS TYPE 1 REPLICATION IN HUMAN BLOOD-DERIVED MACROPHAGES THROUGH MODULATION OF CORECEPTOR EXPRESSION

Sharon M. Harrold,[1] Joanna M. Dragic, Sarah L. Brown, and Cristian L. Achim

Department of Pathology
Division of Neuropathology
University of Pittsburgh
School of Medicine
Pittsburgh, PA 15213

[1] Corresponding Author:
Phone: (412) 647-7127
Fax: (412) 647-5468
Email: harrold@np.awing.upmc.edu

INTRODUCTION

Macrophages are a major target for human immunodeficiency virus type-1 (HIV-1) infection in vivo and represent a reservoir of chronically infected cells in patients receiving highly active retroviral therapy [1,2]. The entry of virus into cells is mediated by binding of the viral glycoprotein gp120 with CD4 and chemokine receptors. The T-cell-tropic strains of HIV-1 utilize the CXC chemokine receptor 4 (CXCR4), while the CC chemokine receptor 5 (CCR5) predominantly binds macrophage-tropic (M-tropic) HIV-1 strains (reviewed in Rucker [3]). Increased cell surface expression of CCR5 during monocyte maturation to macrophages correlates with susceptibility to HIV-1 infection [4]. However, efficient usage of CCR5 and CXCR4 as CD4-independent HIV-1 receptors was recently shown [5,6]. Interestingly, primary isolates and dual-tropic stains of HIV-1 can utilize a plethora of additional G protein coupled coreceptors including CXCR4, CCR5, and the β-chemokine receptors, CCR3 or CCR2b, among others, to gain access into primary T lymphocytes and monocytic lineage cells [3].

Nerve growth factor (NGF) and brain derived neurotrophic factor (BDNF) are members of the neurotrophic factor (NTF) family that are known to influence neuronal growth, differentiation and survival. NGF modulates the expression of enzymes that remove free radicals, while BDNF regulates calcium concentration via CREB [7]. Expression of neurotrophic factors and their receptors have been observed in *in vitro* cultured microglia and macrophages. Our laboratory recently determined that HIV-1 infection induced an increased expression of NTF mRNA in cultured human macrophages [8]. Additionally, we observed increased BDNF reactivity on striatal activated brain macrophages in conjunction with trkB immunoreactive astrocytes in regions of HIVE histopathology [9]. NGF was recently determined to function as an autocrine factor essential for the survival of HIV-1 infected macrophages [10]. Additionally, the role of neurotrophic factors has been expanded to include differentiation and survival of hematopoietic cells [11] and to prevent the initiation and propagation of immune responses by microglia [12]. Thus the aberrant expression of neurotrophic factors and their receptors could have wide-ranging effects on CNS and immunological function.

The hypothesis that neurotrophic factor treatments may alter expression of viral receptors influencing susceptibility of blood-derived macrophages to HIV-1 infection was examined by semiquantitative analyses of viral and cellular RNA expression and by production of viral protein from infected, neurotrophic factor treated cultured macrophages as compared to control cells.

METHODS

Peripheral blood mononuclear cells (PBMC) were obtained by discontinuous gradient centrifugation (Hypaque-Ficol, Sigma) of heparinized blood from healthy HIV-seronegative donors. Monocytes were enriched from allogeneic stimulated PBMC by overnight adherence of 1.8×10^7 cells (resuspended at 1.5×10^6 cells/ml) to plastic 75 cm^2 culture flasks in complete medium (AIM-V (Gibco-BRL) plus 10% human AB serum that was negative for HIV and hepatitis B virus antibodies (Sigma), 100 U penicillin per ml, and 100μg of streptomycin per ml), followed by extensive rinsing with isotonic saline. Monocyte differentiation into macrophages (MDM) was evaluated by morphological criteria and immunofluorescent analysis, which determined <5% contamination with cells that were not immunoreactive with a macrophage specific antibody (anti-CD68).

Cells were cultured for 4 days to allow for optimal cell surface expression of chemokine receptors [4] prior to treatment with 5 or 50 ng/ml of nerve growth factor (NGF, Promega) or brain derived growth factor (BDNF, Promega) in complete medium. Neurotrophic factor treatments were replenished every 2 days. Parallel, untreated cultures served as controls. Half of the cultures were infected after 7 days with the macrophage-tropic isolate $HIV_{ADA\text{-}M}$, which was obtained from the National Institutes of Health AIDS Research and Reference Reagent Program. Viral stocks were produced in MDM and stored in aliquots at -80°C. The viral titer was quantified by p24 antigen capture enzyme-linked immunosorbent assay (ELISA, Coulter). The limit of detection of HIV p24 antigen was 10 pg/ml. MDM cells were inoculated with a minimum of 3 ng of p24 cell-free virus in M-SFM (Gibco-BRL) in the presence of neurotrophic factors. Independent uninfected cultures served as controls. Virus was adsorbed overnight, followed by thorough washing and addition of neurotrophic factor containing medium. Culture supernatants were stored for quantification by viral p24 antigen capture ELISA.

To determine the levels of RNA expression of GAPDH, CD4, CCR5, CXCR4, HIV-1 and HLA-DR in neurotrophic factor treated or untreated MDM, RT-PCR was performed with oligonucleotide primers specific for each RNA species. Briefly, total RNA was isolated with Trizol (Gibco-BRL) according to the manufacture's instructions and 1μg was used for cDNA synthesis with Superscript II (Gibco-BRL) and random hexamers (Gibco-BRL) at 42°C for 1 hr, followed by inactivation at 85°C for 5 min. For PCR amplification, preliminary experiments were performed to determine the amount of cDNA that was within the linear range of the amplification reaction. A master mix consisting of 1X PCR buffer

(ABI), 0.2μM of each primer, 200μM dNTP (Pharmacia), water, 0.025U/μl Taq DNA polymerase (Gibco-BRL) and 0.05μCi/μl of α-^{33}P dATP (2000Ci/mmol, NEN) was added to each amplification reaction. The thermal cycle profile was denaturation at 95°C for 30 sec., annealing at 50-55°C for 1 min., extension at 72°C for 45 sec for 30-35 cycles, followed by a 7 min, 72°C extension. Following PCR amplification, the products were separated on an 8% polyacrylamide gel, dried and the radiolabel incorporated was determined by phosphor imaging technology (Molecular Dynamics, Sunnyvale, CA). The respective primer sequences were: GAPDH, 5' TGGTATCGTGGAAGGACTCATGAC, 3' ATGCCAGTGAGCTTCCCGTTCAGC; CD4, 5' GCTAGGCATCTTCTTCTGTG, 3' CTGCTACATTCATCTGGTCC; CCR5, 5' CCTGTTTTTCCTTCTTACTG, 3' GTGGCT-CTTCTTCTCATTTC; CXCR4, 5' AAATGGGCTCAGGGGACTAT, 3' GGAACACA-ACCACCCACAAG; HIV-1, SK38 ATAATCCACCTATCCCAGTAGGAGAAAT, SK39 TTTGGTCCTTGTCTTATGTCCAGAATGC. RT-PCR analyses of glyceraldehyde-3-phosphate dehydrogenase (GAPDH) mRNA were used for normalization. The amount of radiolabel incorporated during amplification of the target gene was calculated and used to directly compare the levels of RNA expression. Each measurement was performed at least in triplicate and the statistical significance was determined by two-tailed, unpaired Student t test of three independent experiments.

RESULTS

Effects of NGF and BDNF on HIV replication in macrophages

To determine the effects of NGF and BDNF treatment on the replication of HIV-1, MDM cultures were inoculated with the macrophage-tropic strain, HIV_{ADA-M}. The level of viral RNA in infected cells was assayed at 4 days post-infection by semiquantitative RT-PCR. Viral RNA was increased about twofold in cultures treated with low doses of BDNF ($P < 0.03$), while MDM grown in the presence of 50 ng/ml BDNF harbored HIV RNA comparable to untreated cultures (Fig. 1). In contrast, NGF treatment resulted in an approximately twofold decrease in HIV viral RNA ($P < 0.01$ and < 0.001 for 5 and 50 ng NGF per ml, respectively) (Fig. 1). These results indicated discrete modulation of viral replication by NGF and BDNF in macrophages.

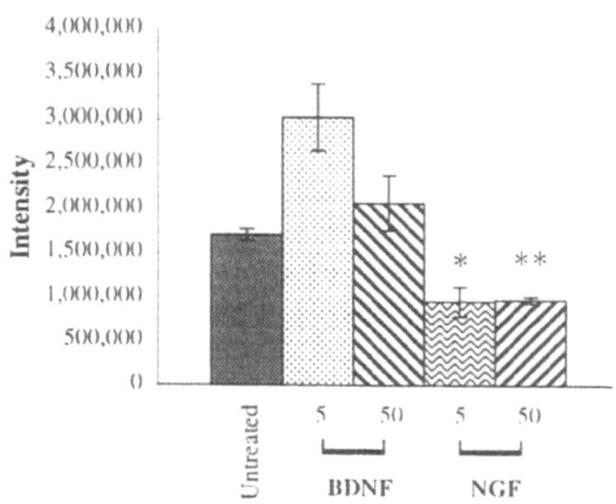

Figure 1 Modulation of HIV-1 RNA expression in cultured MDM treated with neurotrophins. Relative expression of viral RNA levels were analyzed by semiquantitative RT-PCR analyses normalized for GAPDH RNA levels. MDM (4 day cultures) were treated with 5 ( ) or 50 ng/ml ( ) of BDNF and 5 ( ) or 50 ng/ml ( ) NGF for 2 days prior to overnight HIV_{ADA-M} infection. RNA was isolated from NTF treated and control cultures ( ) at 4dpi. All data are represented as the mean and SEM from three independent experiments, assayed at least twice. *, $p < 0.01$; **, $p < 0.001$ compared to control cultures.

To determine whether viral RNA levels are indicative of viral replication, the levels of p24 antigen in culture supernatants was assayed at 1 and 4 days post-infection. At 1 day post-infection, very low amounts of virus were produced (Fig. 2). Interestingly at 4 days post-infection, prior to amplification of viral replication, similar levels of p24 were determined in all treatment conditions.

Relationship between viral replication and expression of chemokine receptors following neurotrophic factor treatments

The susceptibility of monocytes and MDM to HIV-1 infection has been shown to correlate with the level of receptor and coreceptor expression. To determine the effects of neurotrophic factor treatment on the expression of HIV-1 receptors in macrophages, MDM cultures were maintained in 5 or 50ng/ml of BDNF or NGF for 7 days. Total cellular RNA was isolated and analyzed by semiquantitative RT-PCR. Transcripts encoding CD4, CXCR4 and CCR5 were detected (Fig. 3), confirming reports from other laboratories [4].

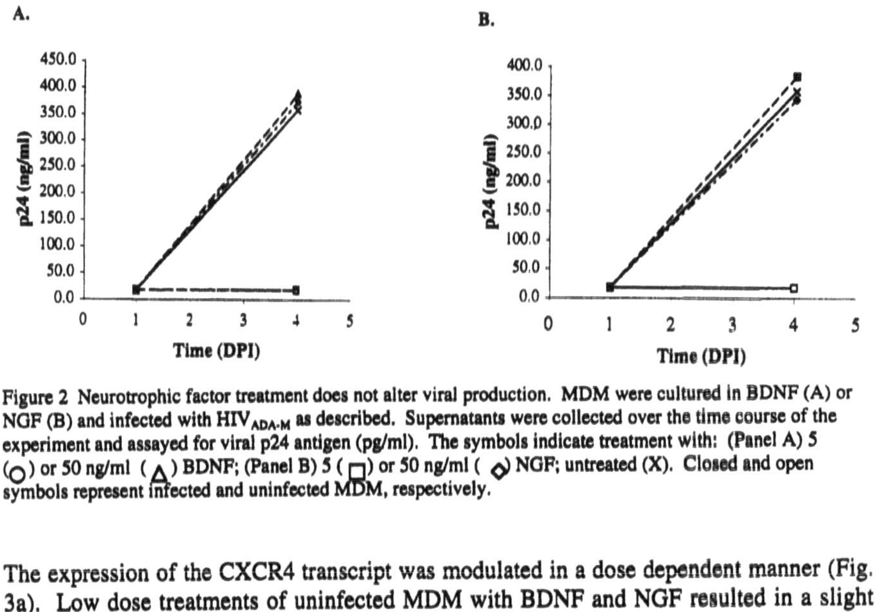

Figure 2 Neurotrophic factor treatment does not alter viral production. MDM were cultured in BDNF (A) or NGF (B) and infected with HIV_{ADA-M} as described. Supernatants were collected over the time course of the experiment and assayed for viral p24 antigen (pg/ml). The symbols indicate treatment with: (Panel A) 5 (○) or 50 ng/ml (△) BDNF; (Panel B) 5 (□) or 50 ng/ml (◇) NGF; untreated (X). Closed and open symbols represent infected and uninfected MDM, respectively.

The expression of the CXCR4 transcript was modulated in a dose dependent manner (Fig. 3a). Low dose treatments of uninfected MDM with BDNF and NGF resulted in a slight increase in CXCR4 mRNA, which was reduced twofold in cells cultured in 50ng/ml neurotrophic factor. In contrast, CCR5 mRNA levels were increased by up to 6-fold in 5 ng/ml BDNF and decreased 2-fold by NGF treatments compared to control untreated cultures (Fig. 3b, $P < 0.001$). Interestingly, CD4 RNA levels in HIV-1 infected cells were not significantly effected by NGF and BDNF treatments (Fig 3c), similar to HLA-DR RNA levels (data not shown). These results suggested that neurotrophic factors specifically modulated chemokine receptors CXCR4 and CCR5 at the transcription level.

Neurotrophic factor pretreated MDM cultures were subsequently infected with the macrophage-tropic strain, HIV_{ADA-M}, and analyzed for the abundance of viral receptor and cellular mRNA levels. While the absolute level of RNA expression was typically different, the general mRNA patterns in virally infected cells recapitulated the uninfected transcript levels (Fig. 3), indicating persistent activation of distinct BDNF and NGF cellular

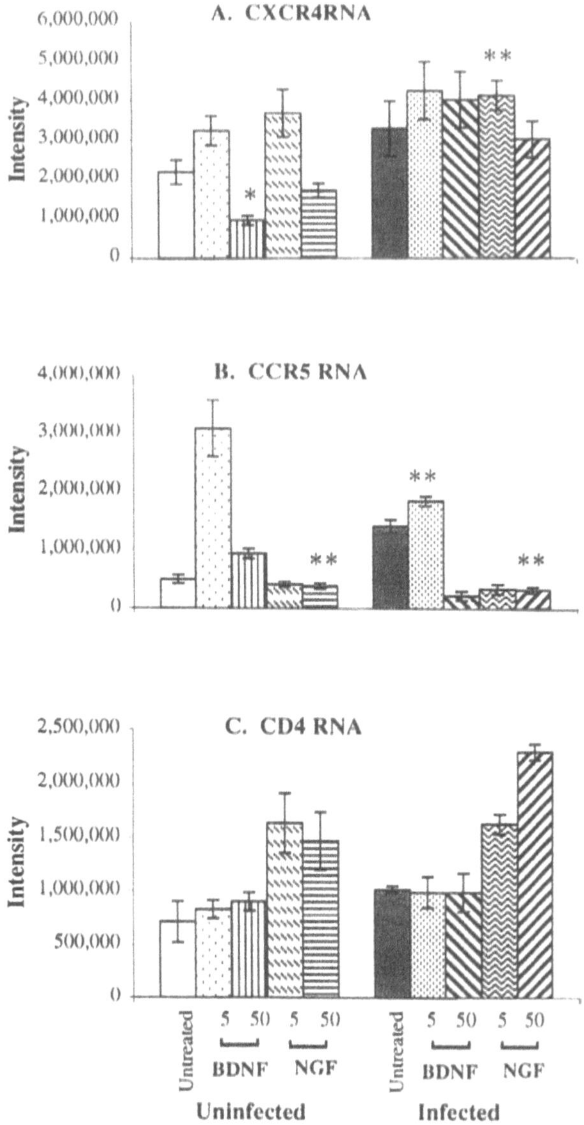

Figure 3. Neurotrophic factor treatment specifically alters chemokine RNA levels. Relative expression of (A) CXCR4, (B) CCR5 and (C) CD4 RNA was analyzed by semiquantitative RT-PCR expression as described above. Uninfected MDM (control (□), treated with 5 (⊡) or 50 ng/ml (▥) BDNF and 5 (▧) or 50 ng/ml (⊟) NGF) and HIV_{ADA}- infected NTF treated cultures (symbols as in Figure 1) are indicated by the left and right histograms, respectively.

pathways. Comparison of the viral RNA and receptor expression revealed that the pattern of expression of CCR5 paralleled that of HIV-1.

DISCUSSION

The results of these experiments supports the hypothesis that neurotrophic factor treatments specifically modulate the expression of HIV-1 and chemokine receptors transcripts (Fig. 1 and 3), suggesting that the local neurotrophic factor milieu may influence the susceptibility of blood-derived macrophages to viral infection. Interestingly, viral production as measured by antigen capture ELISA was insensitive to neurotrophic factor treatments, which could be due to differences in the sensitivity of RNA and protein assays or alternatively to differences in the stability of each species. Examination of the protein levels and cell surface expression of viral receptors and immunomodulatory molecules is currently being investigated.

NGF was recently shown to prevent the initiation and propagation of immune responses [12], and to downregulate costimulatory molecules [13] in brain macrophages in addition to amplification of Th2 responses [14]. Interestingly, inflammatory responses lead to increased NGF secretion [15], suggesting autocrine / paracrine regulation of immune stimulation by NGF. Recently, NGF has been determined to function as an autocrine factor essential for the survival of HIV-1 infected macrophages [10]. Taken together with the current results, this suggests that NGF mediated decrease of virus production may modulate immune mediated clearance of infected cells or production of inflammatory cytokines in sites with active immunological surveillance, such as the spleen and lymph nodes. In contrast, the decreased viral transcription in the central nervous system could limit local immune activation in perivascular areas, contributing to establishment of the brain as a reservoir of infected cells.

CD4 independent primate lentiviral isolates were recently shown to efficiently use CCR5 and CXCR4 as CD4-independent HIV-1 receptors [5,6]. Here we report that viral RNA levels in neurotrophic factor treated MDM vary independently of CD4 and CXCR4 transcripts, suggesting that CCR5 may function as a CD4-independent receptor. Our laboratory has previously shown increased expression of BDNF in macrophages in regions of pathology in HIV brain tissues [9]. The application of low doses of BDNF to macrophages resulted in a significantly increased expression of CCR5 RNA (Fig. 3) accompanied by an increase in viral RNA (Fig 1). Thus increased expression of BDNF in the local CNS microenvironment would result in increased CCR5 expression mediating spread of viral infection in a CCR5-dependent, CD4-independent manner.

Interestingly, while the absolute level of RNA expression varied, the patterns of RNA expression are generally recapitulated in infected and uninfected cells, suggesting persistent activation of distinct BDNF and NGF signal transduction pathways. A recent report determined that all members of the neurotrophin family activated NF-κB in microglia [11]; however, discrete regulation of chemokine and cellular transcripts by BDNF and NGF (Fig. 3) suggests additional signal transduction substrate targets. Results from these experiments have therapeutic implications for HIV infection. A better understanding of the distinct mechanisms of neurotrophic factor regulation of chemokine and viral RNA metabolism is required to effectively target therapies. The present report establishes neurotrophic factors as potential modulators of HIV infection in macrophages.

ACKNOWLEDGEMENTS

This work was supported by National Institutes of Health grant R01 MH 58528. S.M.H. was supported by NIMH training grant 5T32 MH 18273.

REFERENCES

1. Fauci, A. S. Host factors and the pathogenesis of HIV-induced disease. *Nature* **384**, 529-534 (1996).

2. Chun, T. W. and Fauci, A. S. Latent reservoirs of HIV: obstacles to the eradication of virus. *Proc. Natl. Acad. Sci. U S A* **96**, 10958-10961 (1999).

3. Rucker, J. and Doms, R. W. Chemokine receptors as HIV coreceptors: implications and interactions. *AIDS Res. Hum. Retrovir.* **14 Suppl 3**, S241-246 (1998).

4. Naif, H. M., Li, S., Alali, M., Sloane, A., Wu, L., Kelly, M., Lynch, G., Lloyd, A. and Cunningham, A. L. CCR5 expression correlates with susceptibility of maturing monocytes to human immunodeficiency virus type 1 infection. *J. Virol.* **72**, 830-836 (1998).

5. LaBranche, C. C., Hoffman, T. L., Romano, J., Haggarty, B. S., Edwards, T. G., Matthews, T. J., Doms, R. W. and Hoxie, J. A. Determinants of CD4 independence for a human immunodeficiency virus type 1 variant map outside regions required for coreceptor specificity. *J. Virol.* **73**, 10310-10319 (1999).

6. Kolchinsky, P., Mirzabekov, T., Farzan, M., Kiprilov, E., Cayabyab, M., Mooney, L. J., Choe, H. and Sodroski, J. Adaptation of a CCR5-using, primary human immunodeficiency virus type 1 isolate for CD4-independent replication. *J. Virol.* **73**, 8120-8126 (1999).

7. Finkbeiner, S., Tavazoie, S. F., Maloratsky, A., Jacobs, K. M., Harris, K. M. and Greenberg, M. E. CREB: a major mediator of neuronal neurotrophin responses. *Neuron* **19**, 1031-1047 (1997).

8. Achim, C. L. and Wiley, C. A. Inflammation in AIDS and the role of the macrophage in brain pathology. *Curr. Opin. Neurol.* **9**, 221-225 (1996).

9. Soontornniyomkij, V., Wang, G., Pittman, C. A., Wiley, C. A. and Achim, C. L. Expression of brain-derived neurotrophic factor protein in activated microglia of human immunodeficiency virus type 1 encephalitis. *Neuropathol. Appl. Neurobiol.* **24**, 453-460 (1998).

10. Garaci, E. *et al.* Nerve growth factor is an autocrine factor essential for the survival of macrophages infected with HIV. *Proc. Natl. Acad. Sci. U S A* **96**, 14013-14018 (1999).

11. Nakajima, K., Kikuchi, Y., Ikoma, E., Honda, S., Ishikawa, M., Liu, Y. and Kohsaka, S. Neurotrophins regulate the function of cultured microglia. *Glia* **24**, 272-289 (1998).

12. Neumann, H., Misgeld, T., Matsumuro, K. and Wekerle, H. Neurotrophins inhibit major histocompatibility class II inducibility of microglia: involvement of p75 neurotrophin receptor. *Proc. Natl. Acad. Sci. USA* **95**, 5779-5784 (1998).

13. Wei, R. and Jonakait, G. M. Neurotrophins and the anti-inflammatory agents interleukin-4 (IL-4), IL-10, IL-11 and transforming growth factor-beta1 (TGF-beta1) down-regulate T cell costimulatory molecules B7 and CD40 on cultured rat microglia. *J. Neuroimmunol.* **95**, 8-18 (1999).

14. Braun, A., E., E. A., Baruch, R., Herz, U., Botchkarev, V., Paus, R., Brodie, C. and Renz, H. Role of nerve growth factor in a mouse model of allergic airway inflammation and asthma. *Eur. J. Immunol.* **28**, 3240-3251 (1998).

15. Heese, K., Hock, C. and Otten, U. Inflammatory signals induce neurotrophin expression in human microglial cells. *J. Neurochem.* **70**, 699-707 (1998).

DIRECT VS. INDIRECT MODULATION OF COMPLEX *IN VITRO* HUMAN RETROVIRAL INFECTIONS BY MORPHINE

Susan Bell Nyland, Steven Specter, and Kenneth E. Ugen

Dept. of Medical Microbiology and Immunology
University of South Florida College of Medicine
Tampa, Fl 33612

BACKGROUND

The human T-cell leukemia virus type I (HTLV-I) was first characterized by Gallo in 1981 [1]. This type C oncoretrovirus was thought to be confined to well-delineated regions, passed primarily by long-term exposure (i.e., mother to infant, long-term marital relationships) or by frequent exposure to large numbers of infected cells. No recent large-scale epidemiological studies showing the incidence of HTLV-I infection in the United States are available, however, it is estimated that current infection rates are somewhat higher than previously believed. A significant contribution to the estimated increase in HTLV-I infection in the U.S., as well as to the confirmed increases worldwide, is associated with injecting drug users (IDU) [1, 2]. This route of infection is more efficiently exploited by human immunodeficiency virus type 1 (HIV-1). Both retroviruses target the same CD4+ lymphocytes, however an opposite set of outcomes is observed with HTLV-I versus HIV-1 infection. The decimating influence of HIV-1 on CD4+ T cell populations has been well documented and is thought to be responsible for the onset of AIDS. In contrast, a hallmark of active HTLV-I infection is the inappropriate proliferation of CD4+ T cells, expressed as HLTV-I associated myelopathy (TSP/HAM) or as adult T cell lymphoma (ATL). Additional disorders related to HTLV-I infection are also dependent upon uncontrolled T cell proliferation.

Although the characteristics of infection are different, the lentivirus (HIV-1) and the oncoretrovirus (HTLV-I) produce somewhat analogous proteins that have been reported to act in a complementary fashion *in vitro*. Thus, the possibility that superinfection with HIV-1 and HTLV-I in an individual might enhance the progression of HIV-1 infection was investigated during several large-scale epidemiological studies. Earlier case studies using small numbers of coinfected patients suggested that symptoms of complex retroviral infections (dual infection with HIV-1 and HTLV-I) were different from simple HIV-1 infection. However, the larger studies failed to demonstrate a consistent reduction in the time to progression to AIDS [3]. Thus, it appeared that HTLV-I did not enhance the progression of HIV-1 in the body.

A curious feature of dual infection was noted by several researchers, yet was not pursued. Patients coinfected with HIV-1 and HTLV-I were able to maintain higher CD4+ T cell counts, as well as a higher CD4: CD8 ratio, even though progression to AIDS continued at a normal pace [4]. Beilke et al [5] first publicly postulated that this quality of complex infection was due to enhancement of HTLV-I infection. He then demonstrated how HTLV-1-infected cells grew faster in HIV-infected PBMC than in non-infected cultures [6].

Because of opposing cytopathic characteristics attributed to HIV-1 and HTLV-I, the finding that one virus and not the other might be enhanced *in vivo* may impact the treatment of coinfected patients. For example, immunotherapies that seek to restore Type I cytokine levels (such as interleukin 2 and interferon gamma) during later HIV-1 infection might exacerbate the cytopathic effects of HTLV-I [1, 7]. Currently, these effects remain unknown.

Populations with complex retroviral infections are generally centered around regions previously endemic for HTLV-I, and in the IDU and their contacts. The increasing rate of complex infections in IDU in the United States compelled us to examine whether the presence of an opiate (morphine) might also alter the characteristics of *in vitro* HIV-1/ HTLV-I coinfection. Our *in vitro* studies suggest that like Beilke's examination of coinfected PBMC, HTLV-I infection was enhanced in the presence of HIV-1 infection, with or without the addition of morphine. The *in vitro* effect of morphine exposure on either virus alone may be greatly affected by the presence of absence of the host cell, as the following experiments with HIV-1 infection demonstrate.

MATERIALS AND METHODS

Cell Lines. Cell lines were obtained from the American Tissue Culture Collection (www.ATCC.com) and from the AIDS Research and Reference Reagent Repository at NIH (Rockville, MD). H-9, Sup-T1, and Jurkat cells (ATCC) were used as non-infected human T cell lines. H-9MN and MT2 human T cell lines (NIH) were used as HIV-1MN-infected cells (H9-MN) and as HTLV-I-infected cells (MT2) respectively. In addition, the H-9MN cell line was also used to propagate cell-free HIV-1MN, as described below. All cell cultures were grown and maintained in R10 medium: RPMI-1640 (Cell-Gro by Fisher Scientific Co. (www.fishersci.com)) with 1% antibiotic solution (Sigma Chemical Co, St. Louis, MO) and 10% Hyclone fetal bovine serum (Logan, UT).

Propagation and Use of Cell-Free HIV-1MN. In order to maintain consistency in the characteristics of the virus used in during both cell-associated and cell-free *in vitro* HIV-1 challenge, it was decided to propagate the cell-free virus directly from the infected cell line used in the study. HIV-1MN was propagated as described previously [8]. Infectivity titers were determined by the method of Dulbecco [9]. The final $TCID_{50}$ /ml was calculated match the final number of cells/ ml, thus making the M.O.I. 1:1.

Opiates. Morphine Sulfate (Sigma) was measured and diluted according to molecular weight in R10 to achieve the final desired molar concentrations of the drug after further dilution by cells, virus cultures, and antagonist. Naloxone (Sigma) was used as the antagonist due to its ability to interact with multiple μ receptors. The highest non-toxic level of naloxone was determined by cytotoxicity assays of the cell lines as previously described [8]. The highest final molar concentration of morphine was 1 X 10^{-6} M and was reduced by log amounts (1 X 10^{-7}M, 1 X 10^{-8}M) for a total of three molar concentrations. The final concentration of the antagonist naloxone was 2 X 10^{-5}M and remained consistent throughout. It should be noted that naloxone was used as an antagonist by addition to morphine-treated cultures. Naloxone alone was added to the naloxone-treated controls, which did not contain morphine. This was done in order to monitor any effects, cytotoxic or otherwise, that the antagonist might have had on the cells used in the following experiments.

Simple (Single Virus) Infections. Cell-associated simple retrovirus infections were accomplished by
co-culturing infected cells with non-infected targets. Equivalent numbers of cells were combined to a total final concentration of 2 X 10^5 cells/ml, and immediately exposed to morphine-treated medium or to medium containing both morphine and naloxone. Non-treated controls were also maintained in order to observe the progression of infection. Additional single cultures were left untreated and used as overall viability controls for the respective cell lines.

Simple HIV-1 cell-associated infection was continued for three days, using H-9MN cells as the infected line, and Sup-T1 cells as the targets. Simple HTLV-I infection was continued for five days, using Sup-T1 cells as targets.

HIV-1MN was propagated as described above for use as a cell-free virus in simple and complex infections. For simple infections, the cell-free virus was incubated with target cells for two hours, then non-attached virions were removed by centrifugation followed by washing [10]. Cell pellets were resuspended in opiate-treated medium. Pretreatment of the virus involved incubation for one hour in opiate-treated medium or control medium. Target cells were then added and the non-attached virus removed one to two hours later. Infected cultures were resuspended in non-treated medium and infection was allowed to proceed for the above-indicated periods. Because HTLV-I is rarely transmitted as a cell-free virus, all experiments using HTLV-I were performed with the chronically infected MT2 cell line.

Complex (Dual Virus) Infections. As with the simple infections, complex infections were divided into two main groups. The first group was an *in vitro* representation of cell-associated dual infection. To accomplish this, H-9MN and MT2 cells were co-incubated. The experimental design was the same as for the simple infections. The second group was an *in vitro* representation of HIV-1 superinfection during chronic HTLV-I infection. For this group, MT2 cells were challenged with cell-free HIV-1MN for a period of three days.

The use of cell-free HTLV-I is not germane to the natural transmission characteristics of this virus, thus cell-free HTLV-I was not used in the following studies. The levels of both simple and complex infections were determined by the measurement of viral core antigens as well as by the measurement of reverse transcriptase activity.

Measurement of Viral Core Antigens. Core antigen capture assays were utilized to determine the presence of HIV-1MN and of HTLV-I in the test cultures [8]. Standards containing a known amount of the desired viral antigen were used to allow for quantification of the antigen levels in the test samples. HIV-1 standards, pre-coated plates and antibodies needed to measure viral p24 levels were obtained from the AIDS Vaccine Program (Bethesda, MD). The assay was performed according to the recommendations supplied with the kit. The reagents needed for the measurement of HTLV-I p19 core antigen levels were purchased in bulk and as kits from Zeptometrix Corporation (Buffalo, NY).

Measurement of Reverse Transcriptase (RT) Activity. An adaptation of previously described techniques [11, 12] was used to determine the *in situ* activity levels of HIV-1 and/or HTLV-I RT. In order to accomplish this, a poly-A template (Fisher) was immobilized onto microtiter plates with Pierce Reacti-Bind (Pierce Chemical Co., Rockford, IL) and briefly blocked with a 4% casein TBS blocking solution (reagents purchased from Fisher). A reaction mix consisting of oligo dT 18-25, poly T (Pierce), RT buffer (Worthington Biochemical Corp., Lakewood, NJ), and digoxigenin (DIG) -dUTP (NEN Life Science Products, Boston, MA) was prepared. HIV-1 RT standards (Worthington) were diluted in a non-infected cell lysate solution. The reaction mix and whole culture lysate samples or standards were added to the wells of the microtiter plate and the RT reaction was allowed to proceed at 37°C for 20 minutes. RT activity resulted in the incorporation of the DIG-labeled dUTP to the immobilized template. Anti-DIG conjugated antibody (NEN) was then added, and was finally reacted with avidin-HRP (Endogen, Woburn, MA). The entire reaction complex was revealed spectrophotometrically using TMB as a substrate. The color intensity was assumed as directly proportional to the activity of the reverse transcriptase. Controls included non-infected lysates, or singly infected lysates, as dictated by the grouping of the test samples.

Cell Counts. Counts and viability were determined with trypan blue (Sigma) dye exclusion on a hematocytometer. Cytotoxicity was defined as producing a final viability below 90%, and conditions producing cytotoxicity were not tested further for this report. Cell counts were sampled immediately prior to the culture preparations described above. The counts were used to calculate the relative levels of antigen or RT activity per cell.

Statistics. All reported results were from triplicate experiments. The student t test was used to determine the statistical differences between the different treatment groups with p values equivalent to or less than 0.05.

RESULTS

Effect of Morphine on p19 Levels During *In Vitro* HTLV-I Infection

MT2 and Sup-T1 cells were co-cultured with or without morphine for five days. As a control, matching cultures were incubated with morphine plus an excess molar amount of naloxone, an opioid receptor antagonist. The affinity of naloxone is greatest for μ (mu) opioid receptors, followed by δ (delta) antagonism at higher molar concentrations. Two sets of negative controls were used. Non-infected Sup-T1 cells were used to check for cross-reactive proteins. Naloxone-treated Sup-T1 cultures were incubated without morphine in order to ensure that the levels of naloxone used were not cytotoxic to this cell line. The virus controls consisted of MT2/Sup-T1 co-cultures incubated without morphine. One set of virus controls was grown as a non-treated control for any possible morphine effect, and the other set of virus controls contained naloxone. As with the naloxone-treated Sup-T1 controls the viability of the cells was tested. This control also made the observation of opioid antagonist effects on the levels of p19 without the addition of agonist (morphine) possible. On day three of the *in vitro* challenge, the cells were centrifuged and the supernatant removed. This was performed to refresh the test and challenge cultures with new medium. In each case, the medium was replaced in order to maintain the same levels of morphine, naloxone, or both. It was found that replacing the medium in this manner did not affect the ratio of p19 levels to any significant degree (data not shown). However, all test and control cultures demonstrated higher viability levels with lower variations from one well to the next. At the end of the five-day challenge, p19 levels per cell, as well as overall levels, were consistently higher in the morphine-treated cultures. Although the combination of naloxone and morphine appeared to lead to a somewhat lower production of p19, this was significant only when the lower concentration of morphine was used. Figure 1A is an example of the results obtained from a set of three such experiments, corrected for cell counts. Although there does not appear to be a simple dose-related effect, the presence of morphine was associated with a greater level of viral antigen in the test cultures.

The counts (not shown) indicated that at five days post challenge, morphine treatment alone did not significantly affect the number of cells. The combination of morphine plus naloxone led to an increase in the number of cells when compared to naloxone-treated virus controls, yet this change was not significantly different from the non-treated virus controls. So, although the non-treated controls and the morphine-treated samples contained equivalent population sizes, the p19 levels were elevated in the morphine-treated cultures. While the combination of morphine and naloxone moderately enhanced the size of the combined treatment cell populations, p19 production was reduced in comparison to the morphine-treated test samples. It appeared that the increased amounts of viral antigen found in association with morphine treatment was not due simply to an increased number of infected cells per sample.

Effect of Morphine on Reverse Transcriptase Activity During *In Vitro* HTLV-I Infection

Without an increased number of infected cells, it was questioned whether increased viral activity would occur in the presence of morphine, thus leading to the enhanced p19 levels observed in the first set of experiments. In order to address this question, a RT activity assay was devised as described which allowed for the comparison of viral activity using whole culture lysates rather than measuring the activity of centrifuged virions. As demonstrated by the experimental results in (Figure 1B) naloxone enhanced per cell RT activity as well as in whole culture. Importantly, a moderate but consistent increase in the biological activity of the virus was also observed with the addition of morphine. The combination of naloxone plus morphine did not significantly alter the enhanced activity of the virus. The results were consistent with an increased level of syncytia formation observed during opioid treatment of MT2/Sup-T1 co-cultures, which were reported earlier as part of a preliminary study [8].

Effect of HIV-1 on HTLV-I p19 Levels During Morphine Treatment

Results reported by Beilke et al [6] indicated that coinfection of HIV-positive PBMC with MT2 cells led to enhanced HTLV-I infection. However, the effects of opiates on HTLV-I/HIV-1 infection were unknown, even though the IDU are more likely to be coinfected with both retroviruses than any other risk group in the United States. A preliminary study indicated that co-cultures of HIV-infected cells with Sup-T1 targets exhibited moderately increased cytopathic effects in the presence of opioid agonists, especially when μ and δ agonists were used [10]. A co-culture of HIV-1-infected and HTLV-I-infected cells was used to test whether dual infection with cell-associated retroviruses would be affected in the presence of morphine. As demonstrated in Figure 2A, both HTLV-I and HIV-1 antigen levels were enhanced in the presence of morphine, again showing a

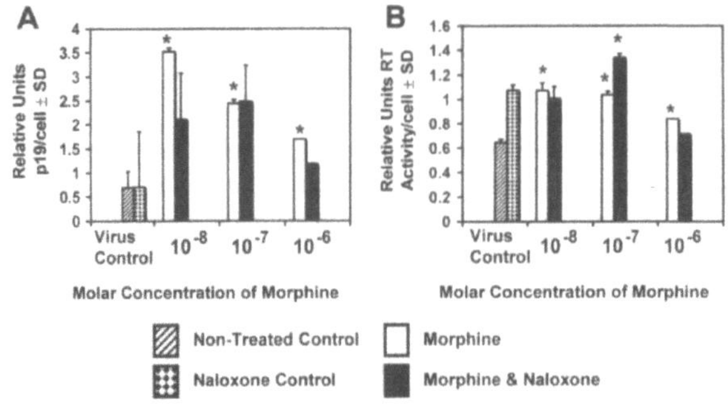

Figure 1. Effect of Morphine on *In Vitro* HTLV-I Infection. Chronically infected MT2 cells were co-cultured with non-infected Sup-T1 cells for five days. As indicated by the graphs, cultures contained morphine, morphine plus naloxone, naloxone (Naloxone Control), or non-treated medium (Non-Treated Control). (A) Relative units of p19 per cell. HTLV-I core antigen (p19) levels were first determined by core antigen capture assays. Relative units were calculated as the picograms (pg) of p19 per 10,000 cells. (B) Relative units of RT activity per cell. The RT activity of the whole culture lysate was determined from the assay described in Materials and Methods. Relative units of activity per cell were calculated as mU RT activity per 10,000 cells. The results shown are from one of three experiments. Values that were significantly different from virus controls are indicated with an asterisk (*).

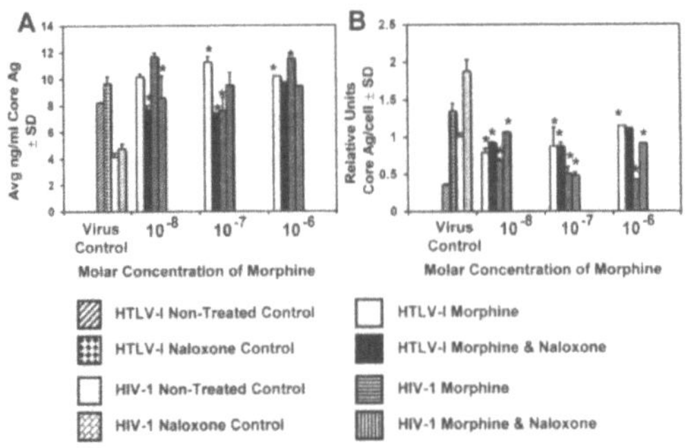

Figure 2. Effect of Morphine and Cell Association on *In Vitro* HTLV-I/HIV-1 Infection. (A) Increased p19 and p24 Production during Dual *In Vitro* HTLV-I/HIV-1 Cell-Associated Infection with Morphine Treatment. HTLV-I-infected MT2 cells and HIV-1-infected H9MN cells were co-cultured with morphine, morphine plus naloxone, naloxone or with non-treated medium for three days. Core antigen levels for each virus were determined by antigen capture assays and were displayed in units of ng/ml. The results shown are from one of three such experiments. (B) Effect of Morphine on Increased p19 versus Decreased p24 Production during *In Vitro* Co-Infection with Cell-Free HIV-1MN. Chronically infected MT2 cells were challenged with cell-free HIV-1MN as described in Materials and Methods. The *in vitro* infection proceeded for three days prior to testing for cell counts and viral core antigen levels. Control and test cultures were tested for both p19 (HTLV-I) and p24 (HIV-1) production. The pg/ml of each core antigen was then compared to the cell number/ml. The relative units of activity were calculated as the pg of p19 or p24 per 10,000 cells. The results shown are from one of three such experiments. Values that were significantly different from virus controls are indicated with an asterisk (*).

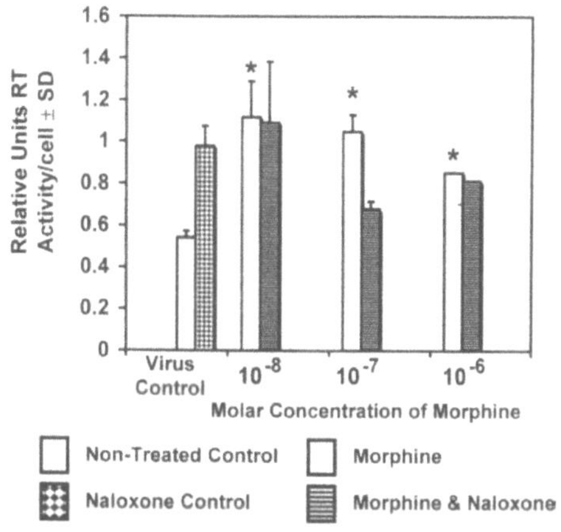

Figure 3. Effect of Morphine and Cell-Free HIV on Relative RT Activity during Dual *In Vitro* HTLV-I/HIV-1 Infection. Graph of the relative RT activity from the co-infection described in Figure 2B. Relative units of activity were derived from the determination of total RT activity and were calculated as mU RT activity per 10,000 cells. The results shown are from one of three such experiments. Values that were significantly different from virus controls are indicated with an asterisk (*).

positive effect on cell-bound viruses. Cell counts were not obtained; therefore, the effect per cell remains to be determined.

However, HIV-1 infection *in vivo* also expands by the budding of cell-free virions, as well as by the release of infectious particles when host cells undergo apoptosis. Cell-free HIV-1MN was obtained from infected cell lines, and then used to challenge HTLV-I-infected MT2 cells in the presence of morphine. HTLV-I antigen levels were enhanced in the presence of morphine, and as with the single infection, this effect was not abrogated by the combined presence of naloxone. However, HIV-1 antigen levels were reduced in the presence of morphine in a dose-related manner, an effect that was somewhat reversed by the addition of naloxone (Figure 2B). In contrast to the stable cell counts during single HTLV-I infection, challenge with cell-free HIV-1 led to a reduction in cell numbers when morphine or combined morphine and naloxone treatments were used. As a control, naloxone negatively affected the population size during dual infection, while the combination of naloxone with morphine had the same effect as morphine alone. Total RT activity per cell was enhanced (Figure 3). Although the RT activity assay cannot distinguish which virus was responsible for the elevated enzyme activity, the increase in p19 levels versus a decrease in p24 antigen suggests that the heightened activity benefited HTLV-I and not HIV-1. Again, a simple relationship between opiate dose and activity was not presented, but the greatest enhancement of activity was observed at the lower morphine concentrations.

The effect of morphine on single HIV-1MN infection using cell-free virus was similar to that seen with dual infection. We previously demonstrated that HIV-1 antigen levels were reduced during early cell-free infection of Sup-T1 targets [8]. This time, we compared the levels of p24 to the numbers of cells per culture. Figure 4A illustrates that p24 levels were reduced not only per volume unit but also per cell during *in vitro* morphine treatment. In addition, RT activity levels remained stable with increasing cell counts during morphine treatment, which indicated a reduced level of viral activity per cell (Figure 4B).

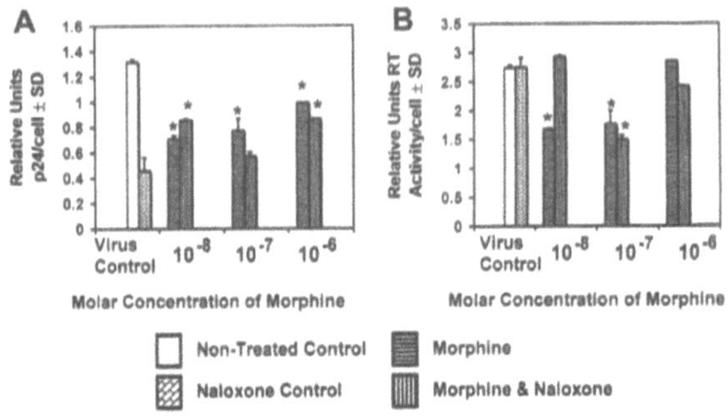

Figure 4. Morphine-Associated Inhibition of *In Vitro* Infection with Cell-Free HIV-1MN. Cell-free HIV-1MN was incubated with morphine, morphine plus naloxone, naloxone, or R10 for one hour prior to challenge of non-infected Sup-T1 cultures. The cells were resuspended in R10 after virus attachment as described in Materials and Methods, and the *in vitro* infection was continued for three days. The cultures were then tested for cell numbers, pg p24, and mU RT activity per ml. (A) Relative units of p24 per cell were based on pg p24 per 10,000 cells. (B) Relative units of RT activity were based on mU RT activity per 10,000 cells. The results shown are from one of three such experiments. Values that were significantly different from virus controls are indicated with an asterisk (*).

DISCUSSION

HTLV-I, HIV-1, and concurrent HTLV-I/HIV-1 infections in the USA are more common in IDU than in the general population. The nature of HTLV-I infection suggests that the mere mechanics of drug injection may not sufficiently account for the greatly increased seropositive rate of this virus in chronic injecting drug users. In order to see of the presence of an opiate could alter the nature of *in vitro* HTLV-I infection morphine was tested as a more stable metabolite of injected heroin. In addition, the *in vitro* influence of attendant HIV-1 infection on some properties of HTLV-I infection was tested with concomitant *in vitro* HTLV-I/HIV-1 challenges. In experimental infections involving only a single virus (simple infections), morphine modestly enhanced p19 production and RT activity in HTLV-I-infected cultures without increasing population sizes. Since HTLV-I is almost exclusively associated with its host cell, the greater levels of viral activity and production might be derived from a greater number of functioning particles per cell, and not with a greater number of infected cells alone. It is tempting to suppose that the close association of the virus with its host cell would mean that opiate effects would be due to classical opioid receptor mechanisms. However, the lack of reversal of the observed effects by an opioid receptor antagonist argues against such an assumption. Combining the data from this and previous work, we have found that μ and δ opiates and opioids alike tend to enhance *in vitro* HTLV-I infection, and that enhancement is dose-related but is not consistently abrogated by opioid receptor antagonists. This does not mean that the effects of opiates on HTLV-I infection are non-specific, although the involvement of a classical opioid receptor is less likely. The greater enhancement of p19 observed at the lower dose range could begin with the involvement of a signal from a high affinity receptor present at low density on the host cell. If such a receptor were identified, it would probably interact with conserved ligand conformations in agonists and antagonists for both μ and δ opioid receptors, thus naloxone alone could enhance aspects of HTLV-I infection, if the dose was low enough.

The combination of HIV-1 and opiates also enhanced HTLV-I infection, whether HIV-1 was added as a cell-free virus, or as a cell-associated virus. Cell counts obtained when the addition of cell-free HIV-1 was being tested showed that the combined cytotoxic effect of HIV-1 plus morphine on the host cells was greater than that of morphine alone. Opiate-treated, co-infected populations were lower in number than non-treated or singly infected populations versus controls, at least for the short duration of the *in vitro* experiments. Nevertheless, both overall RT activity and HTLV-I p19 levels were higher than virus controls that were not treated with morphine. In contrast, a stable RT activity, reduced p24 levels and increased cell counts were found during single infection experiments using cell-free HIV-1. In addition to the results of single infection experiments, the above-described co-infections also yielded lower p24 levels. The findings suggest that HIV-1 is able to contribute to the enhancement of HTLV-I infection during *in vitro* retroviral challenges. Importantly, while the combined presence of HTLV-I and an opiate does not appear to enhance experimental infections with cell-free HIV-1, the concurrent influences of HIV-1 during opiate treatment are evidently beneficial to *in vitro* HTLV-I infection. This finding not only confirms the published data of Beilke et al [6] but also infers that the IDU may be more susceptible to the effects of such a combination. Concurrent HIV-1 infection can mask the presence of HTLV-I in standard clinical tests, suggesting that HTLV-I could be more prevalent in the U.S. than previously reported. Understanding the effects of co-infection on the host cell as well as on the viruses may lead to a more effective protocol for determining risk, identifying dually infected patients, and prescribing the appropriate treatment.

ACKNOWLEDGEMENTS

This work was supported in part with funding provided by National Institutes of Health - National Institute of Drug Abuse grants DA05913 for S. Nyland and DA10161 for K. Ugen and S. Specter.

REFERENCES

1. A. Manns, M. Hisada & L.L. Genade. Human T-lymphotropic virus type I infection. *Lancet* 353, 1951 (1999).
2. A. Gessain. Epidemiology of HTLV-I and associated diseases., in: *Human T-cell Lymphotropic Virus Type I*. Wiley and Sons, New York, (1996).
3. F.R. Cleghorn & W.A. Blattner. Does human T cell lymphotropic virus type I and human immunodeficiency virus type 1 coinfection accelerate acquired immune deficiency syndrome? The jury is still out. *Archives of Internal Medicine* 152, 1372 (1992).
4. M. Schechter, L.H. Moulton & L.H. Harrison. HIV viral load and CD4+ lymphocyte counts in subjects coinfected with HTLV-I and HIV-1. *Journal of Acquired Immune Deficiency Syndromes and Human Retrovirology* 15, 308 (1997).
5. M.A. Beilke, D.L. Greenspan & A. Impey. Laboratory study of HIV-1 and HTLV-I/II coinfection. *Journal of Medical Virology* 44, 132 (1994).
6. M.A. Beilke, J. Shanker & D.G. Vinson. HTLV-I and HTLV -II Virus Expression Increase With HIV-1 Coinfection. *Journal of Acquired Immune Deficiency Syndromes and Human Retrovirology* 17, 391 (1998).
7. J.I. Sin & S. Specter. The role of interferon-gamma in antiretroviral activity of methionine enkephalin and AZT in a murine cell culture. *Journal of Pharmacology and Experimental Therapeutics* 279, 1268 (1996).

8. S.B. Nyland, S. Specter & K.E. Ugen. Morphine effects on HTLV-I infection in the presence or absence of concurrent HIV-1 infection. *DNA and Cell Biology* 18, 285 (1999).
9. R. Dulbecco. Endpoint method-measurement of the infectious titer of a viral sample., in: *Virology*. Lippincott, Philadelphia, (1988).
10. S.B. Nyland, S. Specter, J. Im-Sin & K.E. Ugen. Opiate effects on in vitro human retroviral infection. *Adv Exp Med Biol* 437, 91 (1998).
11. N. Tomita, M. Miyahara, H. Satoh, K. Suzuki, K. Kitajima & K. Miyamoto. Detection of reverse transcriptase activity by enzyme-linked immunosorbent assay in human immunodeficiency virus type 1. *Acta Medica Okayama* 49, 69 (1995).
12. D.H. Ekstrand, R.J. Awad, C.F. Kallander & J.S. Gronowitz. A sensitive assay for the quantification of reverse transcriptase activity based on the use of carrier-bound template and non-radioactive-product detection, with special reference to human immunodeficiency virus isolation. *Biotechnology and Applied Biochemistry* 23, 95 (1996).

GENETIC FACTORS INVOLVED IN CENTRAL NERVOUS SYSTEM/IMMUNE INTERACTIONS

Ronald L. Wilder, Marie M. Griffiths, Grant W. Cannon, Rachel Caspi, Percio S. Gulko, and Elaine F. Remmers

Drs. Wilder and Remmers, Inflammatory Joint Diseases Section, Arthritis and Rheumatism Branch, National Institute of Arthritis and Musculoskeletal and Skin Diseases, National Institutes of Health, Bethesda, Maryland 20892

Drs. Griffiths and Cannon, Research Service, Veteran Affairs Medical Center, Department of Medicine, University of Utah School of Medicine, Salt Lake City, Utah 84132

Dr. Caspi, Laboratory of Immunology, National Eye Institute, National Institutes of Health, Bethesda, Maryland 20892

Dr. Gulko, Columbia University College of Physicians and Surgeons, New York, NY 10032

ABSTRACT

Analysis of several inbred rat strains has led us to hypothesize that HPA axis abnormalities may contribute, in part, to susceptibility to both autoimmune disease and addiction. In this article we review the evidence for this hypothesis and describe our ongoing efforts to genetically characterize these traits. We have mapped the locations of 23 loci that regulate autoimmune disease in rats, and are currently constructing QTL congenic lines in which a genomic region from the resistant strain is transferred to the susceptible strain or vice versa. These QTL congenic lines will be valuable to test whether genes encoding autoimmune regulation also control neuroendocrine traits. Further genetic dissection and identification of the underlying genes will be necessary to infer a mechanistic link between autoimmune and neuroendocrine traits.

INTRODUCTION

Family and twin studies indicate that susceptibility to autoimmune diseases is partially controlled by genetic factors[1]. Other studies support the view that neuroendocrine

Neuroimmune Circuits, Drugs of Abuse, and Infectious Diseases
Edited by Herman Friedman *et al.*, Kluwer Academic/Plenum Publishers, 2001

hormonal and stress response factors can regulate the expression of autoimmune diseases in humans[2-4]. It is possible that some genetic factors that control neuroendocrine hormonal response to various stimuli may also contribute to autoimmune disease expression. However, defining the genetic factors that control these complex phenotypes in humans is extremely difficult. Because significant genetic heterogeneity exists in outbred human populations, different combinations of genes may be responsible for disease expression and complex neuroendocrine/stress responses in different patients. Furthermore, these complex traits are likely to be dependent upon the interaction of genetic factors with non-genetic factors, including environmental factors (e.g. infection or diet) and reproductive factors, as well as age and gender.

A complementary/alternative approach for defining genetic factors that control complex traits is to use genetically defined experimental animal model systems. In particular, inbred rat strains are providing a powerful approach to dissect genetic factors that control autoimmune diseases[5, 6]. Neuroendocrine hormonal factors also appear to regulate inflammatory responses and experimental autoimmune disease in animals[2, 7]. Interestingly, several genetically defined rat strains that are susceptible to experimentally induced autoimmune diseases also display aberrant basal and stimulated hypothalamic-pituitary-adrenal (HPA) axis responses, relative resistance to effects of opioids, low levels of endogenous opioid peptides, and also exhibit traits related to sensitivity to addiction. These observations have led us to hypothesize that some mechanistic factors related to abnormalities in the HPA axis and stress response may be common to both addiction and autoimmune disease susceptibility. In this review, we will discuss the evidence for these relationships and describe our ongoing efforts to address this hypothesis.

SUSCEPTIBILITY OF INBRED RATS TO EXPERIMENTAL AUTOIMMUNE DISEASE

There is remarkable variation in susceptibility to experimentally induced autoimmune diseases among inbred rat strains, indicating significant genetic contributions to these disease models. Two inbred rat strains, DA and LEW, are widely used for autoimmune and autoinflammatory studies because of their sensitivity to autoimmune disease models. Conversely, F344 and BN rats are resistant to most, but not all, forms of experimentally induced autoimmune disease.

DA rats develop inflammatory arthritis in response to immunization with either heterologous (non-rat) type II collagen or homologous (rat) type II collagen, as well as another cartilage constituent, cartilage oligomeric matrix protein[5, 8]. They are also susceptible to arthritis induced in the absence of immunogenic joint constituents, by a variety of adjuvant oils, including complete Freund's adjuvant (which includes mycobacterial immunogens), as well as pristane, avridine, and incomplete Freund's adjuvant (oils without protein immunogens)[5, 8]. DA rats are also highly susceptible to experimental autoimmune encephalomyelitis induced by immunization with whole spinal cord homogenate, myelin basic protein, or myelin oligodendrocyte glycoprotein[9-12], and are susceptible to interphotoreceptor retinoid binding protein- and S retinal antigen-induced experimental autoimmune uveitis (R.Caspi, unpublished data). Furthermore, DA rats are resistant to antigen tolerization protocols[10] and have an inherent propensity to develop autoreactive T cells that exhibit a Th1-like phenotype, i.e., they produce proinflammatory cytokines such as interferon-gamma and tumor necrosis factor alpha.

LEW rats are also used widely as models for autoimmune disease. They are extremely susceptible to arthritis induced by complete Freund's adjuvant, pristane, avridine, and streptococcal cell walls. They develop arthritis following immunization with heterologous type II collagen, but unlike DA rats, they do not develop arthritis following immunization with rat type II collagen or incomplete Freund's adjuvant[5, 8]. They also develop experimental autoimmune encephalomyelitis induced with myelin basic protein and experimental autoimmune uveitis induced with S retinal antigen or interphotoreceptor binding protein[13], but LEW rats develop only mild experimental autoimmune encephalomyelitis after immunization with myelin oligodendrocyte glycoprotein[14]. Unlike DA rats, LEW rats can be tolerized by vaccination with myelin basic protein[10]. Thus, although DA and LEW rats are both susceptible to the development of autoimmune disease, in some ways LEW rats are less susceptible than DA rats.

In contrast to DA and LEW rats, F344 and BN rats are relatively resistant to most, but not all, forms of experimentally induced autoimmune disease. F344 rats are resistant to arthritis induced with collagen, pristane, avridine, incomplete Freund's adjuvant, and streptococcal cell walls, and are only weakly susceptible to complete Freund's adjuvant-induced arthritis[5]. They also do not develop encephalomyelitis in response to myelin basic protein, or uveitis in response to either S retinal antigen or interphotoreceptor binding protein. Although generally resistant, if F344 rats are rendered germ-free, they will then develop severe arthritis in response to complete Freund's adjuvant or streptococcal cell walls[5]. BN rats are also resistant to collagen-, pristane-, and incomplete Freund's adjuvant-induced arthritis, uveitis induced with S retinal antigen, and encephalomyelitis induced with myelin basic protein. They are somewhat susceptible to streptococcal cell wall-induced arthritis and interphotoreceptor binding protein-induced uveitis (R. Caspi, unpublished observation). Surprisingly, BN rats are highly susceptible to myelin oligodendrocyte glycoprotein-induced encephalomyelitis[15]. The observation that BN rats develop encephalomyelitis in response to myelin oligodendrocyte glycoprotein but not myelin basic protein appears to reflect different pathogenic mechanisms for these two models. The myelin basic protein autoimmune response in DA and LEW rats is primarily Th1-dependent, whereas the myelin oligodendrocyte glycoprotein response in BN rats is characterized by antibody-dependent Th2 mechanisms. Furthermore, BN rats in contrast to LEW rats develop severe mercuric chloride-induced autoimmune disease, including arthritis, vasculitis, and glomerulonephritis and this response is also characterized by high levels of the Th2 cytokine IL4[5]. In contrast to DA and LEW rats, it appears that when observed, autoimmune disease pathology in BN rats occurs in the context of Th2 cytokines and pathogenic antibodies. Autoimmune disease susceptibility profiles of the DA, LEW, F344 and BN rats are shown in Table 1.

Table 1. Strain susceptibility/resistance to autoimmune disease and addiction-related phenotypes.

	Rat Strain			
Model: agent	DA	LEW	F344	BN
BII-CIA: Bovine Type II Collagen in IFA	++++	++	-	-
RII-CIA: Rat Type II Collagen in IFA	++++	-	-	-
Mtb-AIA: Mycobacterium spp. in IFA	++++	++++	+/-	+/-
AVIA: Avridine	++++	++++	-	-
PIA: Pristane	++++	++++	-	-
SCWA: Streptococcal cell walls	-	++++	-	+
IFA: Incomplete Freund's adjuvant	++	-	-	-
MBP-EAE: Myelin basic protein	++++	++++	-	-
MOG-EAE: Myelin oligodendrocyte glycoprotein	++++	++	NT	++++
SAG-EAU: S-retinal antigen	+++	++++	-	+
IRBP-EAU: Interphotoreceptor binding protein	++	++++	-	-
Models of alcoholism and/or drug addiction	++++	++++	-	-

HPA AXIS ABNORMALITIES IN DA AND LEW RAT STRAINS

Both of the autoimmune disease-prone/addiction-prone rat strains described above, DA and LEW, exhibit genetically determined HPA axis abnormalities compared with F344, BN and most other rat strains. LEW and DA rats show minimal or insignificant variation in plasma corticosterone throughout the 24 hour circadian cycle, whereas most other rat strains exhibit a daily transient rise in corticosterone levels in the late dark and early light phase[16-20]. Furthermore, a wide variety of stressful stimuli that result in marked increases in plasma corticosterone levels in F344 and most other rat strains either fail to do so in LEW rats or require a greater intensity stimulus[19, 20]. In other words, the HPA axis response is blunted in LEW relative to F344 rats. The molecular basis for the observed differences in HPA axis function among inbred rat strains is unknown.

OPIOID RESISTANCE AND ENDOGENOUS OPIOIDS IN LEW RATS

In several experimental paradigms, it appears that LEW rats are resistant to opioid effects. Profound differences to the antinociceptive effects of mu opioids have been found in various rat strains. Among F344, Sprague Dawley, Long Evans and LEW rat strains, mu-opioids are most potent in F344 and least potent in LEW rats[21]. In addition, kappa-opioid agonists have been shown to suppress adjuvant-induced arthritis in rats[22-24]. This suppression, however, varies among different rat strains. Although the kappa-opioid receptor agonist, MR2034, significantly decreases incidence and severity of adjuvant-arthritis in Wistar rats and reduces arthritis severity in DA rats, it fails to produce any suppressive effect in LEW rats[24]. Sufficient doses of another kappa-opioid receptor agonist, U50488H, can, however, suppress adjuvant-induced arthritis in LEW rats[22]. A high dose of exogenously administered met-enkephalin, a naturally occurring opioid, significantly inhibits guinea pig spinal cord-induced experimental allergic encephalomyelitis in LEW rats. In contrast, a low dose potentiates the neurological and histopathological features of the disease[25]. Additionally, chronic morphine treatment increases the levels of adenylate cyclase and cAMP-dependent protein kinase activities in the nucleus accumbens of F344 rats, but this effect is not seen in LEW rats[26]. These data suggest a degree of opioid resistance or insensitivity in the autoimmune disease-prone and addiction model-sensitive LEW rat strain.

There is also evidence that endogenous opioid peptide levels differ among inbred rat strains. Basal levels of proenkephalin gene expression are lower in the nucleus accumbens and the dorsal striatum, and hypothalamus of LEW rats than F344 rats[27, 28].

Enkephalins and β-endorphin are peptide products of the proopiomelanocortin pathway. Adrenocorticotropin hormone (ACTH) is also derived from the same propeptide. As part of the HPA axis/stress response, ACTH is secreted into the bloodstream and stimulates the adrenal cortex to secrete corticosterone. In a feedback loop, corticosterone inhibits further ACTH production from the proopiomelanocortin pathway (Figure 1). LEW rats produce lower levels of endogenous opioid peptides and are less sensitive than other strains to the effects of exogenously administered opioids. As described above, LEW rats also exhibit HPA axis abnormalities with low corticosterone production. It is interesting to speculate that these hyporesponses are mechanistically linked.

ADDICTION SUSCEPTIBILITY IN LEW, DA AND F344 INBRED RAT STRAINS

LEW and F344, and more recently, DA rats have been used to identify genetically determined traits related to addiction. Interestingly, the two autoimmune disease-prone strains, LEW and DA exhibit behaviors and/or brain biochemical markers that are thought to be involved in susceptibility to addiction. Compared with F344 rats, LEW rats exhibit a higher preference for, or a greater capacity to self-administer or develop place preference associated with cocaine, morphine, ethanol, and nicotine[27-32]. Drug naïve LEW and F344 rats exhibit many differences in brain biochemistry that are suspected to be involved in susceptibility to addiction, particularly in the mesolimbic dopamine pathways [26,27, 34-45]. Similarly, DA rats have also been shown to exhibit behaviors that are related to susceptibility to addiction. DA rats become sensitized to morphine-induced locomotor activity more readily than F344 rats, and these strains also display significant biochemical differences in the ventral tegmental area of the brain[17].

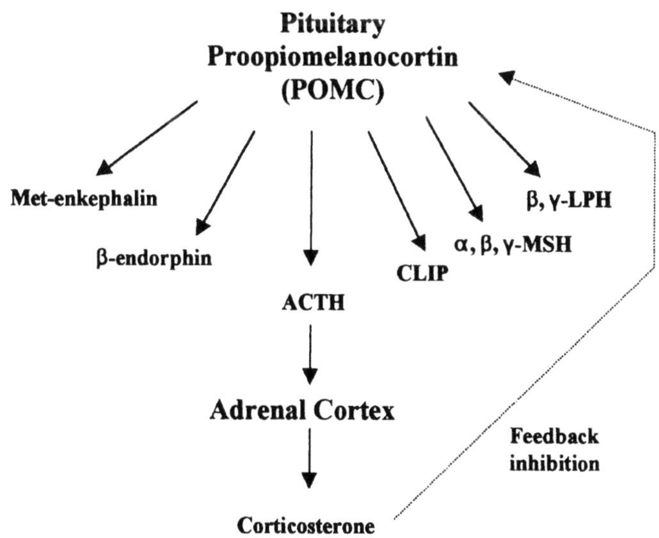

Figure 1: Proopiomelanocortin pathway. The propeptide, proopiomelanocortin is the precursor of a variety of neuropeptides. In a feedback loop, corticosterone prevents further production of piuitary ACTH. ACTH = adrenocorticotropic hormone, CLIP = corticotropin-like intermediary peptide, MSH = melanotropin, LPH = lipotropin.

HPA-AXIS FUNCTION AND SUSCEPTIBILITY TO AUTOIMMUNE DISEASE AND DRUG ADDICTION

Corticosteroids can clearly influence autoimmune disease expression in humans and animals[2, 4, 7]. We have previously suggested that genetically determined abnormal HPA axis function resulting in a defect in central corticotropin releasing factor production may contribute to autoimmune disease susceptibility [2, 46]. Corticosteroids also appear to be capable of modulating susceptibility to drug addiction. For example, cocaine induces elevated plasma corticosterone levels in rats, and LEW rats are more sensitive than F344 rats to the suppressive effect (by feedback inhibition) of centrally administered glucocoticoid agonist, dexamethasone[45, 47]. Furthermore, withdrawal of cocaine, marijuana, or ethanol from rats, results in a rise in centrally produced corticotropin releasing factor which may be responsible for anxiety-like behavior related to drug withdrawal[48-50] and to depressed peripheral corticosterone production. Therefore, it is possible that susceptibility to autoimmune disease and drug addiction, and abnormalities in basal and stress-induced HPA-axis function may share common underlying genetic mechanisms. Although this hypothesis is based on association in a limited number of inbred rat strains and it is therefore possible that the association is merely coincidental, it clearly warrants further investigation.

GENETIC STUDIES INVOLVING LEW, DA, F344 AND BN INBRED RATS

One way to address whether autoimmune diseases and drug addiction are mechanistically linked to abnormalities in basal and stress-activated HPA axis function is to use a genetic approach. Identification of genomic loci containing regulatory genes that control two or more of these traits would provide evidence for an underlying biochemical pathway that may causally link the traits. We have focused our efforts on identifying genomic intervals that contain loci that regulate a number of experimentally induced autoimmune diseases (collagen-induced arthritis, complete Freund's adjuvant-induced arthritis, and experimental autoimmune uveitis) in progeny of crosses between susceptible (DA and LEW) and resistant (F344 and BN) inbred rat strains. For collagen-induced arthritis, we have identified 14 loci including the MHC-associated locus, *Cia1*, and the non-MHC-associated loci, *Cia2-14*, in F2 progeny of DA x F344, DA x ACI and DA x BN crosses (Figure 2). We have also mapped four loci that control adjuvant-induced arthritis in DA x F344 progeny and two loci that control experimental autoimmune uveitis in LEW x F344 progeny (Figure 2).

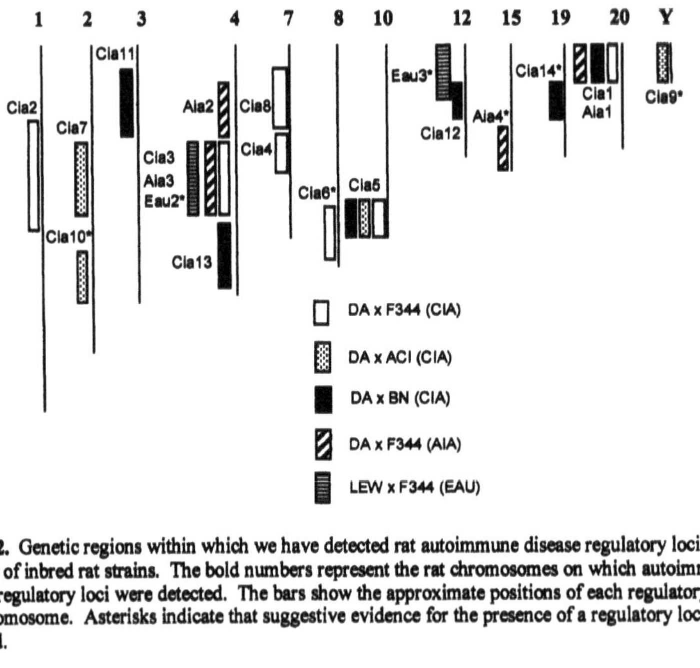

Figure 2. Genetic regions within which we have detected rat autoimmune disease regulatory loci in F2 progeny of inbred rat strains. The bold numbers represent the rat chromosomes on which autoimmune disease regulatory loci were detected. The bars show the approximate positions of each regulatory locus on that chromosome. Asterisks indicate that suggestive evidence for the presence of a regulatory locus was obtained.

Interestingly, Bice and coworkers[51] mapped a QTL that regulates alcohol consumption in selectively bred P and NP rat lines to a region on chromosome 4 that overlaps Cia3/Aia3/Eau2* and Aia2. To confirm the arthritis QTLs and to create a resource that can be used to fine map the loci and characterize the phenotypes controlled by the loci, we are currently constructing QTL congenic lines. In these lines we are transferring genomic segments encoding arthritis and/or uveitis resistance (F344) alleles to the arthritis and/or uveitis susceptible (DA or LEW) genome. Once constructed, we will test whether introgression of the F344 alleles protects these rats from experimental autoimmune disease. We also intend to test these lines for alterations in basal and stress-activated HPA axis function and addiction-related phenotypes compared with DA controls. These congenic lines should allow us to determine whether genetic control of these neuroendocrine and autoimmune phenotypes are encoded in the same DNA segment, whether they are separable, and ultimately, whether a single gene variant is responsible for the observed phentoypic differences among control and QTL congenic lines.

REFERENCES

1. D.A. Carson, Genetic factors in the etiology and pathogenesis of autoimmunity, *FASEB J.* 6:2800 (1992).
2. R.L. Wilder, Neuroendocrine-immune system interactions and autoimmunity, *Ann. Rev. Immunol.* 13:307 (1995).
3. R.L. Wilder, Adrenal and gonadal steroid hormone deficiency in the pathogenesis of rheumatoid arthritis, *J. Rheumato.l Suppl.* 44:10 (1996).
4. R.L.Wilder, R.L. and I.J. Elenkov, Hormonal regulation of tumor necrosis factor-alpha, interleukin-12 and nterleukin-10 production by activated macrophages. A disease-modifying mechanism in rheumatoid arthritis and systemic lupus erythematosus? *Ann. N.Y. Acad. Sci.* 876:14 (1999).
5. R.L Wilder, E.F. Remmers, Y. Kawahito, P.S. Gulko, G.W. Cannon, and M.M. Griffiths, Genetic factors regulating experimental arthritis in mice and rats., in: *Current Directions in Autoimmunity; Vol 1: Genes and Genetics of Autoimmunity.* A.N. Theophilopoulos, ed., Karger, Basel (1999).
6. M.M. Griffiths, J.A. Encinas, E.F. Remmers, V.K. Kuchroo, and R.L. Wilder, Mapping autoimmunity genes, *Curr. Opin. Immunol.* 11:689 (1999).
7. R.L. Wilder, R.L., Hormones and autoimmunity: animal models of arthritis. *Baillieres Clin. Rheumatol.* 10:259 (1996).
8. B. Joe and R.L. Wilder, Animal models of rheumatoid arthritis, *Molecular Medicine Today*, 5:367 (1999).
9. I. Dahlman, L. Jacobsson, A. Glaser, J.C. Lorentzen, M. Andersson, H. Luthman and T. Olsson, Genome-wide linkage analysis of chronic relapsing experimental autoimmune encephalomyelitis in the rat identifies a major susceptibility locus on chromosome 9. *J. Immunol.* 162:2581 (1999).
10. D.C. Lenz, N.A. Wolf, and R.H. Swanborg, Strain variation in autoimmunity: attempted tolerization of DA rats results in the induction of experimental autoimmune encephalomyelitis. *J. Immunol.* 163:1763 (1999).
11. I. Dahlman, J.C. Lorentzen, K.L. de Graaf, A. Stefferl, C. Linington, H. Luthman, and T. Olsson, Quantitative trait loci disposing for both experimental arthritis and encephalomyelitis in the DA rat; impact on severity of myelin oligodendrocyte glycoprotein-induced experimental autoimmune encephalomyelitis and antibody isotype pattern . *Eur. J. Immunol.* 28:2188 (1998).
12. M.K. Storch, A. Stefferl, U. Brehm, R. Weissert, E. Wallstrom, M. Kerschensteiner, T. Olsson, C. Linington, and H. Lassmann, Autoimmunity to myelin oligodendrocyte glycoprotein in rats mimics the spectrum of multiple sclerosis pathology. *Brain Pathol.* 8:681 (1998).
13. R.R. Caspi, C.C. Chan, Y. Fujino, S. Oddo, F. Najafian, S. Bahmanyar, H. Heremans, R.L. Wilder, and B. Wiggert, Genetic factors in susceptibility and resistance to experimental autoimmune uveoretinitis, *Curr. Eye Res.* 11:81 (1992).
14. R. Weissert, E. Wallstrom, M.K. Storch, A. Stefferl, J. Lorentzen, H. Lassmann, C. Linington, and T. Olsson, MHC haplotype-dependent regulation of MOG-induced EAE in rats, *J. Clin. Invest.* 102:1265 (1998).
15. A. Stefferl, U. Brehm, M. Storch, D. Lambracht-Washington, C. Bourquin, K. Wonigeit, H. Lassmann, and C. Linington, Myelin oligodendrocyte glycoprotein induces experimental autoimmune encephalomyelitis in the "resistant" Brown Norway rat: disease susceptibility is determined by MHC and MHC-linked effects on the B cell response. *J. Immunol.* 163:40 (1999).
16. E.S. Brodkin, W.A. Carlezon, Jr., C.N. Haile, T.A. Kosten, G.R. Heninger, and E.J. Nestler, Genetic analysis of behavioral, neuroendocrine, and biochemical parameters in inbred rodents: initial studies in Lewis and Fischer 344 rats and in A/J and C57BL/6J mice, *Brain Res.* 805:55 (1998).
17. E.S. Brodkin, T.A. Kosten, C.N. Haile, G.R. Heninger, W.A. Carlezon, Jr., P. Jatlow, E.F. Remmers, R.L. Wilder & E.J. Nestler, Dark Agouti and Fischer 344 rats: differential behavioral responses to morphine and biochemical differences in the ventral tegmental area. *Neuroscience.* 88:1307 (1999).

18. A.C. Griffin and C.C. Whitacre, Sex and strain differences in the circadian rhythm fluctuation of endocrine and immune function in the rat: implications for rodent models of autoimmune disease, *J. Neuroimmunol.* 35:53 (1991).
19. F.S. Dhabhar, B.S. McEwen, andR.L. Spencer, Stress response, adrenal steroid receptor levels and corticosteroid- binding globulin levels--a comparison between Sprague-Dawley, Fischer 344 and Lewis rats, *Brain Res.* 616:89 (1993).
20. M.S. Oitzl, A.D. van Haarst, W. Sutanto, and E.R. de Kloet, Corticosterone, brain mineralocorticoid receptors (MRs) and the activity of the hypothalamic-pituitary-adrenal (HPA) axis: the Lewis rat as an example of increased central MR capacity and a hyporesponsive HPA axis, *Psychoneuroendocrinology* 20:655 (1995).
21. D. Morgan, C.D, Cook, and M.J. Picker, Sensitivity to the discriminative stimulus and antinociceptive effects of mu opioids: role of strain of rat, stimulus intensity, and intrinsic efficacy at the mu opioid receptor, *J. Pharmacol. Exp. Ther.* 289:965 (1999).
22. J.S. Walker, C.R. Howlett, and V. Nayanar Anti-inflammatory effects of kappa-opioids in adjuvant arthritis, *Life Sci.* 57:371 (1995).
23. W. Binder and J.S. Walker, Effect of the peripherally selective kappa-opioid agonist, asimadoline, on adjuvant arthritis, *Br. J. Pharmacol.* 124:647 (1998).
24. J. Antic, T, Vasiljevic, S. Stanojevic, V, Vujic, V. Kovacevic-Jovanovic, D. Djergovic, C. Miljevic, B.M. Markovic, and J. Radulovic, Suppression of adjuvant arthritis by kappa-opioid receptor agonist: effect of route of administration and strain differences, *Immunopharmacology* 34:105 (1996).
25. B.D. Jankovic and D. Maric, Enkephalins and autoimmunity: differential effect of methionine-enkephalin on experimental allergic encephalomyelitis in Wistar and Lewis rats, *J. Neurosci. Res.* 18:88 (1987).
26. X. Guitart, J.H. Kogan, M. Berhow, R.Z. Terwilliger, G.K. Aghajanian, and E.J. Nestler, Lewis and Fischer at strains display differences in biochemical, electrophysiological and behavioral parameters: studies in the nucleus accumbens and locus coeruleus of drug naive and morphine-treated animals, *Brain Res.* 611:7 (1993).
27. S. Martin, J. Manzanares, J. Corchero, C. Garcia-Lecumberri, J.A. Crespo, J.A. Fuentes, and E. Ambrosio, Differential basal proenkephalin gene expression in dorsal striatum and nucleus accumbens, and vulnerability to morphine self-administration in Fischer 344 and Lewis rats. *Brain Res.* 821:350 (1999).
28. E.M. Sternberg, W.S. Young 3d, R. Bernardini , A.E. Calogero, G.P. Chrousos, P.W. Gold, and R.L. Wilder, A central nervous system defect in biosynthesis of corticotropin-releasing hormone is associated with susceptibility to streptococcal cell wall-induced arthritis in Lewis rats. *Proc. Natl. Acad. Sci. U.S.A.* 86:4771 (1989).
29. T. Suzuki, F.R. George, and R.A. Meisch, Differential establishment and maintenance of oral ethanol reinforced behavior in Lewis and Fischer 344 inbred rat strains. *J. Pharmacol. Exp. Ther.* 245:164 (1988).
30. T.A. Kosten, M.J. Miserendino, S. Chi, and E.J. Nestler, Fischer and Lewis rat strains show differential ocaine effects in conditioned place preference and behavioral sensitization but not in locomotor activity or conditioned taste aversion. J. *Pharmacol. Exp. Ther.* 269:137 (1994).
31. T.A. Kosten, M.J. Miserendino, C.N. Haile, J.L. DeCaprio, P.I. Jatlow, and E.J. Nestler, Acquisition and maintenance of intravenous cocaine self-administration in Lewis and Fischer inbred rat strains. *Brain Res.* 778:418 (1997).
32. B. Horan, M. Smith, E.L. Gardner, M. Lepore, and C.R. Ashby, Jr., (-)-Nicotine produces conditioned place preference in Lewis, but not Fischer 344 rats. *Synapse* 26:93 (1997).
33. T. Suzuki, Y. Ise, J. Maeda, and M. Misawa. Mecamylamine-precipitated nicotine-withdrawal aversion in Lewis and Fischer 344 inbred rat strains. *Eur. J. Pharmacol.* 369:159 (1999).
34. D. Beitner-Johnson, X. Guitart, and E.J. Nestler, Dopaminergic brain reward regions of Lewis and Fischer rats display different levels of tyrosine hydroxylase and other morphine- and cocaine-regulated phosphoproteins, *Brain Res.* 561:147 (1991).
35. X. Guitart, D. Beitner-Johnson, D.W. Marby, T.A. Kosten, and E.J. Nestler, Fischer and Lewis rat strains differ in basal levels of neurofilament proteins and their regulation by chronic morphine in the mesolimbic dopamine system, *Synapse* 12:242 (1992).
36. P.W. Burnet, I.N. Mefford, C.C. Smith, P.W. Gold, and E.M. Sternberg, Hippocampal 8-[3H]hydroxy-2-(di-n-propylamino) tetralin binding site densities, serotonin receptor (5-HT1A) messenger ribonucleic acid abundance, and serotonin levels parallel the activity of the hypothalamopituitary-adrenal axis in rat, *J. Neurochem.* 59:1062 (1992).
37. D.M. Camp, K.E. Browman, and T.E. Robinson, The effects of methamphetamine and cocaine on motor behavior and extracellular dopamine in the ventral striatum of Lewis versus Fischer 344 rats, *Brain Res.* 668:180 (1994).
38. I. Nylander, M. Vlaskovska, and L. Terenius, Brain dynorphin and enkephalin systems in Fischer and Lewis rats: effects of morphine tolerance and withdrawal, *Brain Res.* 683:25 (1995).

39. F. Chaouloff, A. Kulikov, A. Sarrieau, N. Castanon, and P. Mormede, Male Fischer 344 and Lewis rats display differences in locomotor reactivity, but not in anxiety-related behaviours: relationship with the hippocampal serotonergic system, *Brain Res.* 693:169 (1995).
40. P.W. Burnet, I.N. Mefford, C.C. Smith, P.W. Gold, and E.M. Sternberg, Hippocampal 5-HT1A receptor binding site densities, 5-HT1A receptor messenger ribonucleic acid abundance and serotonin levels parallel the activity of the hypothalamo-pituitary-adrenal axis in rats, *Behav. Brain Res.* 73:365 (1996).
41. G. Flores, G.K. Wood, D. Barbeau, R. Quirion, and L.K. Srivastava, Lewis and Fischer rats: a comparison of dopamine transporter and receptors levels, *Brain Res.* 814:34 (1998).
42. Y. Minabe, E.L. Gardner, and C.R. Ashby, Jr., Differential effects of chronic haloperidol administration on midbrain dopamine neurons in Sprague-Dawley, Fischer 344, and Lewis rats: an in vivo electrophysiological study, *Synapse 29*:269 (1998).
43. M. Werme, P. Thoren, L. Olson, and S. Brene, Addiction-prone Lewis but not Fischer rats develop compulsive running that coincides with downregulation of nerve growth factor inducible-B and neuron-derived orphan receptor 1. *J. Neurosci.* 19:6169 (1999).
44. S.E. Lindley, T.G. Bengoechea, D.L. Wong, and A..F. Schatzberg, Strain differences in mesotelencephalic dopaminergic neuronal regulation between Fischer 344 and Lewis rats. *Brain Res.* 832:152 (1999).
45. J. Ortiz, J.L. DeCaprio, T.A. Kosten, and E.J. Nestler, Strain-selective effects of corticosterone on locomotor sensitization to cocaine and on levels of tyrosine hydroxylase and glucocorticoid receptor in the ventral tegmental area. *Neuroscience.* 67:383 (1995).
46. E.M. Sternberg, J.M. Hill, G.P. Chrousos, T. Kamilaris, S.J. Listwak, P.W. Gold, and R.L. Wilder, Inflammatory mediator-induced hypothalamic-pituitary-adrenal axis activation is defective in streptococcal cell wall arthritis-susceptible Lewis rats, *Proc. Natl. Acad. Sci. U.S.A.* 86:2374 (1989).
47. M.R. Simar, D. Saphier, and N.E. Goeders, Dexamethasone suppression of the effects of cocaine on adrenocortical secretion in Lewis and Fischer rats, *Psychoneuroendocrinology* 22:141(1997).
48. Z. Sarnyai, E. Biro, J. Gardi, M. Vecsernyes, J. Julesz, and G. Telegdy G, Brain corticotropin-releasing factor mediates 'anxiety-like' behavior induced by cocaine withdrawal in rats, *Brain Res.* 675:89 (1995).
49. F. Rodríguez de Fonseca, M. Rocío, A. Carrera, M. Navarro, G.F. Koob, and F. Weiss, Activation of corticotropin-releasing factor in the limbic system during cannabinoid withdrawal, *Science* 276:2050 (1997).
50. R.F. Service, Probing alcoholism's 'dark side'. *Science* 285:1473 (1999).
51. P. Bice, T. Foroud, R. Bo, P. Castelluccio, L. Lumeng, T.K. Li, and L.G. Carr, Genomic screen for QTLs underlying alcohol consumption in the P and NP rat lines, *Mamm. Genome* 9:949 (1998).

INTERACTIONS OF OPIOID RECEPTORS, CHEMOKINES, AND CHEMOKINE RECEPTORS

Imre Szabo,[1,2] Michele Wetzel,[1] Lois McCarthy,[1] Amber Steele,[1] Earl E. Henderson,[1] O. M. Zack Howard,[3] Joost J. Oppenheim,[3] and Thomas J. Rogers[1]

[1]Department of Microbiology and Immunology
Center for Substance Abuse Research
and the Fels Institute for Cancer Research and Molecular Biology
Temple University School of Medicine
Philadelphia, PA 19140
[2]Department of Dermatology
University Medical School of Debrecen
Debrecen, Hungary H-4012
[3]Laboratory of Molecular Immunoregulation
Division of Basic Sciences
National Cancer Institute
Frederick Cancer Research and Development Center
Frederick, MD 21702-1201

INTRODUCTION

The endogenous opioid peptide ligands and opioid receptors are widely distributed in brain tissue and the periphery. Three classes of receptors have been identified for the opioids, designated μ, κ, and δ, and each of the opioid receptor genes expressed in brain tissue has been cloned and sequenced.[1-3] Results from radiolabeled-agonist- and antagonist-binding assay experiments suggest that these receptors also exist on cells of the immune system,[4,5] and μ-, κ-, and δ-opioid receptors have been cloned from cells of the immune system.[6-8]

Endogenous opioid compounds are primarily selective for the μ- and δ-opioid receptors β-endorphin, met-enkephalin (MetEnk), and leu-enkephalin, and the exogenous opioid agonist morphine, and have been shown to suppress antibody responses,[9,10] delayed-type hypersensitivity,[11] and natural killer cell activity.[12] Our recent studies have shown that resting macrophage phagocytic activity is depressed by treatment with μ-opioid-, δ-opioid-, and κ-opioid-selective agonists. The inhibitory activities of these compounds can be reversed by the corresponding opioid-type-selective antagonist.[13]

While the phagocytic activity of leukocytes is inhibited following opioid administration, both morphine and endogenous μ-, κ-, and δ-opioids induce chemotaxis of human monocytes and neutrophils.[14-16] It is significant that pre-treatment with opioids, including morphine, heroin, MetEnk, the more selective μ-agonist [D-ala^2, *N*-Me-Phe^4, Gly-ol^5]enkephalin (DAMGO), or the selective δ-agonist [D-Pen^2, D-Pen^5]enkephalin (DPDPE), leads to the inhibition of leukocyte chemotaxis in response to complement-derived chemotactic factors[17] and in response to the chemokines macrophage inflammatory protein (MIP)-1α, regulated on activation normal T cell expressed and secreted (RANTES), monocyte chemotactic protein (MCP)-1, and IL-8.[16] The results of Grimm *et al.*[16] suggest that this may be attributable to the activation of the μ- and δ-opioid receptors resulting in the desensitization of the chemokine receptors CCR1, CCR2, CXCR1, and CXCR2.

The first aim of our studies was to determine the impact of opioids on the expression of chemokines and chemokine receptors. We have studied the impact of the μ-opioid agonist DAMGO on the production of the chemokine MCP-1 and the chemokine receptor CCR5 by human peripheral blood leukocytes and murine thymocytes. Furthermore, studies were conducted to evaluate the impact of opioid administration on chemokine receptor function, as well as the converse effect of selected chemokines on the function of μ-opioid-induced chemotaxis.

EFFECT OF DAMGO ON HUMAN MONOCYTE CHEMOKINE EXPRESSION

Our previous studies showed that the production of the pro-inflammatory cytokines IL-1, IL-6 and TNF-α by both primary macrophages and the macrophage cell line P388D$_1$ is inhibited by κ-opioid agonist administration.[18,19] On the other hand, LPS-stimulated murine peritoneal macrophages pretreated with a low dose of morphine (50 nM) exhibited a naloxone-reversible increase in TNFα and IL-6.[20] In contrast, those receiving a high dose of morphine (50 μM) exhibited a decrease in IL-6 and TNFα which was not reversible with naloxone. It was determined that low doses of morphine augment LPS-induced NF-κB levels whereas high doses of morphine reduce NF-κB levels, suggesting that opioids may be affecting cytokine production at the transcriptional level.[21] Based on the established capacity of μ-opioids to modulate cytokine expression, we wished to determine the capacity of human peripheral blood mononuclear cells (PBMCs) to produce MCP-1 following administration of the μ-opioid agonist DAMGO. Our results (Table 1) show that DAMGO costimulates a 3.5-fold increase in the PHA-induced expression of this chemokine. Additional work (data not shown) has established that DAMGO administration also costimulates an increase in the expression of the chemokines RANTES and interferon-γ-inducible protein-10 by both phytohemagglutinin-stimulated and non-stimulated leukocytes. The alteration in the expression of MCP-1 may have significant consequences for the function of the immune system in a number of disease states. For example, it is becoming clear that MCP-1 is associated with the development of several inflammatory diseases, including multiple sclerosis, inflammatory bowel disease, and AIDS dementia.[22] The primary receptor for MCP-1 is CCR2b. Interestingly, the human CCR2b chemokine receptor is considered to be a minor co-receptor for HIV, but serves as a major monocyte co-receptor for certain HIV strains.[23]

IMPACT OF DAMGO ON CHEMOKINE RECEPTOR EXPRESSION

There has been little attention given to the impact of opioids on the surface expression of chemokine receptors. We wished to determine the effect of the μ-opioid DAMGO on the expression of CCR5 by peripheral blood leukocytes. Studies were carried out by treating human PBMCs for 48 hr with DAMGO (10 nM), and the level of CCR5 expression on gated lymphocytes and monocytes was determined by flow cytometry. The results (Table 2) show a twofold increase in the percentage of lymphocytes positive for CCR5 following treatment with DAMGO. It is well established that CCR5 is a major co-receptor for the monocyte-tropic strains of HIV. Recent studies have shown that IL-10 induces an elevation in the expression of CCR1, CCR2, and CCR5 in human monocytes.[24] Additional work will be necessary to determine whether the DAMGO-induced increase in CCR5 expression is mediated by an elevation in IL-10 expression.

Table 1. Increase in MCP-1 production following DAMGO administration.[1]

DAMGO concentration	MCP-1 level (pg/ml)
0	5,534 ± 37
1 μM	19,282 ± 2131

[1] 2×10^6 cells were cultured with PHA (5 μg/ml) with the designated concentrations of DAMGO for 72 hr, and the level of MCP-1 in the supernatant was determined by ELISA.

Table 2. Increase in CCR5 expression following DAMGO administration.[1]

DAMGO concentration	CCR5 expression
0	4.1 ± 1.1
10 nM	9.9 ± 1.8

[1] 3×10^6 cells were cultured with designated concentrations of DAMGO for 48 hr, and the level of CCR5 expression was determined by flow cytometry. The results are expressed as the percentage of positively stained gated lymphocytes.

RECIPROCAL REGULATION OF CHEMOKINE AND OPIOID RECEPTOR FUNCTION

Previous studies have established that pre-treatment with opioids leads to the inhibition of chemotaxis of neutrophils and monocytes both to complement-derived chemotactic factors[17] and to the chemokines MIP-1α, RANTES, MCP-1 and IL-8.[16] These results showed that the administration of morphine, heroin, MetEnk, the more selective μ-agonist DAMGO, or the selective δ-agonist DPDPE inhibited the subsequent chemotaxis of human peripheral blood neutrophils and monocytes to chemokine ligands for CCR1, CCR2, CXCR1, and CXCR2. Furthermore, this opioid-induced desensitization appears to be due to phosphorylation of the chemokine receptor.

We wished to extend these studies in an effort to more fully define the extent of the cross-talk between opioid and chemokine receptors. Experiments were carried out with a variety of murine and human primary cells and cell lines, and the capacity of various opioids and chemokines to desensitize chemotactic responses was determined. The results (summarized in Table 3) confirm the earlier published data[16] and demonstrate that both μ- and δ-opioids desensitize the response of human monocytes to the chemokine RANTES. In contrast, these results demonstrate that the desensitizing signal(s) delivered by DPDPE, MetEnk, and DAMGO do not alter the response of cells to the chemokine stromal cell-derived factor (SDF)-1α, the ligand for CXCR4. Conversely, the results also show that chemokines inhibit the chemotactic responses of human and murine cells to μ- and δ-opioids. The results show that RANTES, a ligand primarily for CCR1 and CCR5, is able to desensitize the response of human monocytes, keratinocytes, and murine thymocytes to DAMGO. Moreover, SDF-1α is able to induce desensitization of both DPDPE and DAMGO responses, in contrast to the failure of either opioid to desensitize the response to SDF-1. Finally, in contrast to other CC and CXC chemokines tested, the CXCR1 and CXCR2 ligand IL-8 fails to desensitize the response to DAMGO. This result is particularly interesting, since both μ- and δ-opioids have been reported to desensitize the responses of human leukocytes to IL-8.[16]

Table 3. Cross-talk between opioid and chemokine receptors.[1]

Desensitizer (inducer)	Target response	Desensitization	Cell population[2]
DPDPE (DOR[3])	RANTES (CCR1,5[4])	Yes	HM
DPDPE (DOR)	SDF-1α (CXCR4)	No	HM, HJ
MetEnk (DOR)	SDF-1α (CXCR4)	No	HM, HJ
MetEnk (DOR)	RANTES (CCR1,5[4])	Yes	HJ
DAMGO (MOR)	RANTES (CCR1,5[4])	Yes	HM, HK, MT
DAMGO (MOR)	SDF-1α (CXCR4)	No	HM, HJM
RANTES (CCR1,5[4])	DAMGO (MOR)	Yes	HM, HK, MT
SDF-1α (CXCR4)	DPDPE (DOR)	Yes	HM
SDF-1α (CXCR4)	DAMGO (MOR)	Yes	MT
ELC (CCR7)	DAMGO (MOR)	Yes	HK, MT
MCP-1 (CCR2)	DAMGO (MOR)	Yes	MT
IL-8 (CXCR1,2)	DAMGO (MOR)	No	HK

[1] Cells were pre-treated for 15 to 60 min with the designated desensitization agent at the optimal chemotactic concentration. Cells were then washed and tested for the designated chemotactic response (target response) to determine whether the inducing agent abolished the response.
[2] Cellular designations are: human monocytes, HM; human Jurkat T cells, HJ; human Jurkat T cells transfected with MOR, HJM; human HaCaT keratinocytes, HK; murine thymocytes, MT.
[3] Opioid receptor designations are: μ-opioid receptor, MOR; δ-opioid receptor, DOR.
[4] The impact of opioid treatment on the function of CCR5 is not clear at this time.

DISCUSSION

Our collective results suggest an extensive degree of regulation between the opioid and chemokine families both at the level of ligand and receptor expression and at the level of receptor function. We observe a significant elevation of both chemokine and chemokine receptor expression by PBMCs following μ-opioid administration, and RNAse protection analysis suggests that the opioid-mediated effects are at the level of transcription (data not shown). The mechanism of these effects is not clear at this time; however, we cannot rule out the possibility that the increase in chemokine and/or chemokine receptor expression is due to the function of an intervening cytokine. It is also possible that the activation of the μ-opioid receptor mediates the effects on MCP-1 and CCR5 expression through transcriptional activating factor(s). Activation of the μ-opioid receptor has been shown to result in a significant increase in the activity of NF-κB.[21]

The cross-desensitization of opioid and chemokine receptors suggest the possibility of a high level of selective "tuning" of the migratory activities of leukocytes by these two families of chemoattractants. The results also suggest an interesting hierarchy in the selectivity of the cross-desensitization phenomenon. Cross-talk between CC chemokine receptors and the opioid receptors was observed for all receptors tested. However, the opioid receptors fail to desensitize CXCR4, but successfully desensitize CXCR1 and/or CXCR2.[16] On the other hand, the CXCR1 and CXCR2 ligand IL-8 fails to desensitize the μ-opioid receptor, while the CXCR4 ligand SDF-1α desensitizes both μ- and δ-opioid receptors. The signaling cascades which are responsible for these desensitization reactions must possess a high level of target receptor specificity. We suggest that this high level of selectivity through combinations of chemoattractant substances more precisely directs the migration of leukocytes to particular tissue sites. In broader terms, it is possible that the opioids may function to interfere with inflammatory cell infiltration, while the chemokines may inhibit opioid receptor function and heighten pain perception at sites of inflammation.

ACKNOWLEDGMENTS

These studies were supported in part by grants DA06550, DA11130, DA12113, T32DA07237, and F31DA05894 from the National Institute on Drug Abuse.

REFERENCES

1. Y. Chen, A. Mestek, J. Liu, and L. Yu, Molecular cloning of a rat kappa opioid receptor reveals sequence similarities to the mu and delta opioid receptors, *Biochem. J.* 295:625 (1993).
2. S. Li, J. Zhu, C. Chen, Y.-W. Chen, J.K. DeRiel, B. Ashby, and L.-Y. Liu-Chen, Molecular cloning and expression of a rat kappa opioid receptor, *Biochem. J.* 295:629 (1993).
3. C.J. Evans, D. Keith, K. Magendzo, H. Morrison, and R.H. Edwards, Cloning of a delta opioid receptor by functional expression, *Science* 258:1952 (1992).
4. D.J.J. Carr, B.R. DeCosta, C.-H. Kim, A.E. Jacobson, V. Guarcello, K.C. Rice, and J.E. Blalock, Opioid receptors on cells of the immune system: evidence for δ- and κ-classes, *J. Endocrinol.* 122:161 (1989).
5. J.M. Bidlack, L.D. Saripalli, and D.M.P. Lawrence, κ-opioid binding sites on a murine lymphoma cell line, *Eur. J. Pharmacol.* 227:257 (1992).
6. L.F. Chuang, T.K. Chuang, K.F. Killam, Jr., A.J. Chuang, H.-F. Kung, L. Yu, and R.Y. Chuang, Delta opioid receptor gene expression in lymphocytes, *Biochem. Biophys. Res. Commun.* 202:1291 (1994).
7. S.M. Belkowski, J. Zhu, L.-Y. Liu-Chen, T.K. Eisenstein, M.W. Adler, and T.J. Rogers, Sequence of κ-opioid receptor cDNA in the R1.1 thymoma cell line, *J. Neuroimmunol.* 62:113 (1995).

8. M. Sedqi, S. Roy, S. Ramakrishnan, R. Elde, and H.H. Loh, Complementary DNA cloning of a μ-opioid receptor from rat peritoneal macrophages, *Biochem. Biophys. Res. Commun.* 209:563 (1995).
9. H.M. Johnson, E.M. Smith, B.A. Torres, and J.E. Blalock, Regulation of the in vitro antibody response by neuroendocrine hormones, *Proc. Natl. Acad. Sci. USA* 79:4171 (1982).
10. D.D. Taub, T.K. Eisenstein, E.B. Geller, M.W. Adler, and T.J. Rogers, Immunomodulatory activity of μ- and κ-selective opioid agonists, *Proc. Natl. Acad. Sci. USA* 88:360 (1991).
11. N.R. Pellis, C. Harper, and N. Dafny, Suppression of the induction of delayed-type hypersensitivity in rats by repetitive morphine treatments, *Exp. Neurol.* 93:92 (1986).
12. R.J. Weber and A. Pert, The periaqueductal gray matter mediates opiate-induced immunosuppression, *Science* 245:188 (1989).
13. I. Szabo, M. Rojavin, J.L. Bussiere, T.K. Eisenstein, M.W. Adler, and T.J. Rogers, Suppression of peritoneal macrophage phagocytosis of *Candida albicans* by opioids, *J. Pharmacol. Exp. Ther.* 267:703 (1993).
14. M.R. Ruff, S.M. Wahl, S. Mergenhagen, and C.B. Pert, Opiate receptor-mediated chemotaxis of human monocytes, *Neuropeptides* 5:363 (1985).
15. D.E. Van Epps and L. Saland, β-endorphin and met-enkephalin stimulate human peripheral blood mononuclear cell chemotaxis, *J. Immunol.* 132:3046 (1984).
16. M. Grimm, A. Ben-Baruch, D.D. Taub, O.M.Z. Howard, J.H. Resau, J.M. Wang, H. Ali, R. Richardson, R. Snyderman, and J.J. Oppenheim, Opiates transdeactivate chemokine receptors: delta and mu opiate-receptor mediated heterologous desensitization, *J. Exp. Med.* 188:317 (1998).
17. Y. Liu, D.J. Blackbourn, L.F. Chuang, K.F. Killam, and R.Y. Chuang, Effects of in vivo and in vitro administration of morphine sulfate upon Rhesus macaque polymorphonuclear cell phagocytosis and chemotaxis, *J. Pharmacol. Exp. Ther.* 265:533 (1992).
18. C. Alicea, S.M. Belkowski, T.K. Eisenstein, M.W. Adler, and T.J. Rogers, Inhibition of primary murine macrophage cytokine production in vitro following treatment with the κ-opioid agonist U50,488H, *J. Neuroimmunol.* 64:83 (1996).
19. S.M. Belkowski, C. Alicea, T.K. Eisenstein, M.W. Adler, and T.J. Rogers, Inhibition of interleukin-1 and tumor necrosis factor-α synthesis following treatment of macrophages with the kappa opioid agonist U50,488H, *J. Pharmacol. Exp. Ther.* 273:1491 (1995).
20. S. Roy, R.A. Barke, and H.H. Loh, MU-opioid receptor-knockout mice: role of mu-opioid receptor in morphine mediated immune functions, *Brain Res. Mol. Brain Res.* 61:190 (1998).
21. S. Roy, K.J. Cain, R.B. Chapin, and R.G. Charboneau, Morphine modulates NF-κB activation in macrophages, *Biochem. Biophys. Res. Commun.* 245:392 (1998).
22. F. Mennicken, R. Maki, E.B. de Souza, and R. Quirion, Chemokines and chemokine receptors in the CNS: a possible role in neuroinflammation and patterning, *Trends Pharmacological Sci.* 20:73 (1999).
23. B.J. Doranz, J. Rucker, Y. Yi, R.J. Smyth, M. Samson, S.C. Peiper, M. Parmentier, R.G. Collman, and R.W. Doms, A dual-tropic primary HIV-1 isolate that uses fusin and the β-chemokine receptors CKR-5, CKR-3, and CKR-2b as fusion cofactors, *Cell* 85:1149 (1996).
24. S. Sozzani, S. Ghezzi, G. Iannolo, W. Luini, A. Borsatti, N. Polentarutti, A. Sica, M. Locati, C. Mackay, T.N.C. Wells, *et al.*, Interleukin-10 increases CCR5 expression and HIV infection in human monocytes, *J. Exp. Med.* 187:439 (1998).

CROSSTALK BETWEEN CHEMOKINE AND OPIOID RECEPTORS RESULTS IN DOWNMODULATION OF CELL MIGRATION

Imre Szabo[1,2] and Thomas J. Rogers[1]

[1]Department of Microbiology and Immunology
Center for Substance Abuse Research
and the Fels Institute for Cancer Research and Molecular Biology
Temple University School of Medicine
Philadelphia, PA 19140
[2]Department of Dermatology
University Medical School of Debrecen
Debrecen, Hungary H-4012

INTRODUCTION

Opioids have been found to modulate the functional activities of a wide range of immune cells.[1] Recent data show that not only haematopoetic cells, but also epithelial keratinocytes respond to opioids in vitro.[2,3] These cells play an important role in skin immunity.[4,5] Their multipotent functions become manifest during wound healing; however, these cells are able to initiate or maintain inflammatory processes in morphologically intact skin as well.[6-8] Endogenous opioids released by nerve endings or produced during melanin synthesis have been reported to modulate the function of human keratinocytes.[2,3] Opioid ligand-receptor binding in responder immune cells was found to initiate processes responsible for cell migration.[9] Three types of synthetic opioid receptor agonists (DAMGO [μ-receptor], DPDPE [δ-receptor] and U50,488H [κ-receptor]) were used to assess the opioid responsiveness of primary murine thymocytes and human HaCaT keratinocytes.

MATERIALS AND METHODS

Cells and Culture Conditions

Primary murine thymocytes were isolated from 4- to 6-week-old male Balb/c mice, and the freshly prepared cell suspensions were used in migration and adhesion assays. HaCaT cells (human immortalized epithelial keratinocytes) were continuously cultured, and the presheet cultures were used for migration analysis. The DAMGO effect was analyzed in cells preincubated with CTAP at a concentration of 10^{-7} M for 30 min. CTAP was continuously present during the chemotaxis assay. For desensitization of μ-opioid receptors, cells were incubated in the presence of certain chemokines at a concentration of 100 ng/ml for 15 min at 37°C, and were then washed three times.

Migration Assay

The chemotactic migration induced by δ-, μ- and κ-opioid agonists was determined using 48-well microchemotaxis chambers with various pore size polycarbonate membranes (5 μ pore size fibronectin-coated membranes for primary murine thymocytes, and 12 μ pore size uncoated membranes for HaCaT keratinocytes). The migration of primary murine thymocytes was terminated after a 3-h incubation at 37°C; however, HaCaT keratinocytes required 45 min. Migrating cells were stained by hematoxylin-eosin and were counted by light microscopy. Chemotaxis is reported as the percentage change compared to the untreated control. All experiments were done in triplicate and repeated at least three times.

Adhesion Assay

Thymocytes were loaded into the wells of 96-well tissue culture plates coated with extracellular matrix proteins collagen I, collagen IV, fibronectin, laminin and vitronectin (Cytomatrix Screen kit; Chemicon, Temecula, CA), and were incubated in the presence of DAMGO for 1 h. Nonadherent cells were removed by repeated washing, and the adherent cells were stained with crystal violet. Optical density was determined at 570 nm after solubilization of dye in 1% SDS. All measurements were done in triplicate and repeated three times.

RESULTS

We found that primary murine thymocytes expressed strong migration toward DAMGO, but DPDPE and U50,488H induced little significant chemotaxis (Table 1). The HaCaT keratinocyte cells, however, migrated well toward all three opioid agonists. All opioids induced a concentration-dependent cell migration resulting a bell-shaped dose-response curve, with the maximum at 10^{-9}-10^{-10} M. The DAMGO effect was completely blocked by the specific receptor antagonist CTAP at a concentration of 10^{-7} M in both cell types. Pretreatment of primary murine thymocytes with CC chemokines RANTES, MIP-3β and MCP-1 at a concentration of 100 ng/ml completely blocked the DAMGO-induced chemotaxis (Table 2). Pretreatment of HaCaT keratinocytes with CC chemokines RANTES and MIP-3β, but not with the CXC chemokine IL-8, also resulted in an inhibition of the DAMGO-induced chemotaxis. In the presence of DAMGO, thymocytes expressed a significant increase in cell adhesion to collagen I, collagen IV and laminin, more modest adhesion to fibronectin, and failed to adhere to vitronectin (Table 3). The adhesion of highly adherent HaCaT keratinocytes was not significantly changed by DAMGO.

Table 1. Opioid agonist-induced cell migration.

Selective opioid agonists	Response	
	Murine thymocytes	HaCaT keratinocytes
DAMGO (m)	++++[1]	++
DPDPE (d)	-	++
U50,488H (k)	-	++

[1] Maximum chemotactic response achieved: ++++, ≥151% increase over the control; +++, >100% increase over the control; ++, 50-100% increase over the control; +, <50% increase over the control; -, no significant change.

Table 2. The effect of chemokine pretreatment on DAMGO-induced chemotactic migration.

Chemokine pretreatment	Response	
	Murine thymocytes	HaCaT keratinocytes
RANTES	↓[1]	↓
MIP-3b	↓	↓
MCP-1	↓	NT
IL-8	NT	-

[1] Change in DAMGO-induced migration: ↓, complete inhibition; -, no change; NT, not tested.

Table 3. The effect of DAMGO on murine thymocyte adhesion to various extracellular matrix protein-coated surfaces.

Surface coating	DAMGO effect
Uncoated	-[1]
Collage I	++
Collagen IV	++
Fibronectin	+
Laminin	++
Vitronectin	-

[1] ++, significant increase; +, moderate increase; -, no change.

DISCUSSION

Chemotactic migration represents one of the basic mechanisms of immune processes determining the direction and the intensity of immune cell trafficking. However, chemoattractant molecules, like chemokines, complement factors or fMLP, have not been found to promote chemotaxis with somewhat limited cellular specificity. Many chemokines, for example, can bind more than one receptor on distinct cell types, and most chemokine receptors can recognize multiple ligands.[10] Homologous and heterologous desensitization of chemokine receptors has been suggested to result in a more specific mediation of chemotaxis.[11,12] The complexity of chemotactic regulation further widened with the discovery of opioid-induced cell migration. Moreover, it was also demonstrated that interaction between opioid and chemokine receptors might result in the downregulation of certain chemokine responses.[13] We considered that reciprocal regulation of opioid receptors by chemokines might also exist. In order to further analyze the crosstalk between opioid and chemokine receptors, we studied the effect of chemokine pretreatment on opioid-induced migration in two distinct cell types, primary murine thymocytes and human keratinocytes, both expressing strong migration toward DAMGO. Pretreatment of these cells with the CC chemokine RANTES and MIP-3β resulted in a complete inhibition of DAMGO-induced chemotactic migration. Interestingly, the CXC chemokine IL-8, a potent chemoattractant for keratinocytes, was not able to block the DAMGO activity for this cell. As a part of the migration process, adhesion of murine thymocytes to certain extracellular matrix proteins was also increased by DAMGO, while the adhesion properties of the highly adherent keratinocytes were not modulated. These results provide evidence of heterologous desensitization of opioid receptors by certain chemokines in phylogenetically distinct cell types. Our data also suggest that crosstalk between opioid and chemokine receptors, which both belong to the seven transmembrane domain receptor family, might result in propagation of modulatory signals in both directions and provide a more specific regulation of the migration process.

REFERENCES

1. S. Roy and H.H. Loh, Effects of opioids on the immune system, *Neurochemic. Res.* 21:1375 (1996).
2. J.B. Nissen and K. Kragballe, Enkephalins modulate differentiation of normal human keratinocytes in vitro, *Exp. Dermatol.* 6:222 (1997).
3. P.L. Bigliardi, M. Bigliardi-Qi, S. Buechner, and T. Rufli, Expression of mu-opiate receptor in human epidermis and keratinocytes, *J. Invest. Dermatol.* 111:297 (1998).
4. J.D. Bos, ed., *Skin Immune System (SIS)*, CRC Press, Inc., Boca Raton, Florida (1990).
5. T.A. Luger and T. Schwarz, Evidence for epidermal cytokine network, *J. Invest. Dermatol.* 94:100 (1990).
6. M. Fukuoka, Y. Ogino, H. Sato, T. Ohta, K. Komoriya, K. Nishioka, and I. Katayama, RANTES expression in psoriatic skin, and regulation of RANTES and IL-8 production in cultured epidermal keratinocytes by active vitamin D_3 (tacalcitol), *Br. J. Dermatol.* 138:63 (1998).
7. R. Kulke, E. Bornscheuer, C. Schluter, J. Bartels, J. Rowert, M. Sticherling, and E. Christophers, The CXC receptor 2 is overexpressed in psoriatic epidermis, *J. Invest. Dermatol.* 110:90 (1998).
8. K. Nakamura, I.R. Williams, and T.S. Kupper, Keratinocyte-derived monocyte chemoattractant protein 1 (MCP-1): analysis in a transgenic model demonstrates MCP-1 can recruit dendritic and Langerhans cells to skin, *J. Invest. Dermatol.* 105:635 (1995).
9. D.E. Van Epps, and L. Saland, β-Endorphin and Met-enkephalin stimulate human peripheral blood mononuclear cell chemotaxis, *J. Immunol.* 132:3046 (1984).
10. B.J. Rollins, Chemokines, *Blood* 90:909 (1997).
11. R.M. Richardson, H. Ali, E.D. Tomhave, B. Haribabu, and R. Snyderman, Cross-desensitization of chemoattractant receptors occurs at multiple levels - evidence for a role for inhibition of phospholipase C activity, *J. Biol. Chem.* 270:27829 (1995).

12. P.Y. Law and H.H. Loh, Regulation of opioid receptor activities, *J. Pharm. Exp. Ther.* 289:607 (1999).

13. M.C. Grimm, A. Ben-Baruch, D.D. Taub, O.M.Z. Howard, J.H. Reau, J.M. Wang, H. Ali, R. Richardson, R. Snyderman, and J.J. Oppenheim, Opiates transdeactivate chemokine receptors: δ and μ opiate receptor-mediated heterologous desensitization, *J. Exp. Med.* 188:317 (1998).

MORPHINE UPREGULATES KAPPA-OPIOID RECEPTORS OF HUMAN LYMPHOCYTES

Shunji Suzuki[1], Teddy K. Chuang[1], Linda F. Chuang[1]
Roy H. Doi[2], and Ronald Y. Chuang[1]

[1]Department of Medical Pharmacology and Toxicology
[2]Section of Molecular and Cellular Biology
University of California, Davis, CA 95616

ABSTRACT

Opioids such as morphine are potent analgesic and addictive compounds. Chronic morphine use also induces immunomodulatory and immunosuppressive effects, as especially evident in HIV-infected patients. Morphine acts on the immune cells primarily through its binding to *mu*-opioid receptors on the plasma membrane. However, morphine modulation of immune functions still exists in *mu*-opioid receptor knockout mice, suggesting that in addition to the *mu* opioid receptors, morphine may also act by mechanisms mediated by either *delta* or *kappa* opioid receptors. To determine whether morphine activates *kappa* opioid receptors (KOR), a quantitative competitive RT-PCR procedure was utilized to quantify the KOR gene expression of morphine-treated cells. A segment of KOR transcript spanning the second extracellular loop, which has the reported dynorphin specificity, and the seventh transmembrane domain of the receptor was amplified from the total RNA of morphine-treated CEM x174 lymphocytes, along with a competitor molecule. The competitor was constructed by deleting a 33-nucleotide fragment from KOR. The results of the competitive RT/PCR indicated that CEMx174 cells expressed KOR mRNA constitutively, in the order of femto-grams. Treatment of 10 μM of morphine resulted in the up-regulation of KOR gene expression 24 hr post-treatment. The observed morphine effect could be reversed by treating the cells with either naloxone (a KOR-partially selective antagonist) or nor-Binaltorphimine (a KOR-selective antagonist).

Neuroimmune Circuits, Drugs of Abuse, and Infectious Diseases
Edited by Herman Friedman *et al.*, Kluwer Academic/Plenum Publishers, 2001

INTRODUCTION

Immune cells have been shown to express brain-like *kappa* opioid receptors (KOR) both at transcriptional (Chuang et al., 1995) and translational level (Lawrence et al., 1995). However, the immunological functions of KOR remain largely unknown. Like *mu* or *delta* opioid receptors (Law and Loh, 1999), KOR belongs to the G-protein-coupled receptor family with seven transmembrane domains; ligand-binding studies indicate that KOR may selectively dimerize with *delta* but not with *mu* opioid receptors to form a new functional receptor (Jordan and Devi, 1999).

In a recent study we demonstrated that opioids, at micromolar concentrations, suppress the chemokine-mediated migration of both monkey neutrophils and monkey monocytes (Miyagi et al., 1999). Using various opioid receptor agonists and antagonists in the study we found that this inhibition of leukocyte migration by opioids is mediated by opioids binding to *mu* or *kappa* receptors; binding to *delta* opioid receptors was rarely observed (Miyagi et al., 1999). Morphine is a potent analgesic and addictive opioid; it also elicits various immunomodulatory and immunosuppressive effects on rhesus monkeys when used chronically, including suppression of T-cell proliferation response, IL-2 release (Chuang et al., 1993), inhibition of polymorphonuclear cell phagocytosis and chemotaxis (Liu et al., 1992), and alteration of the disease progression of simian immunodeficiency virus (SIV)-infected animals (Chuang et al., 1997). It is generally recognized that morphine induces its immunological actions through binding to its specific receptors, primarily of the *mu* type (Reisine and Pasternak, 1996). In mice lacking the *mu* opioid receptor gene, the morphine-induced immunosuppression was found to be abolished (Gaveriaux-Ruff et al., 1998). However, in another study, it was reported that several morphine-induced immune functions, including morphine reduction of splenic and thymic cell number and mitogen-induced proliferation, and morphine inhibition of IL-1 and IL-6 secretion by macrophages, are not affected in *mu*-opioid receptor-knockout mice, suggesting that morphine may act by a mechanism mediated by either *delta* or *kappa* opioid receptors (Roy et al., 1998). Resorting to an in vitro system, the present study was undertaken to determine whether morphine may indeed activate *kappa* opioid receptors of immune cells.

MATERIALS AND METHODS

Cell Culture

The CEM x174 cell line, a hybrid of human B cell line 721. 174 and human T cell line CEM (Salter et al., 1985), was used as a model system to determine the effect of morphine on KOR gene expression of immune cells.

Treatment with morphine and/or antagonists

Morphine sulfate, naloxone (a KOR receptors-partially selective antagonist, Reisine and Pasternak, 1996), nor-Binaltrophimine (nor-BNI, a KOR receptors-selective antagonist, Reisine and Pasternak, 1996), and naltrindole (NTI, a delta opioid receptors-selective antagonist, Reisine and Pasternak, 1996), were used. CEM x174 cells were incubated in RPMI 1640 medium containing various concentrations of morphine for the indicated times. Antagonists, if used, were added 15 min prior to the morphine treatment.

RNA isolation

Total RNA was isolated from CEM x174 cells using TRIZOL Reagent (LIFE TECHNOLOGIES). Opioid-treated or control CEM x174 cells were collected by centrifugation at 800 rpm for 5 min. Cells were washed twice in PBS. The cells were then lysed by incubating for 5 min at room temperature in 1 ml of TRIZOL Reagent. After phenol-chloroform extraction, the aqueous phase containing RNA was collected. Total RNA was precipitated from the aqueous phase with isopropyl alcohol and then dissolved in RNase-free H_2O.

Quantitative RT-PCR

To quantify KOR gene expression, a quantitative competitive RT-PCR procedure was performed. Briefly, for the construction of a competitor molecule, a segment of KOR transcript was first amplified from total RNA of CEM x174 cells using RT-PCR. A competitor molecule was constructed by deleting a 33-nucleotide from the amplified segment (Fig. 1). The amount of KOR transcripts in the treated or control samples was determined by reverse transcription and then amplification by PCR in the presence of various concentrations of competitors. After 3 % agarose gel electrophoresis, the amplified PCR products were stained by ethidium bromide.

Analysis

The density of each band amplified from KOR cDNA and a competitor was measured by NIH Image TM Software (version 1.61). The amount of kappa mRNA was calculated from a linear regression analysis with log (competitor/kappa cDNA) versus log Competitor (fg).

RESULTS

Amplification of KOR transcript

A segment of KOR transcript spanning the second extracellular loop and the seventh transmembrane domain of the receptor was amplified from the total RNA of CEM x174 using a RT-PCR procedure with a set of K-1 and K-2 primers (Fig.1).

Effect of morphine on KOR gene expression

To determine whether morphine has an effect on KOR, a quantitative RT-PCR procedure was utilized to quantify the KOR gene expression in morphine-treated cells (Fig.2). The results indicated that CEM x174 cells expressed KOR mRNA constitutively, in the order of femto-grams and treatment of morphine from 100 pM to 10 μM resulted in an up-regulation of KOR gene expression 24 hr post-treatment (Fig. 3).

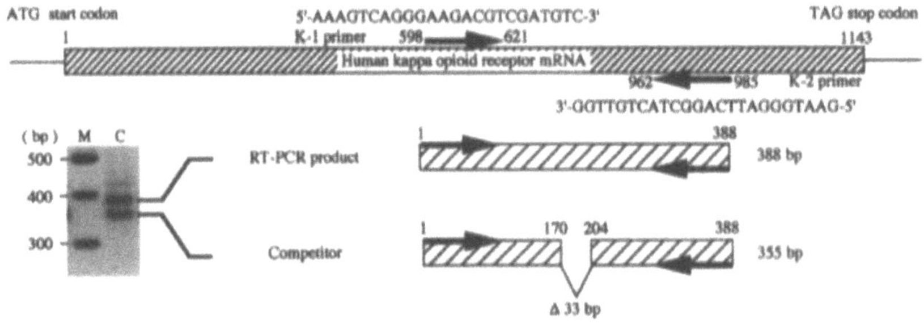

Figure 1. Structure of human KOR mRNA, a KOR RT-PCR product and construction of a competitor. Gel image shows the individual bands amplified from KOR mRNA and a competitor by a competitive RT-PCR. M), 100 bp DNA molecular size marker; C), competitive RT-PCR products.

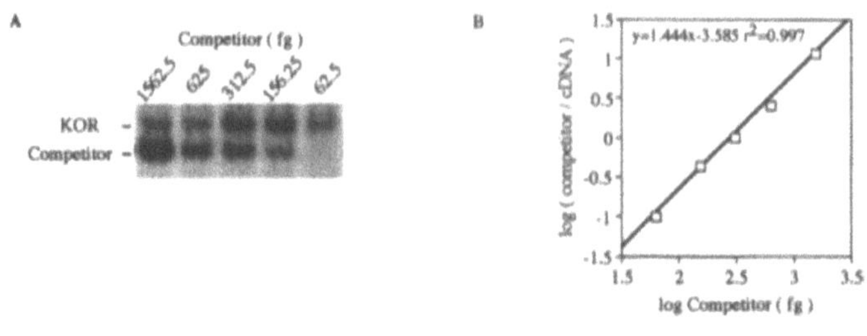

Figure 2. Quantitative RT-PCR of KOR mRNA. (A) Gel images of quantitative RT-PCR of KOR mRNA with a series of competitors. One μg of total RNA isolated from control cells was used. Amplified products were electrophoresed on 3 % agarose gels and visualized by ethidium bromide staining. (B) Data analysis of quantitative RT-PCR. The density of each band was measureded by NIH Image™ Software (version 1.61). The amount of KOR mRNA in 1 μg of total RNA was calculated from a linear regression analysis with log (competitor / KOR cDNA) versus log Competitor (fg).

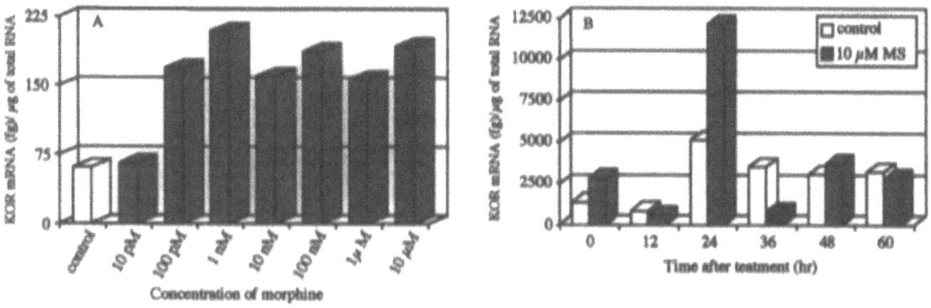

Figure 3. Effect of morphine treatment on KOR gene expression in human CEM x174 cells. A) a dose-response curve. CEM x174 cells were incubated in medium containing various concentrations of morphine, 24 hr post-treatment. B) a time-dependent effect. CEM x174 cells were cultured in medium containing 10 μM morphine (MS) or H_2O (control) for indicated times. After isolation of total RNA, a quantitative RT-PCR was performed and the amounts of KOR mRNA were determined by NIH Image Analysis.

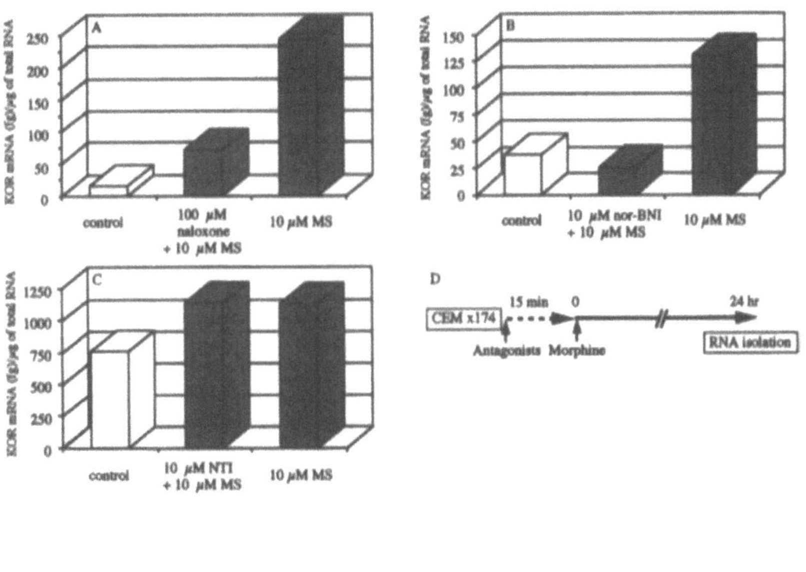

Figure 4. Effects of antagonists on morphine-induced up-regulation of KOR gene expression. CEM x174 cells were treated by A) 100 μM naloxone , B) 10 μM nor-Binaltrophimine (nor-BNI) and C) 10 μM naltrindole (NTI), followed by morphine (10 μM) treatment. D) Schedule of treatment.

Reversibility of morphine effects on kappa gene expression by pretreatment with antagonists

To determine whether morphine induces the up-regulation of KOR gene expression through its binding to opioid receptors, naloxone, nor-BNI and NTI, were used as antagonists. CEM x174 cells were first treated with naloxone (Fig. 4A), nor-BNI (Fig. 4B) or NTI (Fig. 4C) each at 10 μM for 15 min, and then incubated with 10 μM morphine. RNA was isolated for analysis 24 hr post morphine treatment. The results showed that the observed morphine effect could be reversed by pretreatment of cells with KOR antagonists, naloxone or nor-BNI, but not with NTI, a *delta* opioid receptor antagonist (Fig. 4).

Conclusion

The study shows that human CEM x174 lymphocytes express KOR mRNA constitutively, in the order of femto-grams. Treatment of 100 pM to 10 μM morphine resulted in an increase of KOR gene expression 24 hr post-treatment. The observed morphine effect could be reversed by pretreatment of cells with naloxone or nor-BNI, but not with NTI. Together with the studies on *mu*-opioid receptor-knockout mice (Roy et al., 1998), the current study demonstrated that in addition to *mu* opioid receptors, morphine may exert its immunomodulatory effect via activating kappa opioid receptors (KOR).

ACKNOWLEDGMENTS

This work was supported by NIH research grants DA 10433 and DA 05901 from the National Institute on Drug Abuse.

REFERENCES

Chuang, L.F., Chuang, T.K., Killam, K.F.Jr., Qiu, Q., Wang, X.R., Lin, J-j, Kung, H-f., Sheng, W., Chao, C., Yu, L., and Chuang, R.Y., Expression of kappa opioid receptors in human and monkey lymphocytes, *Biochem. Biophys. Res. Commun.* 209:1003-1010 (1995).

Chuang, L.F., Killam, K.F.Jr., and Chuang, R.Y., Opioid dependency and T-helper cell functions in rhesus monkey, *In Vivo* 7:159-166 (1993).

Chuang, L.F., Killam, K.F.Jr., and Chuang, R.Y., SIV infection of macaques: a model for studying AIDS and drug abuse, *Addiction Biol.* 2:421-430 (1997).

Gaveriaux-Ruff, C., Matthes, H.W.D., Peluso, J., and Kieffer, B.L., Abolition of morphine-immunosuppression in mice lacking the μ-opioid receptor gene, *Proc. Natl. Acad. Sci. USA* 95:6326-6330 (1998).

Jordan, B.A., and Devi, L.A., G-protein-coupled receptor heterodimerization modulates receptor function, *Nature* 399:697-700 (1999).

Law, P-Y, and Loh, H.H., Regulation of opioid receptor activities, *J. Pharmacol. Exp. Thera.* 289:607-624 (1999).

Lawrence, D.M., el-Hamouly, W., Archer, S., Leary, J.F., and Bidlack, J.M., Identification of kappa opioid receptors in the immune system by indirect immunofluorescence, *Proc. Natl. Acad. Sci. USA* 92:1062-1066 (1995).

Liu, Y., Blackbourn, D.J., Chuang, L.F., Killam, K.F.Jr., and Chuang, R.Y., Effects of in vivo and in vitro administration of morphine sulfate upon rhesus macaque polymorphonuclear cell phagocytosis and chemotaxis, *J. Pharmacol. Exp. Thera.* 263:533-539 (1992).

Miyagi, T., Chuang, L.F., Lam, K.M., Kung, H-f, Wang, J.M., Osburn, B.I., and Chuang, R.Y., Opioids suppress chemokine-mediated migration of monkey neutrophils and monocytes – an instant response, *Immunopharmacology*, in press (1999).

Reisine, T., and Pasternak, G. Opioid analgesics and antagonists. In: Hardman JG, Limbird LE, Molinoff PB, Ruddon RW and Gilman AG, Eds. The Pharmacological Basis of Therapeutics. McGraw-Hill Co., New York, 521-555 (1996).

Roy, S., Barke, R.A., and Loh, H.H., MU-opioid receptor-knockout mice: role of μ-opioid receptor in morphine mediated immune functions, *Molecular Brain Res.* 61:190-194 (1998).

Salter, R.D., Howell, D.N., and Cresswell, P., Genes regulating HLA class I antigen expression in T-B lymphoblast hybrids, *Immunogenetics* 21:235-246 (1985).

EFFECTS OF MORPHINE ON T-CELL RECIRCULATION IN RHESUS MONKEYS

Robert M. Donahoe,[2*†‡] Larry D, Byrd,[‡] Harold M. McClure,[‡] Mary Brantley,[*] DeLoris Wenzel,[‡] Aftab Ahmed Ansari,[†‡] and Frederick Marsteller[*]

Departments of [*]Psychiatry and Behavioral Sciences, and [†]Pathology and Laboratory Medicine, the School of Medicine, and the [‡]Yerkes Regional Primate Research Center, Emory University, Atlanta GA 30322

ABSTRACT

A 2-yr study on effects of morphine on lymphocyte circulation in rhesus monkeys (*Macaca mulatta*) showed that, over time, a well-maintained morphine-dependency caused biphasic depressive effects on circulating lymphocyte levels. Depression of T cell circulation by opiates actually was a relative effect. Morphine exposure basically stabilized T cell circulation in the context of concurrent increases in controls. Biphasic effects of morphine were attributable to distinctions in circulation kinetics of CD4+/CD62L (+ & -) T cells. That is, levels of CD4+/CD62L+ T cells were selectively depressed by opiates through the first 32wk after initiation of drug, and levels of CD4+/CD62L- T cells were selectively depressed thereafter. Regression analyses also showed that morphine stabilized lymphocyte recirculation. Circulating levels of resting and activated-memory types of T cells were positively correlated in opiate-exposed monkeys during the first 32wk after opiate exposure--an effect not seen with control monkeys. Considerations of changes in the types of experimental stressors extant during the study suggested that temporally differential effects of opiates on T cell recirculation were connected with changes in the stress environment and the ability of morphine to modulate these changes. Thus, morphine, and by inference the endogenous opioid system, are involved in homeostasis of lymphocyte recirculation, probably through effects on central mediation of the stress axis.

Neuroimmune Circuits, Drugs of Abuse, and Infectious Diseases
Edited by Herman Friedman *et al.*, Kluwer Academic/Plenum Publishers, 2001

INTRODUCTION

For over a century there has been sporadic interest in defining the immunological effects of opiates (Cantacuzene, 1898; Plotnikoff et al., 1986; Friedman et al., 1993; Donahoe and Vlahov, 1998). Repeatedly, it has been shown that intravenous opioid abusers are inordinately susceptible to opportunistic infections (Sapira, 1968; Friedman et al., 1993), cancer (Sadeghi et al., 1979) and AIDS (Wormser et al., 1983; Friedman et al. 1993; Donahoe and Vlahov, 1998). From these findings, it has been considered that opioids may directly impact immunocompetence in addicts. Numerous in vitro and in vivo animal-model studies (Sharp, 1998) support this idea by showing that exogenous and endogenous-peptide opioids influence various immunological parameters--including, for morphine, immunocompetence of rodents (Tubaro et al., 1983, Bussiere et al., 1992), swine (Molitor et al., 1991; Risdahl et al., 1993) and monkeys (Chuang et al., 1993; Donahoe et al., 1993) in response to bacterial, fungal and viral infections.

One aspect of immune function investigated for opioid effects with variable results has been the influence of opioids on leukocyte circulation. Some of the earliest and most substantial data indicate that opioids play a definite role in modulating migration of polymorphonuclear leukocytes and mononuclear phagocytes (Cantacuzene, 1898; Ruff et al., 1985, Eisenstein and Hilburger, 1998). The case for lymphocyte recirculation is less certain, however. Depressive opioid effects have been reported with rodents (Flores et al., 1995) but negative results have also been obtained (Fecho et al., 1993). Opioids have also been reported to affect T cell circulation in monkeys (Carr and France, 1993). Still, systematic examinations of chronic effects of opioid exposure on lymphocyte circulation under pharmacological conditions favorable to maintaining homeostasis (i.e., dependencies that largely avoid physiological disruption due to opiate withdrawal) have yet to be reported. We have been studying such a model using rhesus macaques. We report here that morphine delivered 4 times daily by subcutaneous/intramuscular injection for nearly 2 years causes significant changes in circulating T cells and their CD4+/CD62L (+ & -) subsets. Our data suggest that opioids play a homeostatic role in controlling recirculation of lymphocytes and that this role is manifest, at least in part, through central modulation of host stress response systems.

MATERIALS AND METHODS

Animals

Twenty clinically healthy female rhesus monkeys (*Macaca mulatta*) weighing from 5 to 10 kg and ranging in age from 17 to 20 years were used. The monkeys were given psychological enrichments and housed singly in cages in an animal room shared with 4 other monkeys. To maintain behavioral homogeneity and avoid cyclical complications of menses, the monkeys chosen had ovariectomies years prior to this study.

Experimental design

After 8-mo of assay standardization and animal adaptation, baseline-response data were obtained 9 wk prior to the start of morphine and placebo injections. For data analyses, the day injections were initiated was designated day-0. Injections were stopped at wk-82 causing the morphine-dependent monkeys to withdraw from their dependency. Accordingly, immune analyses were performed with blood samples collected 9 wk before day-0 (-9 wk), and at 2, 10, 20, 32, 35, 77 and 92 wk after day-0. To allow sufficient technical time to complete assays, 4 groups of 5 monkeys each (2 or 3 test or controls) were initiated into the study in consecutive weeks. This time-dependent stagger of blood collections was maintained through wk-32 of the study. At wk-32, time time-dependent variations in data between the staggered experimental groups were no longer apparent. Therefore, the blood-collection stagger was stopped. Thus, after wk-32 blood was collected *en masse* from all animals anesthetized as a group on the day of collection.

Reagent injection and blood collection

The morphine-dependency paradigm was after Woods et al. (1984). Injections of morphine sulfate diluted in saline, or a saline placebo were given s.c./i.m. into the upper flank of the lower left leg using sterile, 3/4-inch, 27-gauge needles attached to 1-ml tuberculin syringes. Except as otherwise specified, morphine or placebo were injected every 6 hr during the study. Per injection, the animals received 1mg of morphine per kg body-weight, for the first 3 days; then 2mg/kg for 3 more days; and 3mg/kg, thereafter. Saline placebo was used in proportional volumes relative to animal weights. Animals were anesthetized for blood collection with ketamine hydrochloride (10-20 mg/kg). Blood was withdrawn from the femoral vein by venipuncture between 9:00 and 10:30 AM and collected in EDTA-containing vacutainers.

Differential histological assessment of whole-blood leukocytes

Blood smears were stained with Wright's Geimsa stain. Leukocytes were differentiated by microscopic examination and absolute numbers of lymphocytes determined in relation to whole white blood cell counts obtained with an electronic cell counter (Coulter Electronics, Inc., Hialeah, FL) for each monkey.

Leukocyte subtyping by cytofluorometry

A dual-parameter, whole-blood staining procedure defined by Becton Dickinson (B-D, Mountain View, CA) and a FACScan cytofluorometer from the same manufacturer were used to assess percentages of various leukocyte types as described previously (Ansari et al., 1989). Absolute cell numbers are represented in the data, determined relative to percentages of whole blood cell counts. Briefly, whole blood in EDTA and maintained at room temperature was incubated for 20 min with combinations of monoclonal antibodies conjugated with either fluorescein isothiocyanate (FITC) or phycoerythrin (PE). Red blood cells were then lysed with

B-D lysing reagent and debris was washed from cell pellets after centrifugation. Cells were then suspended in phosphate buffered saline (PBS) with 0.1% NaN_3 and analyzed on the FACScan for forward- and right-angle light scatter characteristics so that the lymphocyte population could be gated for subset analyses. The following antibody (B-D) panel was used [(+) = positive for staining; (-) = negative]: 1. nonspecific mouse FITC-IgG_1 and PE-IgG_2 as background controls; 2. Leu5b-FITC and Leu16-PE (CD2+ pan T cells and CD20+ mature B cells); 3. Leu2a-FITC and Leu3a-PE (CD8+ and CD4+ T cells, respectively); 4. Leu3a-FITC and Leu8-PE [CD4+ T-helper cells for B cells (Leu3a+/Leu8-), and CD4+ T-cells (Leu3a+/Leu8+) that express the peripheral lymph node homing receptor, CD62L]; 5. Leu18-FITC and Leu3a-PE [CD4+ T-helper cells for B cells (Leu3a+/Leu18-, CD45RA-) and CD4+ T-cells (Leu3a+/Leu18+, CD45RA+)]; 6. Leu3a-FITC and DR-framework-PE (CD4 cells that express Class-II mixed histocompatibility complex).

Data Collection and Statistics

Univariate statistics including correlations were computed for cell-count data using SPSS/PC software. For multivariate analyses, substitute scores for missing values due to sample loss or inadequate sample size were imputed using the BMDP AM routine (Dixon, 1992). Less than 3% of values were imputed. Analyses of experimental effects on immunological and behavioral measures were computed using mixed model repeated measures analyses in BMDP 4V (Dixon, 1992). In the repeated-measures analyses, the baseline measurement was the reference for contrasts at subsequent time-points. For the contrast-by-treatment interaction terms (probabilities at each time-point), contrasts at any given time-point when samples were collected corresponded to the null hypothesis that the difference in the average change in a measure from baseline to the time of sample-collection was zero. For the overall time-by-treatment interaction term (run effect), the corresponding null hypothesis was that, at all time points, the difference in average change compared to baseline was zero.

RESULTS

Morphine effects on expression of circulating leukocytes

Effects of morphine exposure on absolute numbers of circulating lymphocytes assessed by differential cell-counting of histologically stained blood smears are shown in Figure 1A. Morphine exposure had significant depressive effects on circulating levels of lymphocytes relative to controls over the course of the study (run effect: $p<0.04$). As such, morphine stabilized circulating levels of lymphocytes relative to variable control levels. Depression was not significant at wk-32, however, nor at wk-92, after opiate withdrawal. Accordingly, the kinetics of opiate effects on lymphocyte circulation was biphasic, a phenomenon seen with all T cell data: CD2+ pan-T cells (Fig. 1B); and the major T cell subtypes, CD4+ (Fig.1C), and, to a lesser extent, CD8+ (Fig.1D) T cells. This type of kinetic effect was not seen for B cells, however (data not shown). Notably, for CD4+ T cells, biphasic kinetics was attributable to temporal distinctions in morphine effects on CD4+ T cell subsets characterized with antibodies to the CD62L lymph node homing receptor (Camerini et al., 1989). Early-phase depression of circulating CD4+ T cells by morphine compared to placebo was restricted, essentially, to CD62L+ cells (Fig. 2A), before wk-32, while late-phase depression was restricted thereafter to CD62L- cells (Fig. 2B).

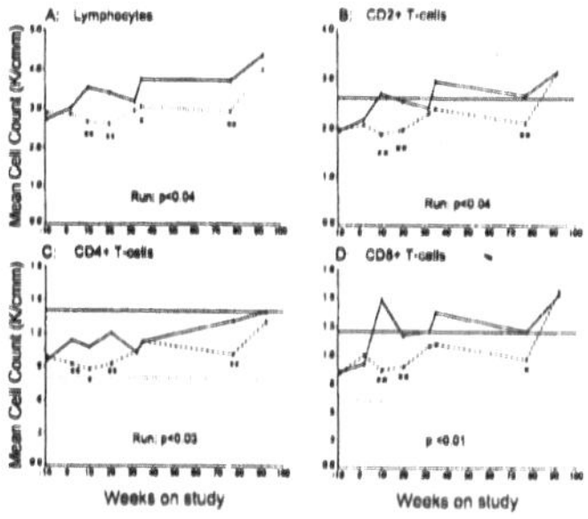

FIGURE 1. Effects of morphine exposure on mean circulating absolute numbers of total lymphocytes (A), and CD2+ (B), CD4+ (C) and CD8+ (D) T cells. Blood for these analyses was collected at -9, 2, 10, 20, 32, 35, 77 and 92 wk. Data collected at -9 wk represent baseline responses of the animals 9 wk before initiation (at wk-0) of 4-times daily injections of either morphine (dashed lines) or saline (solid lines). Solid horizontal lines in panels B, C and D represent norms for rhesus monkeys (Ansari et al., 1989). Dashed horizontal lines represent 1 standard deviation from these norms. Statistical significance for intergroup differences for each time-point of assessment is represented by one ($p \leq 0.05$) or two ($p \leq 0.01$) stars. Statistics concerning comparative differences between the two kinetic curves are presented as run effects. Scales on the ordinate axis vary according to the relative circulating frequency of the cell-type assessed.

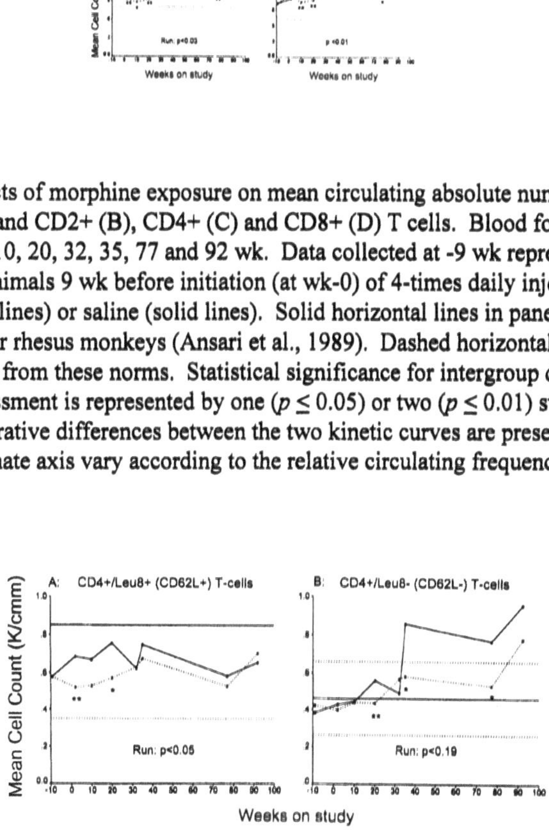

FIGURE 2. Effects of morphine exposure on mean circulating absolute numbers of total CD4+/CD62L+ (A) or CD4+/CD62L- (B) T cells. Collection regimens, statistical analyses and experimental details are the same as described in Figure 1. Morphine data are represented with dashed lines and saline data with solid lines.

Correlative analyses of kinetics of circulating levels of T-cell subset pairs

Stabilizing effects of morphine on lymphocyte circulation were also apparent in regression analyses of circulating levels of activated and inactivated CD4 subset pairs defined by expression of surface activation markers [CD62L, CD45RA, DR and double positive (DP)—CD4+/CD8+ T cells] (Morimoto et al., 1985; Blue et al., 1985; Kanof and James, 1988; Hamann et al., 1988). Morphine, but not saline, exposure caused circulating levels of pairs of non-activated and activated T-cell types [1. CD4+/CD45RA+ and CD4+/CD62L- cells (Fig. 3A), respectively; 2. CD4+/DR- and CD4+/CD62L- cells (Fig. 3B); 3. CD4+/CD45RA+ and DP cells (Fig. 3C)] to be positively correlated throughout the first 32 wk after initiation of morphine or saline injections. After wk-32, correlation coefficients for these pairs of T-cell subsets were generally highly positive for both test and controls. Similar correlative relationships were seen, in a relative sense, for associated expression of CD4+/CD62L+ with CD4+/CD62L- T cells (Fig. 3D), and CD4+/CD62L+ with DP T cells (Fig. 3E). The temporal contrasts in correlative expressions of these various parameters was most apparent when the data for all pairs of T cell subsets were normalized (control = zero) (Fig. 3F).

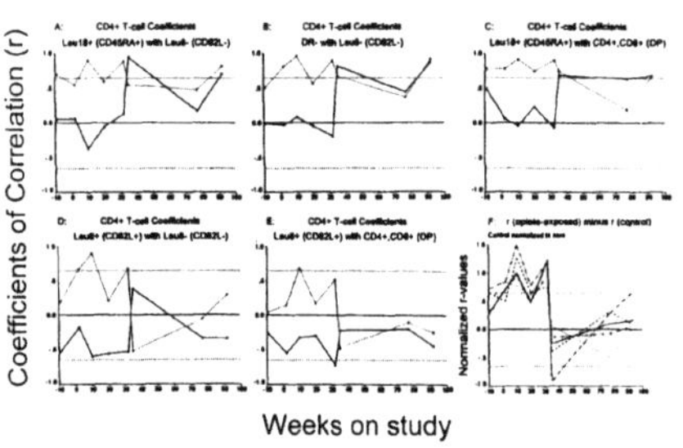

FIGURE 3. Effects of morphine exposure on correlation coefficients for the associated expression of CD4+ T-cell subsets characterized as nonactivated types (CD45RA+, DR- and CD62L+) in comparison with activated types (CD62L- and DP). Regression analyses were run on the following T cell subset pairs: CD4+/CD45RA+ T cells correlated with CD4+/CD62L- T cells (A); CD4+/DR- with CD4+/CD62L- (B); CD4+/CD45RA+ with DP (C); CD4+/CD62L+ with CD4+/CD62L- (D); CD4+/CD62L+ with DP (E). Correlation coefficients for test animals were also normalized relative to controls (F). Dotted horizontal lines represent a measure of statistical significance at $p \leq 0.05$. Experimental conditions are as presented in Figure 1. Morphine data are represented with dashed lines and saline data with solid lines--except for the normalized data in (F), which are delineated accordingly: *dashed line* = normalized data for (A); *dotted line* = normalized data for (B); *line with large and small dots* = normalized data for (C); *dashed and dotted line* = normalized data for (D); *solid line* = normalized data for (E).

DISCUSSION

The repeated-measures design of this study, the statistically significant effects of morphine on circulation of lymphocytes seen herein, and the corroboration of these data with that of other studies (Carr and France, 1993; Flores et al., 1995) leave little doubt that chronic morphine dependency alters lymphocyte circulation. There is also compelling evidence that exogenous and endogenous opioids mediate immune processes through interaction with the endogenous opioid system (Plotnikoff, 1986; Heijnen et al., 1991), and that immune cells produce their own endogenous opioids (Blalock, 1985). Collectively, these findings indicate that the endogenous opioid system is an important mediator of lymphocyte recirculation.

The present data show that morphine exposure depressed circulating levels of lymphocytes and T cells (especially CD4+ T cells) in two distinct phases, before and after the 32nd wk of drug exposure. This depression was relative, however, which relates to the fact that morphine exposure basically stabilized circulating T-cell levels in the context of concurrent increases in controls, in both phases of analyses. That is, until the end of each phase when test and controls equalized—a phenomenon attributable to tolerance to morphine, in the first phase, and to normalization after morphine withdrawal, in the second phase.

The relative depressive effects of morphine on T cell levels were temporally differential in that CD62L (+) and (-), CD4+ T cells were depressed at different stages during the study. This novel observation demonstrates that opiates have differential impact on immune function since relative expression of CD62L distinguishes T cells as having inactivated-naïve (CD62L+) or activated-memory (CD62L-) status (Morimoto et al., 1985; Blue et al., 1985; Kanof and James, 1988; Hamann et al., 1988).

A key to understanding the temporal differential effects of morphine on CD62L (+ or -) T cells lies in the relative nature of these effects. This relativity indicates that adaptive processes in control animals underlay these data, and that morphine effects were essentially counter-adaptive. Accordingly, it follows logically that stress was the underlying process involved in precipitating the adaptations; that experimental manipulation was the source of this stress; and that opiates had a counter-adaptive influence over this stress.

Retrospective evaluation of the stresses associated with manipulation of the animals in this study support the conclusion that opiates affected circulation of CD62L T cell subsets in a temporally differential counter-adaptive way. This consideration fits also with various data that have linked effects of opioids and stress, alone and together, on lymphocyte circulation and other immune processes (Hernandez et al., 1993; Fecho et al., 1993; Baddley et al., 1993; Carr and France, 1993; Flores et al., 1995; Steele et al., 1971; Ernstrom and Sanberg, 1974; Crary et al., 1983; Madden and Livnat, 1991; Ottoway and Husband, 1994).

The main experimental stressors encountered by the animals in this study, through wk-32, were associated with manipulation of the monkeys during administration of 4-times-daily reagent injections and with blood collection. Considering control animals separately, it is logical to presume that the chronic, repeated nature of the saline injections used fostered a relatively rapid adaptation to the novelty of the stresses involved. In this way injection stress was likely to have evolved relatively quickly from an acute, to chronic intermittent type, in accordance with definitions of McCarty (1994). In contrast, adaptation to the stress of blood collections ongoing during this same time was apt to proceed more slowly because these collections were sporadic and because of the unusually stressful nature of the collection-stagger used. This stagger caused 5 different monkeys to be carried about the room during each of the 4 weeks that comprised a single, early-stage, blood-collection episode. Accordingly, cohorts of the animals being bled witnessed these events. Witnessing animals being carried about by staff is a well-known, socially stressful context for monkeys. Thus, through wk-32 wk of the present study, the stress environment evolved from an initially predominantly acute stage to one where chronic intermittent types of stress associated with reagent injections overlapped acute aspects of stress associated with blood collections.

Since kinetics of opiate effects on T cell circulation seen herein were biphasic, demarcated by a phase shift at wk-32, it is especially notable that the stress environment after wk-32 differed dramatically from that before wk-32. After wk-32, blood collections were no longer staggered. All monkeys were bled as a group on a single day. This procedural change essentially eliminated the social context of witnessing cohorts being carried about, thereby effectively eliminating acutely stressful aspects of blood collection. The experimental stresses that remained were associated with the routine reagent injections and other daily handling routines. Essentially, therefore, during the latter stage of the present study, after wk-32, the experimental stresses present can be presumed to have been largely of the chronic intermittent type.

Given the different stress environments that can be surmised to have existed before and after wk-32 of the present study, the temporally differentially counter-adaptive effects of opiates on T cell circulation seen herein can be explained further by considering the probable involvement of norepinephrine (NE) output. It is well known that NE output can modulate lymphocyte circulation (Madden and Livnet, 1991). Also, many immunological effects of opiates are attributable to regulation of sympathetic outflow (Hernandez et al., 1993; Fecho et al., 1993; Baddley et al., 1993). Thus, for the control situation, it is important that our analyses indicate that acute stress, later overlaid with chronic intermittent stress (CIS), prevailed during the early stages of the present study, because these types of stress conditions are known to foster high output of NE (McCarty, 1994). In turn, our analyses that showed that conditions of CIS prevailed later in this study indicate that NE should have been relatively low in controls at this time, because NE is sequestered in neurons under conditions of CIS (McCarty 1994). Therefore, since opioids foster release of NE, directly (Van Loon et al., 1981; Appel et al. 1986), and, indirectly (Harris and Aston-Jones, 1993), it is likely that opioids counteracted circumstances of NE release that otherwise were present in this study in control animals. That is, morphine exposure likely caused a relative reduction in NE release under conditions of acute stress and a relative increase in NE release under conditions of CIS. For these reasons, we surmise that opiate-induced NE tone is the likely cause for the opiate-induced changes in T cell recirculation seen herein. This reasoning fits well, too, with the fact that circulation of T cells overall did not change much in the presence of opiate exposure.

Invoking NE involvement as a causal factor in the outcome of the results of this study also makes sense in the context of known roles of NE in modulating lymphocyte circulation. Catecholamines are known to induce lymphocyte release from the spleen (Steele et al. 1971; Ernstrom and Sandberg, 1974; Crary et al., 1983) and secondary immune organs are prominently innervated with noradrenergic processes (Felten and Felten, 1991). Therefore, the likely elevations in NE outflow described above, in the early stage of the present study, could well have caused bolus release of CD4+/CD62L+ T-cells from lymphocyte reservoirs like the spleen. Such effect would be expected to lead to incremental increases in circulation of these cells, as was seen herein for controls. The fact that NE release is typically rapid and peaks early after exposure to acute stress (McCarty, 1994) does not negate these considerations because NE effects on circulating levels of T cells could be expected to be fairly enduring, requiring more slowly developing immunological counter-adaptations to readjust. It follows, too, that tonic release of

NE by morphine exposure early in the study would be expected to counterbalance release of lymphocytes from the spleen.

Circumstances involving NE effects in the latter stages of the study, on circulation of CD62L- T cells, were likely to have been different because the activated-memory status of these cells, in response to proliferative stimuli, dictates their propensity to recirculate (Kanof and James, 1988; Hamann et al., 1988; Picker, 1994). Therefore, since T cell proliferation is down-regulated by catecholamines (Crary et al., 1983; Madden and Livnat, 1991; Murray et al., 1992), it follows that reduced NE release into secondary lymphoid organs, as we are projecting to have occurred herein for control animals after wk-32, would result in increased fidelity of T cell activation to ambient antigens. Such effect could be the cause for the elevations in circulation of these cells as was seen herein for control animals in the latter part of the study. In contrast, by keeping NE tone relatively high during this same time, morphine would be expected to dampen T cell productivity in response to ambient antigens and cause relative lowering of circulating CD62L- T-cell levels, as was seen herein. Reports that some immune effects of opioids (Lysle et al., 1993) and stress (Keller et al., 1991) are compartmentalized within secondary lymphoid organs of the immune system further support such suppositions.

Obviously, the foregoing speculation will require direct experimental substantiation. It is notable, therefore, that we have been unable to find direct effects of opiates and NE, alone or in combination, on resting or activated CD62L T-cell marker expression in vitro (manuscript in preparation). Also, the present data show that T cell circulation was actually depressed at the outset of this study, to nearly 1 standard deviation below established norms (Ansari et al., 1989). In this context, it is notable that these norms appear valid as a gauge of initial depression in circulating T cell levels because these levels, at the end of the study, after ample time for normalization had occurred, were very near these norms. Therefore, we suspect that initial pre-test depression in T cell circulation was related to the overall stress of animal manipulations prior to the start of the study. In this way, the ability of opiate exposure to stabilize T cell circulation at levels around initially depressed levels, as seen herein, may have been due to the ability of morphine to extend effects mediated by endogenous opioid pathways engaged initially by pre-test stressors. This consideration identifies these circumstances with incidences of pharmacological cross-tolerance (Christie et al., 1982) and further implicates the endogenous opioid system in the opioid effects seen herein.

In conclusion, the net effect of morphine exposure in this study was to balance circulation of lymphocytes, T cells in particular, even though these effects were manifest differentially. This homeostatic effect was evident also in the ability of morphine to maintain positive coefficients of correlation between circulating levels of resting and activated-memory types of T cells. We have found in other studies that morphine administration as done herein, in a homeostatically balancing way, may retard progression of simian AIDS (Donahoe et al., 1993; Donahoe, 1993b). Such effects could well be related to the homeostatic influences of opiates on T cell circulation since a balance in T cell expression is known to be critical to development of AIDS in HIV-1 infected individuals (Balter, 1995; Nowak and McMichael, 1995). These considerations fit well with findings of Capitanio et al. (1998) that progression of AIDS is quicker in animals exposed to high versus low stress environments. Accordingly, we have argued (Donahoe, 1993b; Donahoe and Vlahov, 1998) that the influence of opiates on AIDS progression in populations of street addicts may be conditionally and variably related, largely, to the pharmacologic status of the host relative to whether their opiate-dependency is maintained homeostatically or not. Taken together, our data support the overall supposition that opioids modulate lymphocyte recirculation in vivo in the context of effects on ambient stress. This study suggests the need for further understanding of the role of opioids and the endogenous opioid system in regulating lymphocyte recirculation

ACKNOWLEDEMENTS

This work was supported by NIDA grant DA04400. We thank the clinical veterinary staff and the animal technical staff at the Yerkes Regional Primate Research Center for their dedicated support of the animals during these experiments. The expert technical assistance of Michael Litrel, Jennifer Patera, Protul Shrikant, Granger Sunderland, and Gabriella Thomas, as well as the secretarial assistance of Mrs. Sybil Mashburn and Mrs. Nurit Golan are gratefully acknowledged--as are, too, helpful discussions about the data with Drs. Barbara Bayer, Daniel Carr, John Madden, Philip Peterson, and Burt Sharp.

REFERENCES

Ansari, A.A., Brodie, A.R., Fultz, P.N., Anderson, P.C., Sell, K.W., McClure, H.M., 1989. Flow microfluorometric analysis of peripheral blood mononuclear cells from nonhuman primates: correlation of phenotype with immune function. Amer. J. Primatol. 17, 107-131.

Appel, N.M., Kiritsky-Roy, J.A., VanLoon, G.R., 1986. Mu receptors at discrete hypothalamic and brainstem sites mediate opioid peptide-induced increases in central sympathetic outflow. Brain Res. 378, 8-20.

Baddley, J. W., Paul, D., Carr, D.J.J., 1993. Acute morphine administration alters the expression of β-adrenergic receptors on splenic lymphocytes. Adv. Biosci. 86, 593-597.

Balter, M., 1995. Cytokines move from the margins to the spotlight. Science 268, 205-206.

Blalock, J.E., 1984. The immune system as a sensory organ. J. Immunol. 132:1067-1070.

Blue, M-L., Daley, J.F., Levine, H., Schlossman, S.F., 1985. Coexpression of T4 and T8 on peripheral blood T cells demonstrated by two-color fluorescence flowcytometry. J. Immunol. 134, 2281-1286.

Bussiere, J.L., M.W. Adler, T.J. Rogers, T.K.E. Eisenstein, 1992. Differential effects of morphine and naltrexone on the antibody response in various mouse strains. Immunopharmacol. Immunotoxicol. 14,657-673.

Camerini, D., James, S.P., Stamenkovic, I., Seed B., 1989. Leu8/TQ1 is the human equivalent of the MEL-14 lymph node homing receptor. Nature (Lond.) 342, 78-82.

Cantacuzene, J., 1898. Nouvelles recherches sur le mode de destruction des vibrions dans l'organisme. Ann. Inst. Pasteur 12, 274-300.

Capitanio, J.P., Mendoza, S.P., Lerche, N.W., Mason W.A., 1998. Social stress results in altered glucocorticoid regulation and shorter survival in simian acquired immune deficiency syndrome. Proc. Natl. Acad. Sci., U.S.A. 95, 4714-4719.

Carr, D.J., France, C.P., 1993. Immune alterations in morphine-treated rhesus monkeys. J. Pharmacol. Exp. Ther. 267, 9-15.
Christie, M.J., Trisdikoon, P., Chesher, G.B., 1982. Tolerance and cross tolerance with morphine resulting from physiological release of endogenous opiates. Life Sci. 31:839-845.
Chuang, R.Y., Blackbourn, D.J., Chuang, L.F., Liu, Y., Kilam, Jr., K.F., 1993. Modulation of simian AIDS by opioids. Adv. Biosci. 86, 573-583.
Crary, B., Hauser, S.L., Borysenko, M., Kutz, I., Hoban, C., Haut, K.A., Weiner, H.L., Benson, H., 1983. Epinephrine-induced changes in the distribution of lymphocyte subsets in peripheral blood of humans. J. Immunol. 131, 1178-1181.
Dixon, W. J., 1992. BMDP statistical software manual, second edition. University of California Press, Berkley, CA.
Donahoe, R.M., Byrd, L., McClure, H.M., Fultz, P., Brantley, M., Marsteller, F., Ansari, A.A., Wenzel, D., Aceto, M., 1993. Consequences of opiate-dependency in a monkey model of AIDS. Adv. Exp. Med. Biol. 335, 21-28.
Donahoe, R.M., 1993b. Neuroimmunomodulation by opiates: relationship to HIV-1 infection and AIDS. Adv. Neuroimmunol. 3, 31-46.
Donahoe, R.M., Vlahov, D., 1998. Opiates as potential cofactors in progression of HIV-1 infections to AIDS. J. Neuroimmunol. 83, 77-87.
Eisenstein, T.K., Hilburger, M.E., 1998. Opioid modulation of immune responses: effects on phagocyte and lymphoid cell populations. J. Neuroimmunol. 83, 36-44.
Ernstrom, U., Sandberg, G., 1974. Adrenaline-induced release of lymphocytes and granulocytes from the spleen. Biomedicine 21, 293-296.
Fecho, A., Dykstra, L.A., Lysle, D.T., 1993. Evidence for *beta* adrenergic receptor involvement in the immunomodulatory effects of morphine. J. Pharmacol. Exp. Ther. 365:1079-1087.
Felten, S.Y., Felten, D.L., 1991. Innervation of lymphoid tissue. In: Ader, R., Felten, D.L., Cohen, N. (Eds.), Psychoneuroimmunology (second edition). Academic Press, Inc., New York, pp. 27-69.
Flores, L.R., Wahl, S.M., Bayer, B.M., 1995. Mechanisms of morphine-induced imunosuppression: Effect of acute morphine administration on lymphocyte trafficking. J. Pharmacol. Exp. Ther. 272, 1246-1251.
Friedman, H., Klein, T.W., Spector, S., 1993. Drugs of abuse, immunity and AIDS. Plenum Press, New York, N.Y.
Hamann, A., Jablonski-Westerich, O., Scholz, K-U., Duijvestijn, A., Butcher, E.C., Theile, H-G., 1988. Regulation of lymphocyte homing I: Alterations in homing receptor expression and organ-specific high endothelial venule binding of lymphocytes upon activation. J. Immunol.140, 737-743.
Harris, G.C., Aston-Jones, G., 1993. Beta-adrenergic antagonists attenuate somatic and aversive signs of opiate withdrawal. Neuropsychopharmacology 9, 303-311.
Heijnen, C.J., Kavelaars, A., Ballieux, R.E., 1991. Beta-endorphin: cytokine and neuropeptide. Immunol. Rev. 119, 41-63.
Hernandez, M.C., Flores, L.R., Bayer, B.M., 1993. Immunosuppression by morphine is mediated by central pathways. J. Pharmacol. Exp. Ther. 267, 1336-1341.

Kanof, M.E., James, S.P., 1988. Leu-8 antigen expression is diminished during cell activation but does not correlate with effector function of activated T lymphocytes. J. Immunol. 140, 3701-3706.

Keller, S.E., Schleifer, S.J., Demetrikopoulos, M.K., 1991. Stress-induced changes in immune function in animals: Hypothalamo-pituitary-adrenal influences. In: Ader, R., Felten, D.L., Cohen, N. (Eds.), Psychoneuroimmunology (second edition). Academic Press, Inc., New York, pp. 771-787.

Lysle, D. T., Coussans, M.E., Watts, V.J., Bennet, E.H., Dykstra, L.A., 1993. Morphine-induced alterations of immune status: Dose dependency, compartment specificity and antagonism by naltrexone. J. Pharmacol. Exp. Ther. 265, 1071-1078.

Madden, K.S., Livnat, S., 1991. Catecholamine action and immunological reactivity. In: Ader, R., Felten, D.L., Cohen, N. (Eds.), Psychoneuroimmunology (second edition). Academic Press, Inc., New York, pp. 283-310.

McCarty, R., 1994. Regulation of plasma catecholamine responses to stress. Sem. Neurosci. 6, 197-204.

Molitor, T.W., Morilla, A., Risdahl, J.M., Murtaugh, M.P., Chao, C.C., Peterson, P.K., 1991. Chronic morphine administration impairs cell-mediated immune responses in swine. J. Pharmacol. Exp. Ther. 260, 581-586.

Morimoto, C., Levin, N.L., Distaso, J.A., Aldrich, W.R., Schlossman, S.F., 1985. The isolation and characterization of the human suppressor inducer T cell subset. J. Immunol. 134, 1508-1515.

Murray, D.R., Irwin, M., Reardon, A., Ziegler, M., Motulsky, H., Maisel, A., 1992. Sympathetic and immune interactions during dynamic exercise: Mediation via a β_2-adrenergic dependent mechanism. Circulation 86, 203-213.

Nowak, M.A., McMichael A.J., 1995. How HIV defeats the immune system. Sci. Am. 273, 58-65.

Ottaway, C.A., Husband A.J., 1994. The influence of neuroendocrine pathways on lymphocyte migration. Immunol. Today 15, 511-517.

Picker, L.J., 1994. Control of lymphocyte homing. Curr. Topics Immunol. 6, 394-406.

Plotnikoff, N.P., Faith, R.E., Murgo, A.J., Good, R.A., 1986. Enkephalins and endorphins. Stress and the immune axis. Plenum Press, New York, N.Y.

Risdhal, J.M., Peterson, P.K., Chao, C.C., Pijoan, C., Molitor, T.W., 1993. Effects of morphine dependence on the pathogenesis of swine herpes virus infection. J. Infec. Dis. 167, 1281-1287.

Ruff, M.R., Whal, S.M., Mergenhagen, S., Pert, C.B., 1985. Opiate receptor-mediated chemotaxis of monocytes. Neuropeptides 5, 363-366.

Sadeghi, A., Behmard, S., Vasselinovitch, S., 1979. Opium: a potential urinary bladder carcinogen in man. Cancer 43, 2315-2321.

Sapira, J.D., 1968. The narcotic addict as a medical patient. Am. J. Med. 45, 555-588.

Sharp, C.W., 1998. Pharmaconeuroimmunology, AIDS and other diseases. J. Neuroimmunol. 83, 1-174.

Steel, C.M., French, E.B., Aitchison, W.R.C., 1971. Studies on adrenaline-induced leucocytosis in normal man. I. The role of the spleen and thoracic duct. Br. J. Haematol. 21, 413-421.

Tubaro, E., Borelli, G., Croce, C., Cavallo, G., Santiangeli, G., 1983. Effect of morphine on resistance to infection. J. Infec. Dis. 148, 656-666.

Van Loon, G.R., Appel, N.M., Ho, D., 1981. β-endorphin-induced stimulation of central sympathetic outflow: β-endorphin increases plasma concentrations of epinephrine, norepinephrine, and dopamine in rats. Endocrinology 109, 46-53.
Woods, J.W., Winger, G.D., Medzihradsky, F., Smith, C.B., Gmerek, D., Aceto, M.D., Harris, L.S., May, E.L., Balster, R.L., Slifer, B.L., 1984. Evaluation of new compounds for opioid activity in rhesus monkey, rat, and mouse. NIDA Res. Monogr. 55, 309-393.
Wormser, G.P., Krupp, L.B., Hanrahan, L.P., Gavis, G., Spira, T.J., Cunningham-Rundles, S., 1983. Acquired immunodeficiency syndrome in male prisoners. Ann. Intern. Med. 98, 297-303.

MITOGEN - INDUCED ACTIVATION OF MOUSE T CELLS INCREASES KAPPA OPIOID RECEPTOR EXPRESSION

Jean M. Bidlack and Michael K. Abraham

Department of Pharmacology and Physiology
University of Rochester School of Medicine and Dentistry
601 Elmwood Ave.
Rochester, NY 14642-8711

INTRODUCTION

There is a growing body of evidence demonstrating the influences of opioids on the immune system (Sibinga and Goldstein, 1988; Roy and Loh, 1996). For example, studies find that the κ-selective agonist U50,488 had a suppressive effect on LPS-induced production of TNF-α, IL-1 and IL-6 by peritoneal macrophages (Alicea *et al.,* 1995). Also the treatment of murine splenocytes with the opioid receptor antagonist naltrexone prevented LPS-stimulated inducible nitric oxide synthase production (Lyle and How, 1999). These data strongly suggest the presence of an opioid receptor on mature splenocytes. Previous studies have also shown T-lymphocytes isolated from the thymus of mice express the κ opioid receptor (KOR) on the cell surface (Ignatowski and Bidlack, 1997), validating the presence of a brain-like opioid receptor on normal immune cells. By using the same approach for the identification of receptors, the presence of the KOR on the more mature splenocytes has been identified (Ignatowski and Bidlack, 1999). Immature thymocytes and resting peritoneal macrophages expressed the highest levels of the KOR with 60% or more of the cells expressing the receptor (Ignatowski and Bidlack, 1997, 1999). In contrast, less than 30% of the mature T-helper (CD4+) and T-cytotoxic (CD8+) cells from male 6-8 week-old C57Bl/6By mice expressed the KOR, and only 15% of the B cells (CD45R+) cells expressed the receptor. The goal of the present study was to determine if mitogen activation of CD4+ and CD8+ splenocytes would alter the amount of receptor expression and/or the percentage of cells that expressed the receptor.

A method of measuring T-cell activation using the cell surface antigen CD69 and the flow cytometry has been established (Maino *et al.,* 1994). Multiparameter analysis capabilities of flow cytometry allow for the concurrent detection of the phenotypic markers and the κ opioid receptor. Using the indirect immunofluorescence method to measure KOR expression (Lawrence et al., 1995) , and CD69 expression to measure activation, a possible link between activation of lymphocytes and receptor expression could be demonstrated. Most studies suggest that many immune cells either express an opioid receptor or are modulated by opioid peptides and alkaloids (Sibinga and Goldstein, 1988; Roy and Loh, 1996). A specific increase in KOR expression in relation to the phenotypic markers following mitogen activation would suggest a role for the KOR in regulating immune responses following activation of lymphocytes. In this study, changes in KOR expression are compared to changes in the expression of phenotypic markers to standardize for an increase in general protein expression due to an increase in cell size and number.

MATERIALS AND METHODS

Mice

Male C57BL/6J mice aged 6-8 weeks (Jackson Laboratories, Bar Harbor, ME), were used for all studies.

Murine Splenocyte Preparation

Mice were sacrificed by CO_2 inhalation, and spleens were removed aseptically as described previously (Mishell *et al.*, 1980). Spleens were dissociated, and cell suspensions were passed over glass wool columns to remove dead cells and debris. After centrifugation at 200 *x g* for 10 min at 4°C, cells were treated with an isotonic ammonium chloride solution to lyse contaminating erythrocytes. Unfixed cells, washed twice by centrifugation, were then counted in a hemacytometer using the trypan blue exclusion test.

Cell Culture and Mitogen Activation of Splenocytes

Splenocytes were cultured at a concentration of 3×10^6 cells/ml in RPMI 1640 media, buffered with 12.5 mM HEPES, pH 7.2, and containing 10% (v/v) iron supplemented bovine calf serum, 300 mg/ml L-glutamine, 100 μg/ml streptomycin, 1000 units/ml penicillin, 50 μM 2-mercaptoethanol, and 60 μM 2-aminoethanol. Phytohemagglutinin (PHA) and Concanavilin A (Con A) were both used to activate the splenocyte suspension. The mitogens were titrated for optimal results. Splenocytes were then incubated with 5 μg/ml PHA or 2.5 μg/ml Con A for 0-36 hr.

Detection of Kappa Opioid Receptors on Lymphocytes by Indirect Immunofluorescence

The FITC-AA labeling and amplification procedures were performed as previously described (Lawrence *et al.,* 1995). In a final volume of 200 μl HEPES-BSS, cells were incubated with 30 μM FITC-AA for 30 min at 25°C. The κ-selective antagonist nor-binaltorphimine (nor-BNI) at a final concentration of 500 μM was used to measure nonspecific fluorescence. Samples were chilled on ice and washed three times by centrifugation. After aspirating supernatants, cells were resuspended in 100 μl of HEPES-buffered saline solution (HBSS) containing 10% goat serum. After 10 min, 10 μl of biotinylated goat anti-fluorescein IgG was added to the appropriate tubes. After incubation for 30 min at 4°C in the dark, cells were washed twice. Cells were then suspended in 40 μl HBSS and 10 μl extravidin-R-PE for 15 min at 4°C in the dark. The cells were again washed twice and resuspended with 0.5 μl HBSS, for flow cytometric analysis. Phenotypic characterization of splenocytes possessing the KOR was performed as previously described (Ignatowski and Bidlack, 1998). Quantum Red conjugated rat monoclonal antibodies (mAb) directed against the mouse cell surface markers CD4 (CD4-QR) and CD8 (CD8-QR) were used to measure lymphocyte subset populations. FITC-conjugated CD69 (CD69-FITC) was used to measure cellular activation due the mitogens (Maino *et al.*, 1995). Optimal amounts of the mAb's were added to the appropriate tubes during the second 30-min incubation, followed by the subsequent steps listed above.

Flow Cytometric Analysis

Samples were analyzed on a Becton Dickinson FACScan equipped with a 15mW argon-ion laser for excitation (488nm) of FITC (FL1, green) and PE (FL2, orange) using band pass filters of 530 ± 15 nm and 585 ± 21 nm respectively. The QR fluorescence channel (FL3, red) which emits at 670 nm, uses a long-pass emission filter which transmits long-wavelength light beams above 650 nm. In each sample 25,000 splenocytes were analyzed. Data (forward (FSC) and right angle (SSC) light scatter, as well as FITC, QR, and PE fluorescence) were collected and stored using CELLQuest software. Auto fluorescence of unlabeled cells was used, as a negative control to ensure that fluorescein did not contribute to the PE signal as measured in FL2. Background controls, consisting of cells incubated with only biotinylated-α-FITC IgG and extravidin-R-PE, were used to establish baseline (background) by taking into account any nonspecific staining. Viability discrimination on samples not containing PE or QR were done using 5 μg/ml Propidium Iodide (Loken and Stall, 1982). Cell viabilities were >90% and >70% for control and treated groups, respectively. In samples containing PE or QR, light scatter gating provided live/dead cell discrimination. The cells labeled positively for each particular QR-mAb could either be gated upon in a FITC histogram for CD69 or a PE histogram for phenotypic κ opioid receptor dual labeling. The percent specific fluorescence for labeling of the κ opioid receptor was calculated using the median phycoerythrin fluorescence intensity values as: [((Total fluorescence – background) – (Fluorescence with nor-BNI – background))/(Total fluorescence – background)] X 100.

RESULTS

Activation of Lymphocytes by Con A and PHA

Figure 1 shows the time course and percentage of CD4+ (T-helper) and CD8+ (T-cytotoxic) cells that were activated with the mitogen Con A. Activation of cells was measured using a fluorescent antibody directed against CD69, a cell surface antigen, whose expression increases rapidly upon cell activation. Con A activated both CD4+ and CD8+ cells to an equal amount. Approximately 75% of both CD4+ and CD8+ cells expressed the activation marker CD69 after a 24-hr incubation with 2.5 μg/ml Con A. Conversely, a 24-hr incubation with 5 μg/ml PHA produced a greater activation of T-cytotoxic than T-helper cells. Greater than 80% of the CD8+ cells were activated, compared to less than 35% of the cells expressing the CD4+ phenotype.

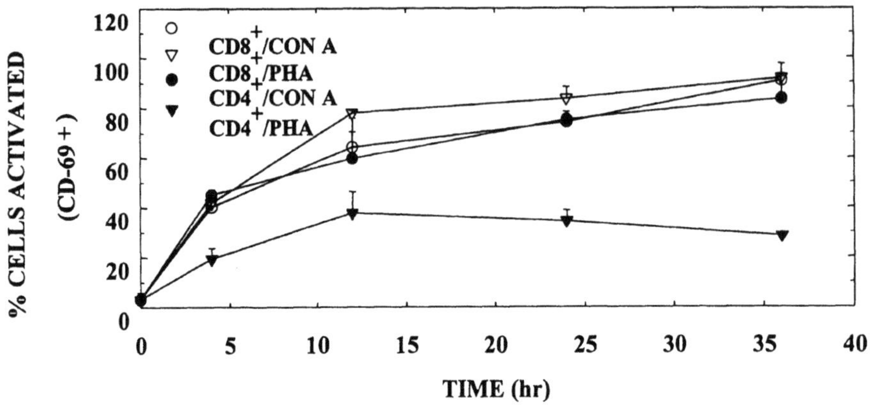

Fig. 1. Percentage of CD4+ and CD8+ cells activated by either Con A or PHA. Mouse splenocytes were cultured in the presence of either 2.5 μg/ml Con A or 5 μg/ml PHA for varying times. Double labeling of cells with either the cell surface marker CD4 (T-helper) or CD8 (T-cytotoxic) and the marker for early activation of cells, CD69 was measured and quantitated using flow cytometry as described in the Materials and Methods.

Figure 2 shows the increase in fluorescence of the κ opioid receptor as a function of time after activation of CD8+ cells with PHA. Splenocytes were treated with 5 μg/ml PHA and were cultured for varying times. CD8+ cells were identified and gated, and then the labeling of the κ opioid receptor was measured. Similar results were obtained with CD4+ cells. Both mitogens increased the expression of the κ opioid receptor, but by differing amounts.

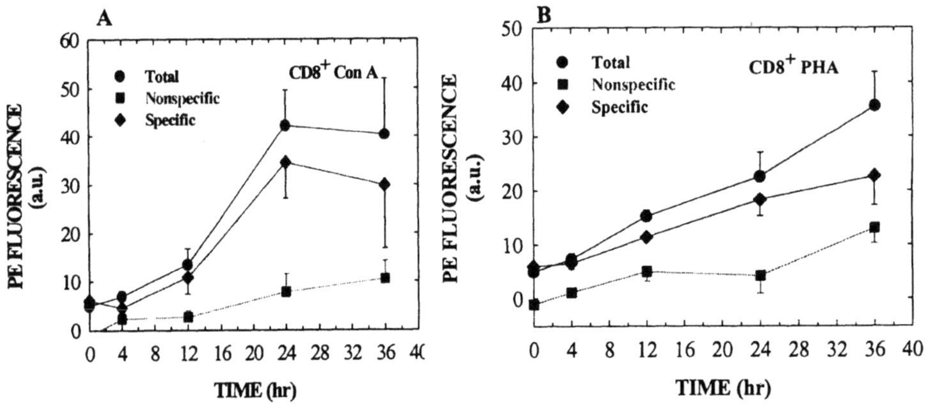

Fig. 2. Kappa Opioid receptor expression on CD8+ cells from splenocytes treated with Con A and PHA. (A) At 24 hr, KOR expression on Con A-activated CD8+ cells reached a limit of 34.6 ± 7.34 a.u. (B) The KOR expression on PHA-treated CD8+ cells increased linearly with a maximal PE fluorescence of 22.7 ± 5.35 a.u.

Studies have addressed the increase in KOR expression relative to other proteins expressed on the lymphocytes. When cells are activated, they increase in size, and thus, their plasma membrane surface area is increased. Therefore, it is possible that the increase in the expression of the κ opioid receptor is due solely to an increase in the surface area of the membrane. To start addressing this point, studies have compared the relative change in the expression of the κ receptor with the change in the expression of either CD4 or CD8 antigens, on the respective populations of cells. Figure 3 shows that the κ opioid receptor is increased approximately 6-fold greater than the CD8 marker in Con A stimulated CD8+ cells. The mitogen PHA increased the receptor 4-fold relative to the CD8 marker. In contrast, Con A and PHA increased the level of the κ opioid receptoron CD4+ cells by 2-fold relative to the cell surface markers.

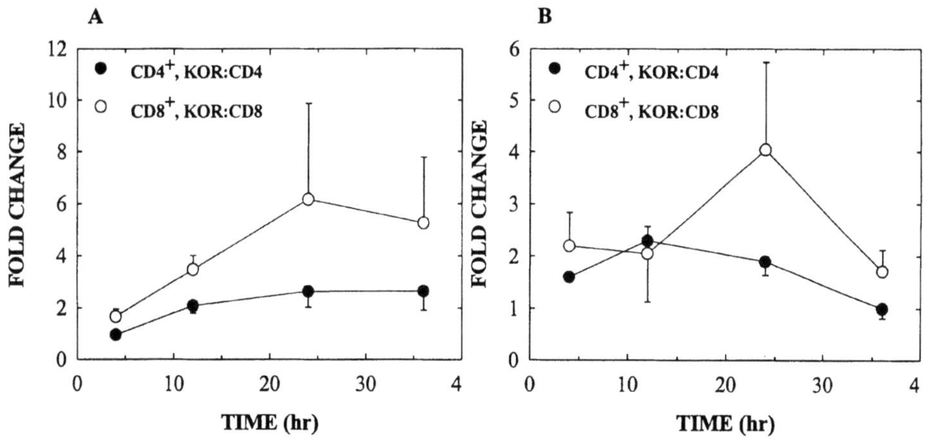

Fig. 3. Relationship between the κ opioid receptor (KOR) and phenotypic marker expression. (A) The ratio of CD8:KOR for Con A stimulated CD8+ cells was greater than a 6-fold increase at 24 hr, while CD4:KOR was only slightly above 2-fold at the same time point. (B) Consistent with the lesser stimulation of splenocytes by PHA, the ratio increase of CD8:KOR for PHA CD8+ cells was approximately 4-fold, while the KOR:CD4 ratio increased only 2-fold at 24 hr.

Table 1 shows that while the amount of κ receptor increased on CD8+ lymphocytes, the percentage of CD8+ cells that expressed the κ receptor did not change as a result of the activation of the lymphocyte, regardless of whether the mitogen Con A or PHA was used.

In summary, this study has shown that mitogen activation of mouse splenocytes increased the expression of the κ opioid receptor on CD4+ and CD8+ cells. Con A increased the receptor expression 6-fold on CD8+ cells relative to the phenotypic marker. While the amount of receptor appears to increase, the percentage of either CD4+ or CD8+ cells that expressed the κ receptor did not change with mitogen activation. Future studies will used purified populations of T-helper and T-cytotoxic cells, and will activate the cells by crosslinking the T-cell receptor.

Table 1. Percentage of CD4+ and CD8+ cells expressing the κ opioid receptor at various times after the addition of either Con A or PHA to mouse splenocytes in culture

	% CD4+/ KOR+ cells ± S. E.		% CD8+/ KOR+ cells ± S. E.	
Time (hr)	**Con A**	**PHA**	**Con A**	**PHA**
0	29 ± 10	29 ± 10	16 ± 1	16 ± 1
4	41 ± 4	26 ± 4	32 ± 1	20 ± 1
12	35 ± 12	35 ± 7	32 ± 8	28 ± 1
24	27 ± 7	32 ± 7	24 ± 9	25 ± 2
36	32 ± 11	22 ± 1	31 ± 6	16 ± 6

CELLQuest software from Becton-Dickinson was used to perform histogram subtractions of the cells labeled for each phenotypic marker and the κ opioid receptor in the presence (nonspecific) and absence (total) of the κ-selective antagonist nor-BNI, on a channel by channel basis. This subtraction yielded the relative percentage of cells specifically labeled for both the phenotypic marker of interest and the κ receptor. The treatment of splenocytes with the mitogens caused no significant increase in the number of cells that were double positive for the phenotypic markers and the κ opioid receptor at any time point.

DISCUSSION

Recent studies have shown that the κ opioid receptor could be detected on mouse thymocytes by using the same labeling procedure used in these studies (Ignatowski and Bidlack, 1998). Thymocytes, isolated from 6-8 week old C57BL/6ByJ mice, incubated with FITC-AA followed by the PE amplification procedure demonstrated specific labeling of the κ opioid receptor. This labeling was inhibited 55 ± 4% above background by excess nor-BNI, a κ-selective antagonist. This κ opioid receptor positive population consisted of 58 ± 2% of all gated thymocytes. Phenotypic characterization determined that not only were 64 ± 3% of the gated thymocytes CD4+/κ opioid receptor positive, but 60 ± 1% of all thymocytes were CD8+/κ opioid receptor positive. Two subpopulations of CD3+ thymocytes, consisting of both mature and immature cells, also displayed labeling for the κ opioid receptor. Double labeling of thymocytes with anti-CD4 and anti-CD8 antibodies demonstrated 82 ± 0.5% of these cells were of the double-positive phenotype. Therefore, these findings demonstrated that thymocytes that express the κ opioid receptor are predominantly of the immature CD4+/CD8+ phenotype. Collectively, these findings not only establish the presence of the κ opioid receptor on immune cells involved in opioid responsiveness, but also further indicate that this technique allows for the identification of distinct lymphocyte subpopulations which express the receptor.

Studies then addressed at determining if mature CD4+ and CD8+ cells expressed the κ opioid receptor to the same level as the double-positive immature T cells (Ignatowski and Bidlack, 1999). Unfixed primary splenocytes from 6-8 week old C57BL/6ByJ male mice were labeled as described previously (Lawrence et al., 1995). In contrast to the earlier report, where we showed that greater than 60% of immature thymocytes (CD4+/CD8+) demonstrated specific κ opioid receptor labeling, the studies examining mature T-cells showed that less than 25% of either T-helper or T-cytotoxic splenic lymphocytes expressed the κ opioid receptor. Likewise, only 16% of all splenic B lymphocytes were labeled for the κ opioid receptor. These findings demonstrate a decrease in κ opioid receptor expression upon maturation of mouse lymphocytes. Interestingly, resident peritoneal macrophages showed a greater magnitude of specific receptor labeling, compared to either thymocytes or splenocytes, and approximately 50% of the resting macrophages expressed the κ opioid receptor. Greater than 80% of human fetal microglial cells have been shown to express the κ opioid receptor (Chao et al., 1996). Taken together, these findings demonstrate the diversity in the expression of the κ opioid receptor on immune cells at varying stages of differentiation.

Because the percentage of cells on mature CD4+ and CD8+ that expressed the κ opioid receptor was considerably lower than the percentage of immature T cells that expressed the receptor, the present study addressed whether the amount of receptor expression or the percentage of cells that expressed the receptor changed upon mitogen activation. Stimulation of mouse splenocytes with either Con A or PHA activated the T-helper and T-cytotoxic cells (Fig. 1). Mitogen activation increased the amount of the KOR expression relative to the CD4 and CD8 cell surface markers. In contrast, the percentage of either CD4+ or CD8+ cells that expressed the κ opioid receptor did not change as a result of mitogen activation. These findings suggest that in the subpopulations of cells that do express the receptor, mitogen activation of these cells increases the KOR expression to a greater degree than the increase expression of other proteins.

Similar findings have been observed with the δ opioid receptor. Sharp et al. (1997) reported that mRNA levels for the δ opioid receptor increased in mouse splenocytes, which had been activated by crosslinking the T-cell receptor. An increase in the mRNA levels for the δ opioid receptor is strongly suggestive of an increase in receptor expression. Collectively, these studies suggest that activation of lymphocytes increases the expression of both the κ and δ opioid receptors.

ACKNOWLEDGEMENTS

This study was supported by grants K05-DA00360 and DA04355 from the National Institute on Drug Abuse.

REFERENCES

Alicea, C., Belkowski, S., Eisenstein, T.K., Adler, M.W., and Rogers, T.J., 1995, Inhibition of primary murine macrophage cytokine production in vitro following treatment with the κ-opioid agonist U50,488H, *J. Neuroimmunol.* 64: 83.

Chao, C.C., Gekker, G., Hu, S., Sheng, W.S., Bu, D.-F., Archer, S., Bidlack, J.M., and Peterson, P.K., 1996, Kappa opioid receptors in human microglia downregulate human immunodeficiency virus-1 expression, *Proc. Natl. Acad. Sci. U.S.A.*, 93: 8051.

Ignatowski, T.A., and Bidlack, J.M., 1998, Detection of kappa opioid receptors on mouse thymocyte phenotypic subpopulations as assessed by flow cytometry, *J. Pharmacol. Exp. Ther.* 284:298.

Ignatowski, T.A., and Bidlack, J.M., 1999, Differential κ-opioid receptor expression on mouse lymphocytes at varying stages of maturation and on mouse macrophages after selective elicitation, *J. Pharmacol. Exp. Ther.*, 290: 863.

Lawrence, D.M.P., El-Hamouly, W., Archer, S., Leary, J.F., and Bidlack, J.M., 1995, Identification of κ opioid receptors in the immune system an indirect immunofluorescence method, *Proc. Natl. Acad. Sci. U.S.A.* 92:1062.

Lysle, D.T. and How, T., 1999, Endogenous opioids regulate the expression of inducible nitric oxide synthase by splenocytes, *J. Pharmacol. Exp. Ther.* 288:502.

Loken, M.R., and Stall, A.M., 1982, Flow cytometry as an analytical and preparative in immunology, *J. Immunol. Meth.* 50:85.

Maino, V.C., Suni, M.A., and Ruitenberg, J.J., 1995, Rapid flow cytometric method for measuring lymphocyte subset activation, *Cytometry* 20: 127.

Mishell, B.B., Shigii, S.M., Henry, C., Chan, E.L., North, J., Gallily, R., Slomich, M., Miller, K., Marbrook, J., Parks, D., and Good, A.H., 1980, I. In Vitro Immune responses. Preparation of mouse cell suspensions, in: *Selected Models in Cellular Immunology*, B.B. Mishell, and S.M. Shigii, S.M. eds., 3, Freeman Press, New York, NY.

Roy, S. and Loh, H.H., 1996, Effect of opioids on the immune system, *Neurochem. Res.* 21: 1375.

Sibinga, N.E.S., Goldstein, A., 1988, Opioid peptides and opioid receptors in cells of the immune system, *Ann. Rev. Immunol.* 6: 219.

Sharp, B. M., Shahabi, N., McKean, D., Li, M. D., and McAllen, K., 1997, Detection of basal levels and induction of delta opioid receptor mRNA in murine splenocytes, *J. Neuroimmunol.* 78: 198.

SELF-ENHANCEMENT OF PHAGOCYTOSIS BY MURINE RESIDENT PERITONEAL MACROPHAGES AND ITS RELATIONSHIP TO MORPHINE EFFECTS ON THE PROCESS

Wanda E. Pagán, Nancy Y. Figueroa, and Fernando L.Renaud

Department of Biology, University of Puerto Rico, Río Piedras Campus
San Juan, PR 00931

INTRODUCTION

Macrophages are an important component of the immune system since they are part of both innate and cellular immune responses (Rooijen and Sanders, 1997). It is well documented that macrophage activation results in enhancement of antimicrobial and antitumoral activities in these cells. For example, upon activation, macrophages can become engaged in the clearance of microorganisms or altered self-material by enhanced oxidative burst, chemotaxis, phagocytosis, intracellular killing and cytokine secretion (Peterson et al., 1998; Dearman et al., 1996, Tachibana et al., 1992, Langermans et al., 1990).

One important factor that can also modulate immune cell functions is exposure to opioids. Enkephalins, endorphins and morphine-like opioids are capable of influencing the immune system through its binding to opioid receptors (Eisenstein et al., 1998). In our laboratory we have shown that *in vitro* morphine can inhibit Fc-mediated phagocytosis by thioglycolate-elicited murine peritoneal macrophages (Casellas et al., 1991; Tomei and Renaud, 1997). In these cells the response to acute exposure to morphine is biphasic, since inhibition is not observed at low or high doses. We have also shown that upon chronic exposure to morphine macrophage phagocytosis becomes insensitive to the drug, suggesting that a state akin to tolerance is developed in these cells. Furthermore, morphine withdrawal from putatively tolerant cells results in inhibition of phagocytosis (Tomei and Renaud, 1997; Lázaro et al., 1999). This latter observation suggests that chronically exposed macrophages may be in a dependent state, since they seem to need morphine to phagocytize at a control level. The tolerant state can be prevented if the cells are simultaneously incubated with both morphine and naloxone, which suggests that development of the tolerant/dependent state is mediated through classic opioid receptors (Lázaro et al., 1999). Other investigators have demonstrated that thioglycolate-elicited cells show an altered response to other stimuli, when compared to resident macrophages. For example, thioglycollate treatment can increase the production of cytokines, such as TNFα (Tachibana et al., 1992). However, no thorough studies have been performed to determine

Neuroimmune Circuits, Drugs of Abuse, and Infectious Diseases
Edited by Herman Friedman *et al.*, Kluwer Academic/Plenum Publishers, 2001

if the effect of morphine on phagocytosis by macrophages is dependent on the state of activation of macrophages. In the present study, we have focused on the effect of the activation state on the process of phagocytosis by macrophages and how this state can affect the response of macrophages to opioids. Specifically we study the process by which resident peritoneal macrophages are activated *in vitro* and its relationship to their response to morphine.

METHODS

Resident Macrophages Cell Culture and Phagocytosis Assay

The phagocytosis assay was previously described (Casellas et al., 1991; Tomei and Renaud, 1997). Briefly, female mice C3Heb/Fej (Jackson Labs.) 8-12 weeks old were used. Animals were kept in the animal house with food and water *ad libitum*. Resident peritoneal macrophages were collected by peritoneal lavage, washed and cultured in RPMI medium (Sigma) supplemented with 10% fetal bovine serum in 8-well chamber slides (Lab-Tek, VWR Co.). Cell concentrations were adjusted to 1.5 x 10^4 per well. After two hours the medium was changed to remove non-adherent cells and they were cultured at 37°/ 5% CO_2. For the phagocytosis assay, opsonized sheep red blood cells (SRBC) were added to the cells at the appropriate time at a ratio of 100 SRBC per macrophage, followed by a gentle centrifugation, and incubated at 37°C for 30 minutes, unless otherwise specified. Cell ingestion was stopped by hypotonic lysis of non-ingested SRBC, followed by methanol fixation and Giemsa staining. Phagocytosis was quantified. The number of cells ingesting at least one SRBC; in 300 cells counted per variable were scored with a light microscope. The results are expressed as the phagocytic factor, which represents the ratio of the experimental results against the control (% phagocytosis in the presence of agonist ÷ %

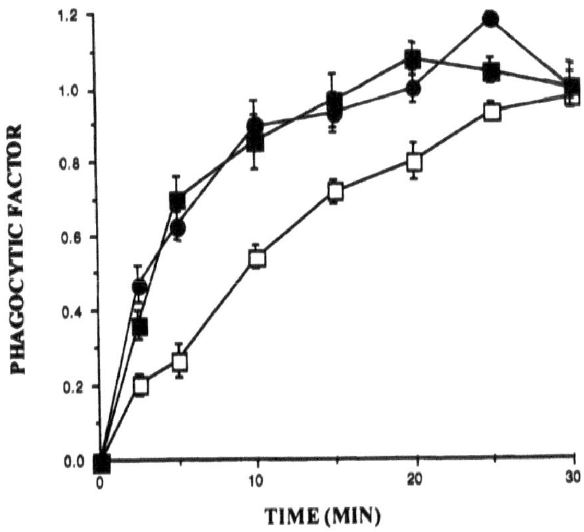

Figure 1. Phagocytosis of SRBC by murine resident peritoneal macrophages. Assay was performed as described in Methods. Phagocytosis was measured at different time intervals for macrophages cultured overnight (open squares), for three days (closed squares) and for macrophages cultured overnight in the presence of conditioned medium from obtained from cultured (3day) macrophages from a three-day culture (closed circles).

phagocytosis in control cells). At least three independent replicates were performed of each experiment. Phagocytosis was measured at different time intervals after macrophages were cultured overnight (24hrs) and for a three-day period (72hrs). We also measured the phagocytic capacity of macrophages cultured overnight in the presence of medium obtained from macrophages cultured for a three-day period (conditioned medium). This conditioned medium was added to fresh cultures at the time of removal of non-adherent cells.

The effect of acute and chronic morphine (Sigma) exposure was determined for cells cultured overnight and for cells cultured for a three day period. For acute morphine exposure experiments, phagocytosis was determined after a 30 min pre-incubation in the presence or absence of 10^{-10} M or 10^{-6} M morphine. To assess the effect of chronic exposure, 10^{-10} M or 10^{-6}M morphine were added after cells had been cultured for 16 hr and the phagocytosis assay was performed after 8 hr later. For macrophages cultured for three days, morphine concentrations were added after 64 hr of culture, and the phagocytosis assay was performed 8 hr later. The morphine concentrations used in this study are physiologically relevant since they are comparable with the concentration of morphine found in the blood from morphine pelleted mice, (Rojavin et al, 1993). These concentrations are also comparable to clinically measured serum morphine concentration found in patients (Peterson et al., 1989, 1992).

RESULTS

The phagocytic activity of resident macrophages cultured for different time intervals is shown in Figure 1. These data show that resident macrophages cultured for a three-day period (closed squares) exhibited of phagocytosis when compared to cells cultured overnight (open squares).

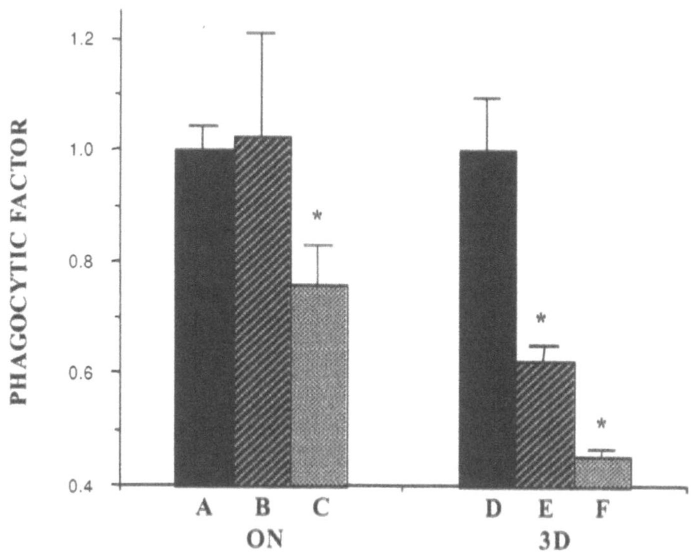

Figure 2. Effect of acute morphine exposure on phagocytosis of opsonized SRBC by resident peritoneal macrophages cultured overnight (ON) or for a three-day period (3D). Assay was performed as described in Methods. Phagocytosis was measured after acute (30 min) exposure to 10^{-10} M (B, E) and 10^{-6} M (C, F) morphine. Phagocytosis assays in the absence of drug serve as controls (A, D) $^*p \leq 0.05$; n=3

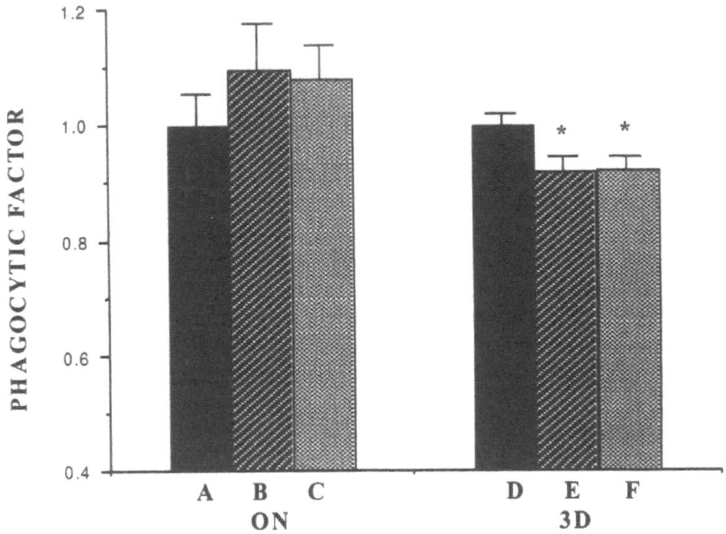

Figure 3. Effect of chronic morphine exposure on phagocytosis of opsonized SRBC by resident peritoneal macrophages cultured overnight (ON) and for a three-day period (3D). Assay was performed as described in Methods. Phagocytosis was measured after chronic (8 hrs) exposure to 10^{-10} M (B, E) and 10^{-6} M (C, F) morphine. Phagocytosis assays in the absence of the drug serve as controls (A, D) *$p \leq 0.05$; n=3

Interestingly, when cells are cultured overnight in the presence of conditioned medium (closed circles) phagocytosis is at the same high level as seen for 3-day cultures. (Figure 1). These data suggest that macrophages secrete self-activating signals that increase phagocytosis

It was previously reported that acute morphine exposure can inhibit Fc-mediated phagocytosis by murine peritoneal macrophages (Tomei and Renaud, 1997; Rojavin et al., 1993; Casellas et al., 1991). We determined effects of acute morphine on phagocytosis by resident macrophages, cultured overnight compared to those cultured for three days (Figure 2). The data show that, when compared to control cells, inhibition of phagocytosis is observed in overnight cultures when cells are acutely exposed to 10^{-6}M morphine, but not at a lower drug concentration (10^{-10}M). In contrast, when cells cultured for a three-day period are acutely exposed to morphine, the data show that, when compared to control cells, phagocytosis is inhibited at both low (10^{-10}M) and high (10^{-6}M) morphine concentrations. We also have evidence which suggests that these types of effects of morphine are mediated through opioid receptors since they can be prevented by simultaneous incubation of morphine with naloxone (data not shown).

We have also studied the effect of chronic opioid treatment on cells cultured overnight and on cells cultured for a three-day period (Figure 3). When compared to control cells, no inhibition of phagocytosis is observed in cells cultured overnight after chronic exposure to 10^{-10}M and 10^{-6}M morphine, suggesting that after chronic exposure macrophages cultured overnight become insensitive to morphine. Interestingly, for cells cultured for a three-day-period, inhibition of phagocytosis at 10^{-10}M and 10^{-6}M morphine concentrations is still observed, suggesting these cells are still sensitive to the drug, although the magnitude of this inhibition is less than that observed in cells after acute exposure (Figure 2).

DISCUSSION

It is well documented that macrophage activation plays an important role in inducing specialized cellular functions, such as phagocytosis. We have evidence which suggests that resident (unactivated) murine peritoneal macrophages can secrete their own activating signals, since after three days in culture they showed a significant enhancement of phagocytosis. Furthermore, conditioned medium obtained from macrophages cultured for three days can enhance the phagocytic capacity of macrophages cultured overnight. The enhanced phagocytic capacity shown by resident macrophages cultured for three days suggests that these cells have become activated *in vitro*. It will be of interest to determine if this enhancement of phagocytosis is due to secretion of cytokines, such as TNF∝ and/or GM-CSF, which have been shown to induce enhancement of phagocytosis by both microglia (von Zan et al., 1997) and peritoneal macrophages (Collins and Baneroft, 1992).

We also find that the observed *in vitro* activation can modulate the response to morphine. Acute morphine treatment of macrophages cultured overnight showed a dose-related inhibition of phagocytosis. However, with cells cultured for three days inhibition of phagocytosis is observed at lower drug concentrations. Furthermore, the results obtained with acute exposure as seen herein differ markedly from the results observed with thioglycolate-elicited cells where inhibition of phagocytosis after acute exposure to morphine, was biphasic with no effects at low or high doses of the drug (Tomei and Renaud, 1997). This increased sensitivity to morphine suggests that there has been a change in the opioid receptor population in resident cells, either in number or affinity, when compared to thioglycolate-elicited cells.

Our results show that resident macrophages cultured overnight become insensitive to morphine after chronic exposure to the drug. This observation is similar to the case with thioglycolate-elicited macrophages that develop a state of putative tolerance after an 8 hr exposure to morphine (Tomei and Renaud, 1997; Lázaro et al., 1999). However, phagocytosis by cells cultured for a three-day period is still inhibited by opiate after chronic exposure, although the inhibitory effect is of lesser magnitude. Again, this observation suggests a difference in the state of the receptor population in resident cells when compared to thioglycolate-elicited cells. How these effects of morphine, upon acute or chronic exposure to the drug, correlate with possible changes in the receptor population is at present unknown.

In conclusion, resident macrophages in culture secrete their own activating signals, and this activation process affects the response to morphine of these cells. Therefore activation should be one of the variables considered when studying the effect of opiates on immune cells.

ACKNOWLEDGEMENTS

This work was supported by NIH Grant SO6 GM08102, the Marc/MBRS Program and the FIPI Program of the University of Puerto Rico.

REFERENCES

Alicea, C., Belkowski, S., Eisenstein,T.K., Adler, M,W. and Rogers, T.J., 1996, Inhibition of primary murine macrophage cytokine production *in vitro* following treatment with the k-opioid agonist U50,488H. J. *Neuroimmunol.* 64: 83.

Casellas, A.M., Guardiola, H. and Renaud, F.L., 1991, Inhibition induced opioids of phagocytosis by peritoneal macrophages. *Neuropeptides* 188: 350.
Collins, H.J. and Baneroft, G.J., 1992, Cytokine enhancement of complement-dependent phagocytosis by macrophages: synergy of tumor necrosis factor ∝ and granulocyte macrophage colony stimulating factor for phagocytosis of *Cryptoccocus neoformans*. *Eur. J. Immunol.* 22: 1447.
Eisenstein, T.K. and Hilburger, M.E., 1998, Opioid modulation of immune responses: effects on phagocyte and lymphoid cell populations. *J. Neuroimmunol.* 83: 36.
Langermans, A.M., Van der Hulst, M.E.B., Nibbering, P.H. and Van Furth, R., 1990, Activation of mouse peritoneal macrophages during infection with *Salmonella thyphimurium* does not result in enhanced intracellular killing. *J. Immunol.* 144: 4340.
Lázaro, M.I., Tomassini, N., González, I. and Renaud, F.L., 1999, Reversibility of morphine effects on phagocytosis by murine macrophages. *Drug Alcohol Depend.* In press.
Peterson, P.K., Molitor, T.W. and Chao, C.C., 1998. The opioid-cytokine connection. *J. Neuroimmunol.* 83: 63.
Peterson, P.K., Sharp, B.M., Gekker, G., Brummitt, C. and Keane, W.F., 1987, Opioid-mediated suppression of cultured peripheral blood mononuclear cell respiratory burst activity. *J. Immunol.*138: 3907.
Rojavin, M., Szabo, I., Bussiere, J.L., Rogers, T.J., Adler, M.W. and Eisenstein, T.K. 1993, Morphine treatment *in vitro* or *in vivo* decreases phagocytic functions of murine macrophages. *Life Sci.* 53: 997.
Rooijen, N.V. and Sanders, A., 1997. Elimination, blocking, and activation of macrophages: three of a kind? *J. Leuk. Biol.* 62: 702.
Tachibana, K., Chen, G., Huang, D.S., Scuderi, P. and Watson, R.R., 1992, Production of tumor necrosis factor α by resident and activated murine macrophages. *J. Leuk. Biol.* 50: 251.
Tomei, E.Z., and Renaud, F.L., 1997, Effect of morphine on Fc-mediated phagocytosis by murine macrophage *in vitro*. *J. Neuroimmunol.* 74: 111.
Von Zan, J. Moller, T., Kettenmann, H. and Nolten, C., 1997, Microglial phagocytosis is modulated by pro- and anti-inflammatory cytokines. *Neuroreport* 8: 18.

ROLE OF MU-OPIOID RECEPTOR IN IMMUNE FUNCTION

Sabita Roy, Richard G. Charboneau, Roderick A. Barke, and Horace H. Loh

Department of Pharmacology and Surgery
University of Minnesota
Minneapolis, MN 55455

INTRODUCTION

Chronic morphine use has been associated with an increased incidence of many diseases (1), and animal studies have demonstrated that morphine can alter a number of immune parameters (2,3,4). Still, though immune effects of morphine are very well established, two important questions remain unresolved: 1) which receptor type (i.e. mu, delta, kappa or naloxone insensitive morphine receptor) mediates morphine's immunosuppressive effects and 2) are central opioid receptors or peripheral opioid receptors responsible for the morphine effects. Gene targeting methodology makes it possible to eliminate or knock-out the mu opioid receptor in mice (MORKO), providing a powerful tool to address this question. In such animals, both direct and indirect effects of morphine on the immune system can no longer be mediated by mu opioid receptors. It is therefore possible to evaluate the role of other classical opioid receptors (delta and kappa) as well as non-classical naloxone-insensitive receptors in immunosuppressive effects of morphine. The objective of the studies carried out in this paper is to determine the contribution of mu-opioid receptors in morphine mediated effects on the immune system.

MATERIALS AND METHODS

Generation of MORKO Animals

The MORKO mice were generated in our laboratory (5,6). The procedure used to generate these animals is as follows: A genomic segment spanning mu-opioid receptor gene exons 2 and 3 was used as the starting material. A double-selection (positive selection by neo^r and negative selection by thymidine kinase) targeting vector was

Neuroimmune Circuits, Drugs of Abuse, and Infectious Diseases
Edited by Herman Friedman *et al.*, Kluwer Academic/Plenum Publishers, 2001

generated by replacing a XhoI/XbaI fragment, which spans the entire exons 2 and 3, with a Neo^r cassette followed by ligation of a thymidine kinase (tk) expression cassette to the 3'-end of this segment. The genotypes of ES clones were characterized by Southern blot hybridization analysis of their genomic DNA, using a neo specific probe and a 5'-flanking sequence probe. MOR-targeted ES clones were injected into blastocysts derived from C57/BL6 mice using a standard microinjection procedure. Live born animals were genotyped by southern blot hybridization analysis of genomic DNA isolated from their tails. Heterozygotes and homozygotes were obtained by a standard breeding program.

Treatment of Animals

***In vivo* Morphine Treatment of Animals**

a) Chronic Morphine Treatment

Both WT and MORKO mice were divided into four groups of at least 10 animals per group. A standard murine model of chronic morphine treatment was used (7). Group 1 received two placebo pellets. Group 2 received one placebo pellet and one morphine (75mg) pellet. Group 3 received one naltrexone (10mg) pellet and one placebo pellet. Group 4 received one morphine pellet and one naltrexone pellet. Animals were treated for 48 hours with the respective pellets before tissue harvest.

b) ICV injection of Mu, Delta and Kappa Antagonist

ICV injections were made directly into the lateral ventricle according to the modified method of Haley and McCormick (8). Mice were lightly anesthetized with Methoxyflurane (Pitman-Moore, Mundelein, IL), an incision made in the scalp and the injection made 1 mm lateral and 0.034 mm caudal to bregma at a depth of 2.35 mm using a Hamilton microliter syringe with a 30 gauge needle.

Analysis of RNA by Reverse Transcriptase Polymerase Chain Reaction

One microgram of total RNA was reverse transcribed to synthesize the first-strand cDNA (42°C, 30 min) using random hexamers (2.5 microM), Moloney murine leukemia virus reverse transcriptase (2.5 Units; Perkin-Elmer) and 1 microM each of dATP, dCTP, dGTP and dTTP in a final reaction volume of 20 microL. Following first-strand synthesis, the reaction mixture was heated (95°C, 5 min) to inactivate the reverse transcriptase. Amplification was performed using upstream and downstream primers specific for mouse mu, delta or kappa specific opioid receptor (9). The mu opioid receptor primers amplify a 785 bp fragment, the delta receptor primers amplify a 697 bp fragment and the kappa opioid receptor primers amplify a 788 bp fragment. The first-strand cDNA reaction mixture was added to PCR buffer containing 2.5 Units of AmpliTaq DNA polymerase (Perkin-Elmer), 2 mM $MgCl_2$, 50 mM KCl, 10 mM tris-HCl, and 0.10 microM of each primer in a total volume of 100 microL. PCR conditions were 94°C for 30 sec (denaturation), 60°C for 30 sec (annealing), and 72°C for 40 sec (extension), followed by a final extension at 72°C for 10 min. Pilot studies showed a good correlation between template input and intensity of amplified fragments at 30 cycles. Control reactions included total RNA without reverse transcriptase or without template. PCR products were analyzed on 1.5% agarose gels and visualized by ethidium bromide staining.

Determination of Plasma Corticosterone

Plasma concentrations of corticosterone were determined using a DA Rat Corticosterone Radioimmunoassay kit purchased from ICN.

Statistical Analysis

Data were tested using analysis of variance followed by hypothesis testing using Tukey's test, where applicable. Significance was accepted at $P<0.05$.

RESULTS

RT-PCR Using Total RNA from Brain

To confirm that the MORKO mice did not express mu-opioid receptors, total RNA was extracted from whole brains of WT and MORKO animals, and RT-PCR was performed with opioid receptor specific primers as described by Gaveriaux et. al.(9). Our results show that, as predicted, in the MORKO animals the mu-opioid receptor primers did not amplify a 785 bp mu-specific band while the expected band was detected in the WT animals (Figure 1). Both kappa and delta specific primers amplified the expected bands (788-kappa and 697-delta) both in the MORKO and WT.

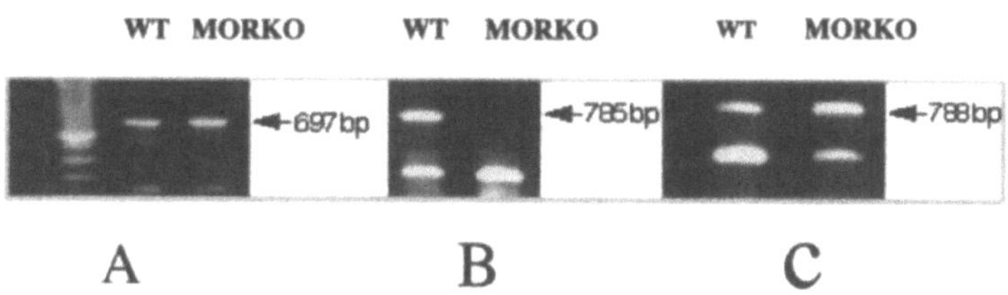

A RT-PCR performed with delta opioid receptor specific primers
B RT-PCR performed with mu opioid receptor specific primers
C RT-PCR performed with kappa opioid receptor specific primers

Figure 1: RT-PCR performed on messenger RNA purified from brain of WT and MORKO animals with mu, delta and kappa specific opioid receptor primers as described above. Arrow indicates position of amplified mu, delta and kappa message.

Effect of Morphine on Plasma Corticosterone in WT and MORKO Mice

It has been postulated that one of the mechanisms by which morphine mediates its immunosuppressive effects is by the activation of the hypothalamic-pituitary-adrenal axis. Such activation results in secretion of corticosterone which then indirectly affects the immune system. We therefore investigated the effect of morphine on plasma corticosterone levels, in the WT and MORKO mice. Our results confirm what has been shown by several investigators (10, 11,12,), that chronic morphine treatment in WT mice results in a 5 fold increase in plasma corticosterone levels around 24 hrs following pellet implantation and remains at that level 48 hr after pellet implantation (Figure 2). In contrast, in the MORKO mice, morphine pellet implantation did not result in any significant increase in plasma corticosterone; levels were similar to the placebo implanted animals (Figure 2). However, the baseline plasma corticosterone level in the MORKO mice given placebo was higher when compared to that in the WT mice.

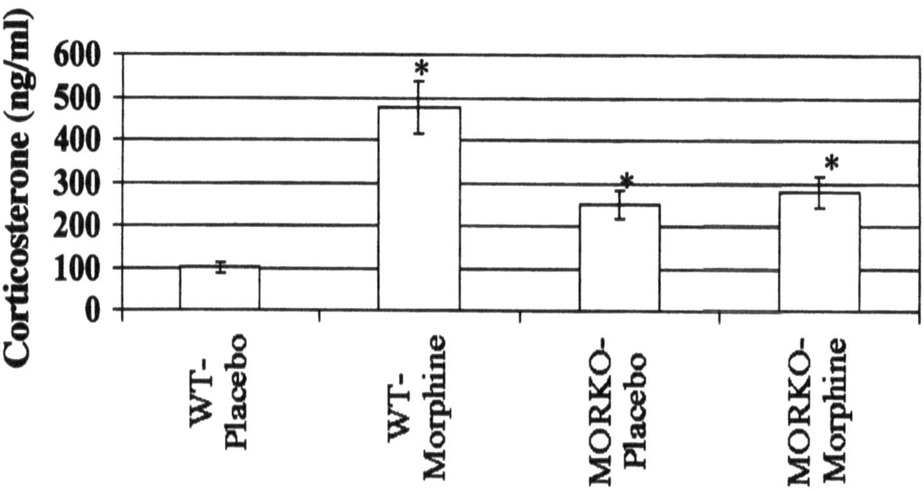

Figure 2: Effect of Morphine on plasma coticosterone in WT and MORKO mice. Plasma corticosterone levels were measured using a rat corticosterone RIA kit. Each bar represents studies carried out in 10 animals. Significance was determined using a student t test. * represents p value< 0.005.

To further verify that the increase in plasma corticosterone in the WT animals was mediated by the mu-opioid receptor, animals were pretreated with a kappa, delta or mu-receptor antagonist before morphine pellet implantation. Our results show that only the mu-receptor antagonist CTOP was able to antagonize the increase in plasma corticosterone following morphine pellet implantation (Fig 3). The delta antagonist NTI and the kappa antagonist norBNI did not antagonize a morphine induced increase in plasma corticosterone (Fig 3).

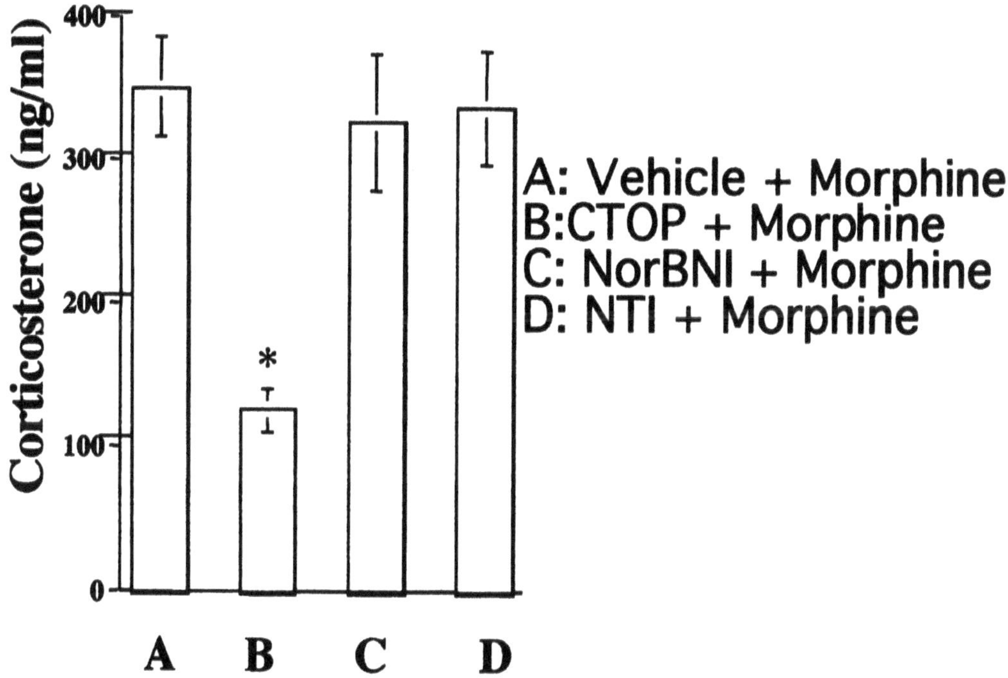

Figure 3: Effect of Opioid antagonist on morphine induced corticosterone release in WT mice. ICV injection of A: Vehicle B: Mu (CTOP), C: delta (NTI) , and D: kappa (norBNI) antagonists were made in the lateral ventricle in a final volume of 2 microliter. Plasma corticosterone levels were measured using a rat corticosterone RIA kit. Each bar represents studies carried out in 10 animals. Significance was determined using a student t test. * represents p value< 0.005.

Effect of Morphine on Thymic weight in WT and MORKO Mice

Thymic weight in the WT and MORKO mice was measured following pellet implantation. Chronic morphine treatment (75 mg implanted morphine pellet) resulted in a significant reduction in thymic and splenic weight in normal mice (Reviewed in 4). A similar reduction, with an 85 % reduction in thymic weight, was observed in the WT mice (Fig 4). This effect of morphine on thymic weight in WT animals was only partially reversed by naltrexone. (Fig 4). In the MORKO mice a significant decrease was still observed (28 %) but not to the same extent as that observed in the WT mice. The inhibition seen in the MORKO mice was not reversed by naltrexone (Fig 4).

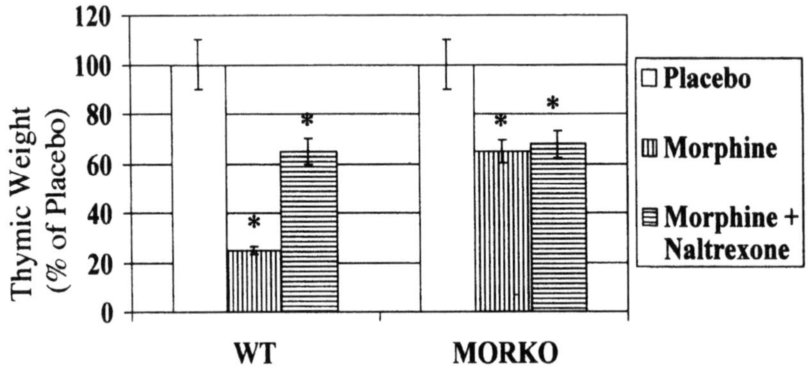

Figure 4: Effect of morphine on thymic weight in WT and MORKO mice. Animals were implanted with either placebo (open bar), morphine (vertical lines) or morphine + naltrexone (horizontal lines) pellets. Thymic weight was measured and adjusted to per gram body weight. Thymic weight in the placebo animals in each group was represented as 100%. Each bar represents studies carried out in 10 animals. Significance was determined using a student t test. * represents p value< 0.005.

Role of Corticosterone in Morphine Induced Reduction of Thymic Weight

Experiments were carried out to investigate if the reduction in thymic weight seen in WT animals was due to increases in plasma corticosterone. In these experiments, animals were implanted with either morphine or placebo pellets received a single dose injection of 10 mg/kg dexamethasone. Thymus were harvested 24 hours after this injection. Another group of animals received both morphine pellets and the dexamethasone. Our results show that in the WT animals, dexamethasone injection alone resulted in 40% reduction in thymic weight but a combination of morphine and dexamethasone resulted in almost a 95% decrease in thymic weight (Figure 5). In the MORKO animals, dexamethasone injection alone resulted in the same decrease (40%) as that observed in the WT animals. However, a combination of dexamethasone and morphine resulted in only a 68% decrease in thymic weight (Figure 5).

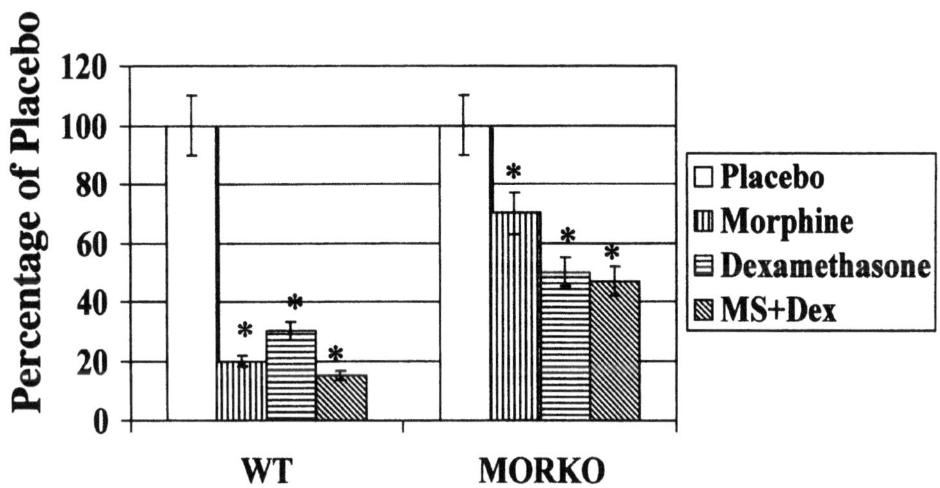

Figure 5: Role of corticosterone in morphine induced reduction of thymic weight. Animals were implanted with either placebo (open bar), morphine pellets (vertical lines). Placebo pelleted (horizontal lines) or morphine pelleted (hatched bar) animals were injected with a single injection of 10mg/kg body weight dexamethasone. Forty-eight hours following dexamethasone injection animals were sacrificed and thymuses removed. Thymic weight was measured and adjusted to per gram body weight. Thymic weight in the placebo animals in each group was represented as 100%. Each bar represents studies carried out in 10 animals. Significance was determined using a student t test. * represents p value< 0.005.

DISCUSSION

Chronic morphine has been shown to alter a number of immune parameters. These include, in addition to a decrease in thymic and splenic weight, the lymphocyte proliferative response to mitogens (13), T-cell rosette formation (14) and the total number of circulating lymphocytes (15). In animal models, morphine treatment has been found to increase mortality rates in experimentally infected mice (16,17,18). Furthermore, lymphocyte proliferative responses (19), NK cell cytotoxicity activity (20) antibody and serum hemolysin formation (21), macrophage phagocytosis (22) and interleukin-2 (IL-2) synthesis (23) are all attenuated after in vivo chronic morphine exposure to animals. While the mechanisms responsible for morphine-induced changes in the immune system are not completely understood, it has been speculated that it may be mediated directly through the opiate receptors present on lymphocytes (4), or indirectly through opiate receptors present in the central nervous system which activate the hypothalamic-pituitary-adrenal axis to release immunosuppressive glucocorticoids (24).

Results from our present study show that there is a significant increase in plasma corticosterone following chronic morphine administration in the WT animals. However this increase in plasma corticosterone was not observed in the MORKO animals. From this result it can be concluded that the increase in plasma corticosterone following morphine treatment is mediated by the mu-opioid receptor. This conclusion is further supported by our observations that the increase in plasma corticosterone in WT animals was only antagonized by the mu-receptor antagonist CTOP but was not antagonized by the delta antagonist NTI or the kappa antagonist NorBNI. Thus, this animal model provides us with a valuable tool to evaluate if the effects of morphine on immune parameters are mediated directly by the mu-opioid receptor or indirectly by corticosterone.

When thymic weight was measured in the WT and MORKO mice following pellet implantation it was found that chronic morphine treatment (75 mg implanted morphine pellet) resulted in significant reduction in thymic weight in the WT mice (Figure 4). This effect of morphine on thymic weight in WT animals was only partially reversed by naltrexone in the WT mice (Figure 4). In the MORKO mice a significant decrease was still observed (28%) but not to the same extent as that observed in the WT mice. The inhibition seen in the MORKO mice was not reversed by naltrexone (Figure 4). From these results it can be concluded that the 28% decrease in thymic weight observed in the MORKO mice was not mediated by the classical mu-opioid receptor and was not corticosterone dependent. We speculate that this decrease in thymic weight by morphine is mediated by the morphine binding site that was previously described by our laboratory (25).

To investigate if the reduction in thymic weight seen in WT animals was due to the increase in plasma corticosterone, in addition to pellet implantation, animals were injected with a single dose of 10 mg/kg dexamethasone. Thymus was harvested 24 hours after the injection. Our results show that in the WT animals, dexamethasone injection alone resulted in a 40% reduction in thymic weight but a combination of morphine and dexamethasone resulted in almost a 95% decrease in thymic weight (Figure 5). In the MORKO animals, dexamethasone injection alone resulted in the same decrease (40%) as that observed in the WT animals. However, a combination of dexamethasone and morphine resulted in only a 68% decrease in thymic weight (Figure 5). From these experiments it can be concluded that in the WT animals, corticosterone was responsible for only a 40% reduction since injection of dexamethasone resulted in only a 40 % decrease in thymic weight in MORKO animals. However to observe the almost 95% decrease in thymic weight following morphine treatment, an intact morphine receptor was essential. Furthermore, it can be concluded that the morphine effect on thymic weight was not totally an indirect effect mediated by corticosterone, but a direct effect of morphine.

ACKNOWLEDGEMENTS

This work was supported by grants from the National Institutes of Health (grant No. R01-DA12104-01A1 and No. P50-DA11806-02) and the Department of Defense/Veterans Affairs.

REFERENCES

1. N.E.S. Sibinga and A. Goldstein, Opioid peptides and opioid receptors in cells of the immune system, *Ann. Rev. Immunol.* 6:219-249 (1988).
2. H.U. Bryant, B.C. Yoburn, C.E. Intrissi, E.W. Bernton, and J.W. Holaday, Immunosuppressive effects of chronic morphine treatment, *Eur. J. Pharm.* 149:165-169 (1988).
3. P.K. Peterson, T.W. Molitor, and C.C. Chao, Mechanisms of morphine induced immunomodulation, *Biochem. Pharmacol.* 46:343-348 (1993).
4. S. Roy and H.H. Loh, Effects of opioids on the immune system, *Neurochem. Resch.* 21(11):1373-1384 (1996).
5. H.H. Loh, H.S. Liu, A. Cavalli, W. Yang, Y.F. Chen, and L.N. Wei, Mu-opioid receptor knockout in mice: effects on ligand-induced analgesia and morphine lethality, *Mol. Brain Res.* 54:321-326 (1998).
6. S. Roy, H.S. Liu, and H.H. Loh, Mu-receptor knockout mice: the role of mu-opioid receptor in gastrointestinal transit, *Mol. Brain Res.* 56:281-283 (1998).
7. E.L. Way, H.H. Loh, and F.H. Shen, Morphine tolerance, physical dependence, and synthesis of brain 5-HT, *Science.* 162:1290 (1968).
8. T.J. Haley and W. G. McCormick, Pharmacological effects produced by intracerebral injections of drugs in the conscious mouse, *Br. J. Pharmacol. Chemother.* 12:12-15 (1957).
9. C. Gaveriaux, J. Peluso, F. Simonin, J. Laforet, and B. Kieffer, Identification of kappa and delta opioid receptor transcripts in immune cells, *FEBS Letters.* 369:272-276 (1995).
10. R.D. Mellon and B.M. Bayer, Evidence for central opioid receptors in the immunomodulatory effects of morphine: review of potential mechanism(s) of action, *J. Neuroimmunol.* 83(1-2):19-28 (1998).
11. R.J. Weber and A. Pert, The periaqueductal grey matter mediates opiate-induced immunosupression, *Science.* 245:188-190 (1989).
12. H.U. Bryant, E.W. Bernton, J. R. Kenner, and J.W. Holaday, Role of the adrenal cortical activation in the immunosuppressive effects of chronic morphine treatment, *Endocrinology.* 128:3253-3258 (1991).
13. S.M. Brown, B. Stimmel, R.N. Taub, S. Kochwa, and R.E. Rosenfeld, Immunological dysfunction in heroin addicts, *Arch. Intern. Med.* 134:1001-1006 (1974).
14. J. Wybran, T. Appelboon, J.P. Famey, and A. Govaerts, Suggestive evidence for receptors for morphine and methionine enkephalin on normal human blood T-lymphocytes, *J. Immunol.* 123:1068-1070 (1979).
15. R.J. McDonough, J.J. Madden, A. Falek, D.A. Shafer, M. Pline, D. Gordon, P. Bokos, J.C. Kuehnle, and J. Mendelson, Alterations of T and null lymphocyte frequencies in the peripheral blood of human opiate addicts, *J. Immunol.* 125:2539 (1980).
16. S. Roy, R. Charboneau, and R.A. Barke, Morphine synergizes with LPS in an experimental sepsis model, *J.Neuroimmunol.* 95:107-114 (1999).

17. E. Tubaro, U. Avico, C. Santiangeli, P. Zuccaro, G. Cavallo, R. Pacifici, C. Croce, and G. Borelli, Morphine and methadone impact of human phagocytic physiology, *Int. J. Immunopharmacol.* 7:865-874 (1985).
18. C.C. Chao, B.M. Sharp, C. Pomeroy, G.A. Filice, and P.K. Peterson, Lethality of morphine in mice infected with Toxoplasma gondii, *J. Pharmacol. Exp. Ther.* 252:605-609 (1990).
19. H.U. Bryant, B.C. Yoburn, C.E Intrissi, E.W. Bernton, and J.W. Holaday, Immunosuppressive effects of chronic morphine treatment, *Eur. J. Pharm.* 149:165-169 (1988).
20. D.T. Lysle, M.E. Coussons, V.J. Watts, E.H. Bennett, and L.A. Dykstra, Morphine-induced alterations of immune status: dose dependency, compartment specificity and antagonism by naltrexone, *J. Pharmacol. Exp. Therap.* 265:1071-1078 (1993).
21. S.S. Lefkowitz and C.Y. Chiang, Effects of certain abused drugs on hemolysin forming cells, *Life Sci.* 17:1763-1768 (1975).
22. I. Szabo, M. Rojavin, J.L. Bussiere, T.K. Eisenstein, M.W. Adler, and T.J. Rojers, Suppression of peritoneal macrophage phagocytosis of *Candida albicans* by opioids, *J. Pharmacol. Expt. Therap.* 267:703-706 (1993).
23. S. Roy, R. Chapin, K. Cain, R. Charboneau, S. Ramakrishnan, and R.A. Barke, Morphine inhibits transcriptional regulation of IL-2 synthesis in thymocytes, *Cell. Immunol.* 179:1-9 (1997).
24. R. George and E.L. Way, Studies on the mechanism of pituitary-adrenal activation by morphine, *Br. J. Pharm.* 10:260-264 (1955).
25. S. Roy, B.L. Ge, N.M. Lee and H.H. Loh, Characterization of 3H-Morphine binding to Interleukin-1 activated thymocytes, *J. of Pharm. And Expt. Ther.* 263,(2):451-456 (1992).

OPIATES PROMOTE T CELL APOPTOSIS THROUGH JNK AND CASPASE PATHWAY

Pravin Singhal, Aditi Kapasi, Krishna Reddy, and Nicholas Franki

Molecular Biology and Experimental Pathology
Division of Kidney Diseases and Hypertension
Long Island Jewish Medical Center
New Hyde Park, NY 11040

ABSTRACT

Opiate addicts are prone to recurrent infections. In the present study we evaluated the molecular mechanism of opiate-induced T cell apoptosis. Both morphine and DAGO ([D-Ala^2,N-Me-Phe^4,Gly^5-ol]enkephalin) enhanced T cell apoptosis. Morphine as well as DAGO activated c-Jun NH_2-terminal kinase (JNK) in T cells. Moreover, opiates increased the expression of ATF-2, a specific substrate for JNK and P38 mitogen activated kinases (MAPK). Furthermore, opiates attenuated extracellular signal related kinase (ERK) in T cells. Both morphine and DAGO cleaved pro-caspases 8, 9, and 10 and generated caspases 8, 9 and 10 (active products). Morphine as well as DAGO also cleaved poly- (ADP-ribose) polymerase (PARP) into 116 and 85 kD proteins indicating the activation of caspase-3. These results suggest that opiate-induced T cell apoptosis may be mediated through the JNK cascade and activation of caspases 8 and 3.

INTRODUCTION

The role of opiates in immune mediated injury has been debated for a long period. Opiate addicts are prone to infections; however, it was predominantly attributed to the use of non-sterile and contaminated needles for the intravenous administration of heroin.[1,2] Nevertheless, immune modulatory cells such as lymphocytes have not only been demonstrated to carry specific opiate receptors but opiates have also been reported to modulate their functions including signal transduction, calcium flux, and generation of cytokines and lymphokines.[3-8] Recently, we and other investigators reported that morphine promotes lymphocyte apoptosis.[9,10] In addition, we demonstrated that morphine accelerates apoptosis of T cells in *in vitro* and in *ex vivo* studies.[10]

Opiate addicts are at increased susceptibility to HIV infection.[11] CD4 cell depletion is an important pathological process associated with HIV-1 infection.[12] CD4 cell count has been reported to be an important denominator for the development of bacterial infections in patients with HIV infection.[12] Previously, other investigators demonstrated that HIV-induced CD4 depletion is mediated through the generation of reactive oxygen species.[13] Recently, we also reported that morphine causes lymphocyte apoptosis through the generation of reactive oxygen species.[10] Thus HIV infection and opiates may have an additive effect in induction of lymphopenia.

T cell activation involves specific steps for their activation i.e. in Th1 cell lines this may include initial induction of c-Jun, c-Fos, and IL-2 receptor α chain followed by IL-2 gene expression and subsequent mitogenic response.[14] Whether a cell will proliferate or die depends on activation of the ERK or JNK pathways.[15] The ERK pathway will augment synthesis of c-Fos whereas the JNK pathway will enhance synthesis of c-Jun to form an AP-1 complex.[15,16] The synthesis of c-Fos may initiate the transcription of genes promoting cell proliferation; whereas the synthesis of c-Jun may trigger a chain of events from gene transcription to cell death.

Caspases have been demonstrated to play an important role in propagation of apoptosis in T cells.[17,18] The caspase family consists of cysteine proteases, which specifically cleave their substrates after an aspartate moiety: hence the name c-asp-ases. They are activated by proteolytic cleavage after aspartate residues themselves. The mature form of the enzyme has the ability to cleave and activate other procaspases, producing a caspase cascade that may induce the demise of a cell.

Recently, Yin et al. reported that morphine enhanced Fas expression and induced apoptosis in lymphocytes.[19] FasL-Fas interaction recruits FADD (Fas –associated death domain containing protein) and procaspase- 8 to form DISC (death inducing signal complex).[20,21] This interaction causes autocatalytic activation of caspase-8. If caspase –8 is in abundance, it can activate other caspases to initiate the effector stage of apoptosis. It has been suggested that amplification loops between caspase-10 and 8, and between caspase-6 and 3 accelerate the caspase cascade. The latter leads to the demise of a cell by caspase-6-induced lamin proteolysis facilitating nuclear collapse as well as by caspase –3 –induced activation of a DNA fragmentation factor causing DNA fragmentation. However, if caspase-8 is generated in small amounts it can cleave Bid, a Bcl-2 interacting protein, into a truncated form that promotes cytochrome c release from the mitochondria. The latter may activate downstream effector caspases including caspase-3.[21,22]

In the present study, we determined the effect of opiates on the activation of T cell JNK and ERK pathways. We also evaluated the role of caspases in the induction of opiate-induced T cell apoptosis.

METHODS

Human T Lymphocytes

Twenty ml aliquots of blood were obtained from healthy volunteers. Lymphocytes were isolated with the use of a T-lymphocyte separating kit (Accurate Chemical & Scientific, Westbury, NY). Lymphocytes were incubated in RPMI 1640 (Life Technologies, Grand Island, NY) containing 10% fetal calf serum (FCS, heat-inactivated), 1 mM L-arginine, 1% Hepes, 0.2% $NaHCO_3$, 50 U/ml penicillin, and 50 □g/ml streptomycin (Life Technologies). To evaluate contamination of the lymphocyte population with monocytes, lymphocytes were labeled with human monoclonal anti-CD14 antibody. Less than 2% of the cells showed positive staining. The purity of lymphocytes in different experiments varied from 97 to 99%.

Jurkat Cells (T-cell line, human)

The human leukemic T cell line, Jurkat (EG-1) was obtained from American Type Culture Collection (ATCC), Rockville, MD. Jurkat cells were cultured in RPMI 1640 medium supplemented with 10% FCS, 1.5 g/L $NaHCO_3$, 1 mM sodium pyruvate, 1% Hepes, 4.5 g/L glucose, 50 U/ml penicillin, and 50 µg/ml streptomycin in a humidified incubator with 5% CO_2 in air at 37°C.

Apoptosis Studies

To evaluate the occurrence of necrosis and apoptosis, we used propidium iodide (Sigma Chemical Co, St. Louis, MO) and Hoechst (H)-33342 (Molecular Probes, Eugene, OR) stains. Propidium iodide stains the necrosed cells; whereas H-33342 stains the nuclei of live cells and identifies apoptotic cells by increased fluorescence.[22] Human T lymphocytes or Jurkat cells were prepared under control and experimental conditions. At the end of the incubation period, aliquots of methanol containing H-33342 (final concentration, 1 µg/ml) were added and incubated for 10 min at 37°C. Cells (without a wash) were placed on ice, and propidium iodide (final concentration, 1 µg/ml) was added to each well. Cells were incubated with dyes for 10 min on ice, protected from light, and then examined under ultraviolet light using a Hoechst filter (Nikon, Melville, NY). The percentage of live, apoptotic, and necrosed cells was recorded in eight random fields by two observers unaware of experimental conditions.

DNA Isolation And Gel Electrophoresis

Equal numbers of subconfluent human T lymphocytes or Jurkat cells were prepared under control and experimental conditions. At the end of the incubation period, cells were washed twice with phosphate buffered saline (PBS) and lysed in DNA lysis buffer. DNA was extracted and run on a 1.8% agarose gel and electrophoresed at 5 V/cm in 0.5 X TE buffer (Tris 10 mM, EDTA 1 mM, pH 8.0) containing 10 µg/ml ethidium bromide.

RNA Extraction And Northern Blotting

To evaluate the effect of morphine on T lymphocyte c-Jun expression, equal numbers of subconfluent lymphocytes (Jurkat cells or isolated human lymphocytes) were incubated under control and experimental (variable concentrations of morphine) conditions. At the end of the incubation period, cells were lysed, and total RNA was extracted by the method of Chomczynski and Sacchi.[23] Aliquots of total RNA were dissolved in 0.5% SDS, electrophoresed in a 1.2% agarose gel, and transferred to Hybond-N, nylon membranes. The gels were stained with ethidium bromide to determine the position of the 28S and 18S ribosomal RNA bands and to assess the integrity of the RNA. cDNA probes specific for c-Jun (gift of Sanjeev Gupta, Albert Einstein College of Medicine, Bronx, NY) were used for hybridization after [^{32}P]dCTP labeling by a random-primed method. Filters were hybridized at 42°C for 16 hours with the labeled cDNA probe. The membranes were washed under highly stringent conditions with 2 to 0.2 X SSC (varied according to probe) and 0.1% SDS at 65°C and then kept in contact with XAR-5 film and an intensifying screen at -70°C before developing. The membranes were stripped to remove the hybridized probe and reprobed with a GAPDH probe (ATCC) to ascertain that similar amounts of RNA were applied to the gel. Three sets of experiments were carried out. Densitometric analysis was performed on each blot.

Protein Extraction And Western Blotting

To evaluate the effect of morphine on activation of the caspase pathway, equal numbers of subconfluent Jurkat cells were incubated under control and experimental (variable concentrations of morphine or [D-Ala^{2},N-Me-Phe^{4},Gly^{5}-ol]enkephalin, DAGO) conditions. At the end of the incubation period, cells were lysed with lysis buffer and protein was assayed using a BCA kit (Pierce, Rockford, IL). Twenty micrograms of protein from each variable was separated on a 4 to 20% gradient polyacrylamide gel and blotted onto a nitrocellulose membrane using a BIO-RAD Western blotting apparatus (Hercules, CA). The nitrocellulose membranes were then processed for JNK, ERK, ATF-2, caspase-8-10 (Santa Cruz Biotechnology Inc., Santa Cruz, CA), and PARP (Upstate Biotechnology, Lake Placid, NY) antibodies. The concentrations of primary antibodies were: 0.2 μg/ml for JNK, ERK and ATF-2, 0.04 μg/ml for Caspase 8-10 and 1 μg/ml for PARP. The membranes were processed using horseradish peroxidase labeled secondary goat anti-rabbit (in case of ATF-2) or donkey anti-goat in case of JNK, ERK, Caspase 8-10 and PARP (Santa Cruz). Blots were developed using enhanced chemiluminescence (ECL, Amersham, and Arlington Heights, IL).

Statistical Analysis

For comparison of mean values between two groups, the unpaired t test was used. To compare values between multiple groups, analysis of variance (ANOVA) was applied and a Newman-Keuls multiple range test was used to calculate a q value. All values are means ± SEM except where otherwise indicated. Statistical significance was defined as $p<0.05$.

RESULTS

To determine the dose response effect of morphine and DAGO, equal numbers of Jurkat cells were incubated in media containing either vehicle or variable concentrations of morphine or DAGO (10^{-8} to 10^{-6}M) for 24 hours. At the end of the incubation period, cells were prepared for H-33342 and propidium iodide staining and DNA fragmentation assay. Both morphine and DAGO enhanced apoptosis of T cells (data not shown). Morphine treated cells also showed integer multiples of 180 base pairs in the form of a ladder pattern (data not shown).

To determine the effect of opiates on JNK/ERK pathways, equal numbers of Jurkat cells were incubated in media containing either vehicle (control) or variable concentrations of morphine or DAGO (10^{-8} to 10^{-6}M) for 12 hours. At the end of the incubation period, cells were prepared for probing for JNK/ATF-2/ERK pathways. As shown in Fig. 1, both morphine (10^{-8} to 10^{-6} M) and DAGO (10^{-6} M) activated JNK.

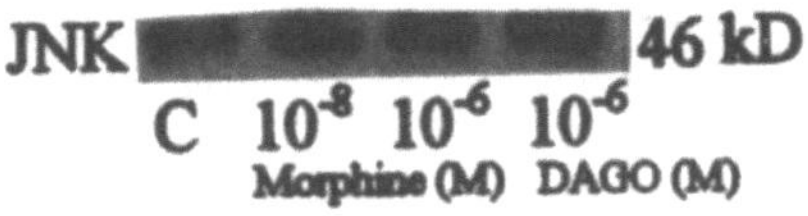

Figure1. Effect of morphine and DAGO on Jurkat cell JNK activation. Both morphine and DAGO activated Jurkat cells when compared to vehicle treated cells (C, control).

Similarly, morphine (10^{-8} to 10^{-6} M) and DAGO (10^{-8} to 10^{-6} M) increased expression of ATF-2, the substrate for JNK/P28 MAPK (Fig. 2).

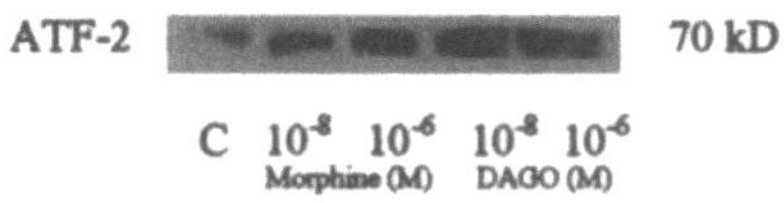

Figure 2. Effect of morphine and DAGO on Jurkat cell ATF-2 production. Both morphine and DAGO enhanced the production of ATF-2 ; whereas control cells © showed minimal production of ATF-2.

On the contrary, both morphine (10^{-8} to 10^{-6} M) and DAGO (10^{-8} to 10^{-6} M) attenuated ERK activation (Fig. 3).

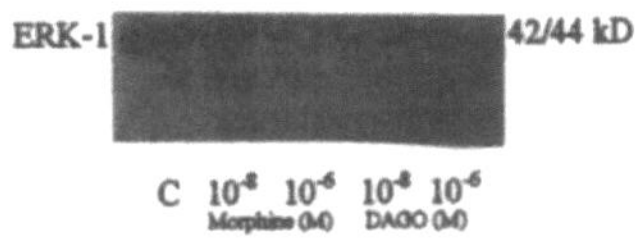

Figure 3. Effect of morphine on ERK activation. Vehicle treated cells © showed activation of ERK. Morphine (10^{-6}M) and DAGO (10^{-8} to 10^{-6} M) decreased activation of ERK.

To evaluate the effect of morphine on Jurkat cell c-Jun expression, equal numbers of Jurkat cells were incubated in media containing either vehicle or morphine (10^{-8} to 10^{-6} M) for 2 h. At the end of the incubation period. Total cellular RNA was extracted and probed for c-Jun mRNA (n=3). As shown in Fig. 4, morphine enhanced Jurkat cell c-Jun mRNA expression.

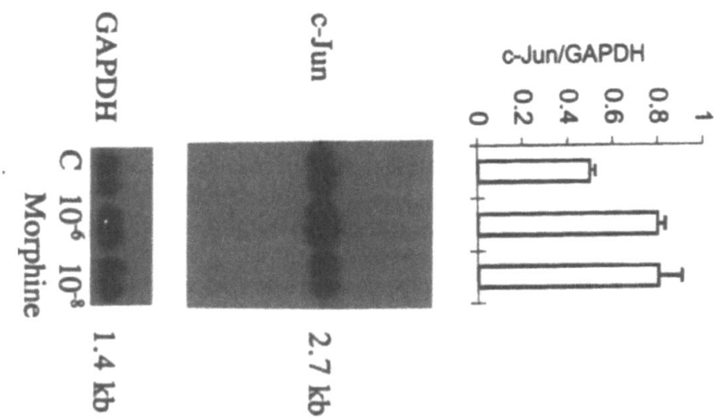

Figure 4. Effect of morphine on Jurkat cell c-Jun expression. Morphine enhanced Jurkat cell mRNA expression of c-Jun .

To determine the role of caspases in opiate-induced T cell apoptosis, equal number of Jurkat cells were incubated in media containing either vehicle (control), morphine (10^{-8} to 10^{-6} M), or DAGO (10^{-8} to 10^{-6} M) for 12 hours. At the end of the incubation period, cells were prepared for probing for PARP and caspases-8, 9, and 10. Both morphine and DAGO not only

prepared for probing for PARP and caspases-8, 9, and 10. Both morphine and DAGO not only cleaved pro-caspases 8, 9 and 10 but also cleaved PARP into 116 and 85 kD proteins suggesting the activation of caspase-3 (Fig. 5).

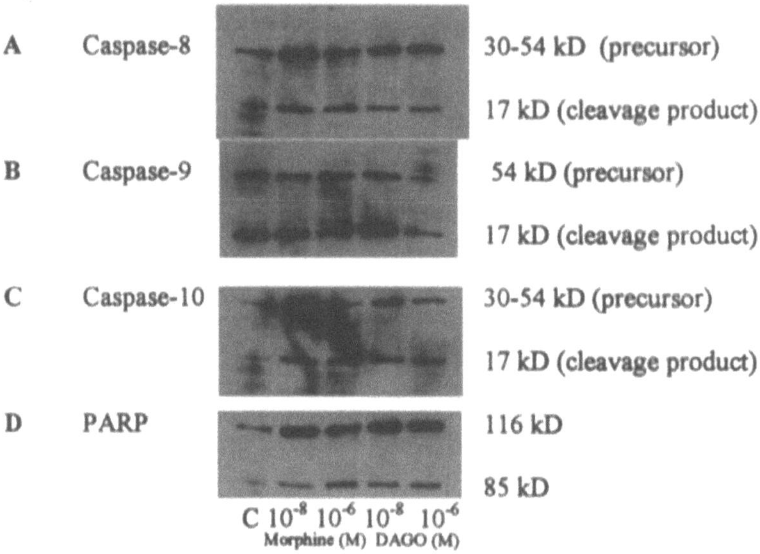

Figure 5. Effect of morphine and DAGO on cleavage of procaspase-8,9, 10 and PARP.
Both morphine and DAGO cleaved procaspase 8 (A),9 (B) and 10 (C) into their cleavage products (active caspases). At levels that were higher than cleavage levels for control cells (C). Morphine as well as DAGO cleaved PARP into 116 and 85 kD proteins indicating the actvation of caspase-3. Control cells (C) showed minimal cleavage of PARP, a specific substrate for caspase-3.

DISCUSSION

In the present study, opiates promoted T cell apoptosis. Opiates also activated caspase-8 and downstream effector caspases. Similarly, opiates not only increased expression of ATF-2 but also stimulated activation of JNK. On the other hand, opiates inhibited ERK activation. These results suggest that opiate-induced T cell apotosis may be mediated through caspase-8 and its downstream effector pathway. In addition, the JNK pathway seemed to mediate the opiate-induced apoptosis.

Lymphocytes are more sensitive to injury, especially radiation, than other cells. It takes thousands of rads to kill fibroblasts or macrophages; whereas as little as 5 rads is lethal to lymphocytes.[24] A damaged lymphocyte usually does not respond by repairing itself but by committing suicide. On the contrary, fibroblasts and macrophages have the ability to repair even severe damage. T cells undergo apoptosis predominantly through TCR (T cell antigen receptor) or activation-induced cell death.[25] Activation-induced apoptosis of T cells occurs through the upregulation of the Fas (CD95) pathway and associated death receptor pathways.[26,27] T cell receptor ligation enhances Fas expression and induces the expression of FasL .[26,27] Ligation of Fas by FasL results in the clustering of Fas and the recruitment of the adopter protein, Fas-associated death domain (FADD) and procaspase-8 to the clustered Fas intracellular domains. This may cause autocleavage of procaspase-8 leading to activation of caspase-8. Activated caspase-8 may mediate apoptosis either by direct activation of downstream effector caspases by direct cleavage, or by indirect activation of downstream effector caspases by inducting the release of cytochrome c from mitochondria.[28-30] Bcl-2 and Bcl-x_L block apoptosis occurring through the mitochondrial pathway rather than through the direct pathway. It appears that some T cells use the direct apoptotic pathway; whereas other T cells use the mitochondrial pathway to undergo apoptosis. Interestingly, co-stimulation by professional antigen presenting cells or by the direct ligation of CD28 attenuates activation-induced apoptosis of previously activated T cells.[31] Similarly, interleukin (IL)-2,by increasing the expression of Bcl-2 inhibits the activation-induced apoptosis.[32] In addition, costimulation of CD28 prevents expression of FasL .[31] Repeated TCR activation enhances the Fas proapoptotic pathway; on the other hand costimulation triggers the Bcl-2-associated anti-apoptotic pathway. Yin et al. in in vitro and in vivo studies demonstrated that opioids enhance lymphocyte Fas expression and induce FasL-mediated apoptosis.[19] Since this effect of opioids was inhibited by naloxone, these investigators suggested the role of opioid receptors in morphine-induced lymphocyte apoptosis.[19]

In the present study morphine and DAGO not only activated JNK (stress kinase activated pathway) but also inhibited ERK in T cells. These results indicate that opiates act as a stress factor or mitogen on these cells. Morphine as well as μ specific agonist DAGO not only enhanced T cell apoptosis but also activated caspases including cleavage of PARP into 116 and 85 kD proteins. Although these studies suggest some role for μ receptors in opiate-induced T cell apoptosis the presence of μ receptors on T cells has not been demonstrated. On the contrary, Wang et al. could not detect μ receptor mRNA by a reverse transcription/polymerase chain reaction assay in human peripheral T cells.[33] These investigators did observe saturable morphine binding to sonicated, activated lymphocytes. However, this binding was not stereospecific and not inhibited by DAMGO. Interestingly, these investigators demonstrated that the microsomal fraction of PMA-activated Jurkat cells bind morphine sterospecifically. Compounds such as (-)- naloxone, levorphanol and DAMGO competed with (-)-morphine for this binding site; whereas (+)-morphine, dextrophan, cocaine and ß-endorphin did not compete for this binding site. This study suggests that the morphine-induced effects on T cells may not be mediated through μ receptors but perhaps mediated through a microsomal binding site. The exact nature of this binding site needs further evaluation.

We conclude that opiates enhance T cell apoptosis. This effect of opiates seems to be mediated through the JNK pathway. Caspases including 8 and 3 play an important role in the propagation of opiate-induced T cell apoptosis.

ACKNOWLEDGEMENT

This work was supported by a grant 12111 from the National Institute on Drug Abuse.

REFERENCES

1. W.F. Luttgens, Endocarditis in "main line " opium addicts, *Arch. Intern. Med.* 83:653 (1949).
2. J.H. Briggs, C.G. McKerron, R.L. Souhami, D.J.E. Taylor, and H. Andrews, Severe systemic infections complicating "main line" heroin addiction, *Lancet* 2:1227 (1967).
3. J.J. Madden, R.M. Donahoe, J. Zwemer-Collins, D.A. Shafer, and A. Falek, Binding of naloxone to human T lymphocytes, *Biochem. Pharmacol.* 36:4103 (1987).

4. R.T. Radulescu, B.R. DeCosta, A.E. Jacobson, K.E. Rice, J.E. Blalock, and D.J.J. Carr, Biochemical and functional characterization of a μ opioid receptor binding site on cells of the immune system, *Progr. Neuroendocrine Immunol.* 4:166 (1991).
5. A. Rao, Signalling mechanisms in T cells, *Crit. Rev. Immunol.* 10:495 (1991).
6. A. Weiss and D.R. Littman, Signal transduction by lymphocyte antigen receptors, *Cell* 76:263 (1994).
7. R.H. Schwartz, Costimulation of T lymphocytes: the role of CD28, CTLA4, and B7/BB1 in interleukin production and immunotherapy, *Cell* 71:1349 (1990).
8. J.D. Fraser, B.A. Irving, G.R. Crabtree, and A. Weiss. Regulation of interleukin-2 gene enhancer activity by the T cell accessory molecule CD28. Science 251:313(1991).
9 M.P.N. Nair, S.A. Schwartz, R. Polasani, J. Hou, A. Swet, and K.C. Chadha, Immunoregulatory effects of morphine on human lymphocytes. Clin. Diagnost. Lab. Immunol. 4:127 (1997).
10 P.C. Singhal, A.A. Kapasi, K. Reddy, N. Franki, N. Gibbons, and G. Ding, Morphine promotes apoptosis in Jurkat cells, *J. Leukocyte Biol.* 66:650 (1999).
11 S.A. Fauci, The human immunodeficiency virus : infectivity and mechanism of pathogenesis, *Science* 239: 617 (1988).
12 A.E. Greenberg, P.A. Thomas, S.H. Landesman, D. Mildvan, M. Seidlin, G.H. Friedland, R. Holzman, J.B. Starrett, E.L. Bryan, and R.F. Evans, The spectrum of HIV-1-related disease among outpatients in New York City, *A.I.D.S.* 6:849 (1992).
13 T.S. Dobmeyer, S. Findhammer, J.M. Dobmeyer, B. Raffel, D. Hoelzer, E.B. Helm, D. Kabelitz, and R. Rossol, Ex vivo induction of apoptosis in lymphocytes is mediated by oxidative stress: Role for lymphocyte loss in HIV infection. *Free Rad. Biol. Med.* 22:775 (1997).
14 G.R. Crabtree, Contingent genetic regulatory events in T lymphocyte activation, *Science* 243:355 (1989).
15 M. Karin, The regulation of AP-1 activity by mitogen-activated protein kinases. *J. Biol. Chem.* 270:16483 (1995).
16 J.M. Kyriakis, and J. Avruch, Sounding the alarm: protein kinase cascades activated by stress and inflammation, *J. Biol. Chem.* 271:24313(1996).
17 M. Enari, R.V. Talanian, W.W. Wong, and S. Nagata, Sequential activation of ICE-like and CPP32-like proteases during Fas-mediated apoptosis, *Nature* 380:723 (1996).
18 S.M. Srinivasula, M. Ahmad, A.T. Fernandes, G. Litwack, and E.S. Alnemri, Molecular ordering of the Fas-apoptotic pathway: The Fas/APO-1 protease Mch5 is CrmA-inhibitable protease that activates multiple Ced-3/ICE-like cystein proteases, *Proc. Natl. Acad. Sci.* USA 93:14486 (1996).
19 D. Yin, A. Mufson, R. Wang, and Y. Shi. Fas-mediated cell death promoted by opioids. *Nature* 397:218(1999).
20 A. Ashkenazi and V.M. Dixit. Death receptors: signaling and modulation. *Science* 281:1305(1998).
21 K. Schulze-Osthoff, D. Ferrari, M. Loss, S. Wesselborg, and M.E. Peter, Apoptosis signaling by death receptors, *Eur. J. Biochem.* 254:439(1998).
22 P.C. Singhal, P. Sharma, A. Kapasi, K. Reddy, N. Franki, and N. Gibbons, Morphine enhances macrophage apoptosis. *J. Immunol.* 160:1886(1998).
23 P. Chomczynski and Sacchi N, Single-step method of RNA isolation by acid guanidine thiocyanate phenol-chloroform extraction, *Anal. Biochem.* 162:156(1987).
24 R.E. Anderson and N.L. Warner, Ionizing radiation and the immune response, *Adv. Immunol.* 24:215(1976).
25 J.J. Cohen, Apoptosis: Mechanisms of life and death in the immune system, *J. Allergy Clin. Immunol.* 103:548(1999).

26 J. Dhein, H. Walczak, C. Baumler, K.M. Debatin, and P.H. Krammer, Autocrine T-cell suicide mediated by APO-1/(Fas/CD95), *Nature* 373:438(1995).
27 M.R. Alderson, T.W. Tough, T. Davis-Smith, Fas ligand mediates activation-induced cell death in human T lymphocytes, *J. Exp. Med.* 181:71(1995).
28 D.A. Martin, R.M. Siegel, L. Zheng, and M.J. Lenardo, Membrane oligomerization and cleavage activates the caspase-8 (FLICE/MACHalpha1) death signal, *J. Biol. Chem.* 273:4345(1998).
29 M. Muzio, B.R. Stockwell, H.R. Stennicke, G.S. Salvesen, and V. M. Dixit, An induced proximity model for caspase-8 activation, *J. Biol. Chem.* 273:2926(1998).
30 H. Li, H. Zhu, C.J. Xu, and J. Yuan, Cleavage of BID by caspase 8 mediates the mitochondrial damage in the Fas pathway of apoptosis, *Cell* 94:491(1998).
31 Y. Collette, A. Benziane, D. Raznajaona, and D. Olive, Distinct regulation of T-cell death by CD28 depending on both its aggregation and T-cell receptor triggering: a role for Fas-FasL, *Blood* 92:1350(1998).
32 A.T. Vella, S. Dow, T.A. Potter, J. Kappler, and P. Marrack, Cytokine-induced survival of activated T cells in vitro and in vivo, *Proc. Natl. Acad. Sci.* USA. 95:3810(3815) 1998.
33 Y.C. Wang, W.L.Whaley, R.M. Donahoe, P. Turner, and J.J. Madden JJ, The interaction of (-)-morphine and peripheral T lymphocytes: if not mu receptors, what? Seventh Annual Conference: Neuroimmune Circuits and Infectious Disease. October 7, 1999, NIDA, Bethesda, MD

ROLE OF BETA-ENDORPHIN IN THE MODULATION OF IMMUNE RESPONSES : PERSPECTIVES IN AUTOIMMUNE DISEASES

Paola Sacerdote, Leda Gaspani, and Alberto E. Panerai

Department of Pharmacology, via Vanvitelli 32, University of Milano, 20129 Milano, Italy

BETA-ENDORPHIN PRODUCTION BY IMMUNE CELLS

Until a decade ago, the opioid peptide beta-endorphin (BE) was considered a "neuropeptide", due to its multiple activities as neurotransmitter and neuromodulator. It is produced as a larger precursor molecule, proopiomelanocortin (POMC), and the POMC precursor molecule is post-translationally processed by a series of enzymatic events, which give rise to a number of peptide hormones e.g. ACTH, βLPH, αMSH and BE. In the CNS, BE is synthesized mainly in the arcuate nucleus of the hypothalamus, and projects in several brain areas. In the periphery the main source of BE is the intermediate pituitary or its vestige (1).

Also it was shown that BE is constitutively synthesized by cells of the immune system, that, upon appropriate stimulation release the opioid peptide (2). The hypothalamic peptide CRH, which is considered the main signal for BE synthesis and release in the brain and in the pituitary, can also induce the synthesis and the secretion of the POMC-derived peptides from Peripheral Blood Mononuclear cells (PBMC) and splenocytes (2,3). Consistently, during stress, which is associated with CRH synthesis and secretion, an increase of immunocyte beta-endorphin is present (4). Our laboratory also demonstrated that BE is similarly controlled in lymphocytes and brain by the same neurotransmitters, in fact BE is under a dopaminergic and GABAergic inhibitory tonic control, and a serotoninergic tonic stimulatory input (5,6). Besides hypothalamic releasing hormones and neurotransmitters, also the cytokine IL-1 has been shown to stimulate BE synthesis and secretion (2,3).

Beta-endorphin synthesis is therefore subject to many regulatory inputs, coming from both the CNS and the immune system.

EFFECTS OF BETA-ENDORPHIN ON THE IMMUNE SYSTEM

The *in vivo* administration of BE to rodents has been shown to decrease cellular immune functions such as mitogen induced lymphoproliferation and natural killer activity

Neuroimmune Circuits, Drugs of Abuse, and Infectious Diseases
Edited by Herman Friedman *et al.*, Kluwer Academic/Plenum Publishers, 2001

(7,8,9). On the contrary, the block of BE activity, either with the administration of the opioid antagonists naloxone and naltrexone, or of immunoglobulins that neutralize the activity of BE, induces an increase of NK and lymphoproliferation within minutes (7,8,9).

The possibility that the effect of BE in the immune system could be the activation of the two types of mature T-helper cells, Th1 and Th2, was explored. The Th1 and Th2 cells produce different patterns of cytokines: Th1 cells produce mainly IL-2, IFN-γ and lymphotoxins, whereas Th2 cells produce IL-4, IL-5, IL-6, IL-10 and IL-13. Th1 cells are mostly involved in cell-mediated reactions, while the Th2 cytokines are commonly found in association with strong antibody and allergic responses. Moreover, the characteristic cytokine products of Th1 and Th2 cells are inhibitory for the differentiation and effector function of the opposite subset (10).

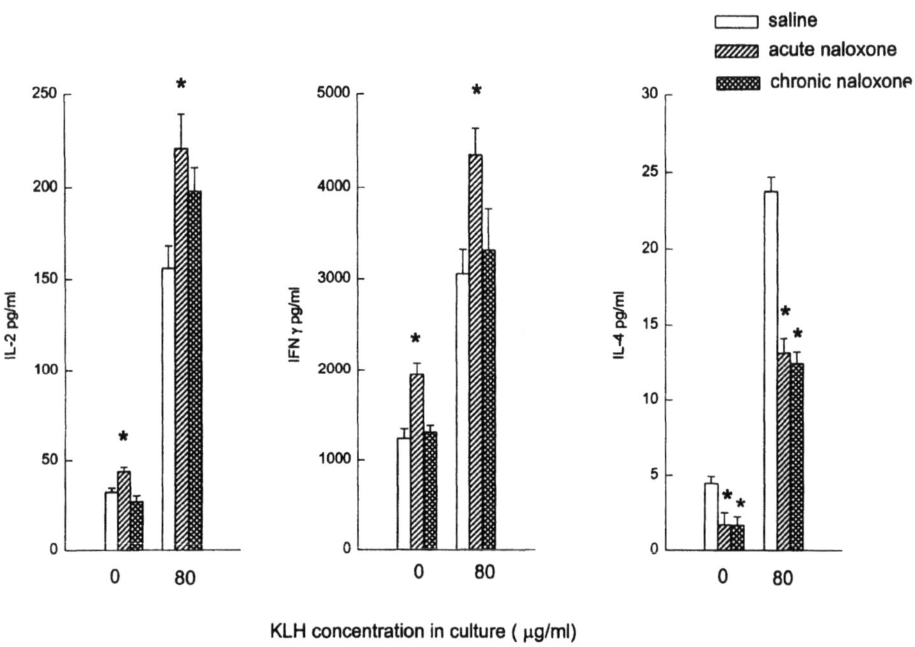

Figure 1. Effect of acute and chronic naloxone treatment on KLH-stimulated cytokine production *in vitro* by splenocytes from Balb/cJ mice. Animals were immunized *in vivo* with 100 μg KLH, and treated acutely (one single injection at the moment of immunization) or chronically (6 days) with naloxone (5 mg/Kg). Animals were killed 6 days after immunization. Spleen cells were cultured with 0 or 80 μg/ml KLH for 48 h for IL-2 and IFN-γ, and 72 h for IL-4. All measurements were performed by ELISA.
* $p < 0.01$ vs saline

Strains of mice in which one type of T cell subsets dominates over the other have been described. Balb/cJ mice are susceptible to infection by intracellular pathogens, have a weak cell-mediated immune response, and consistently present a Th2 dominance (11). In contrast, C57Bl/6 mice, which are resistant to intracellular pathogens and have a highly effective cell-mediated response, exhibit a Th1 dominance (11).

Administration of the opioid antagonist naloxone profoundly affects splenocyte production of the Th1 cytokines,IFN-γ and IL-2, and of the Th2 cytokine, IL-4 in Balb/cJ and C57Bl/6 mice immunized with the protein antigen Keyhole Limpet Hemocyanin (KLH). In fact, acute as well chronic naloxone treatment (5 mg/Kg i.p.), decreased IL-4 production both in Balb/cJ (Figure 1) and C57/Bl6 (Figure 2) mice. On the contrary, IL-2 and IFN-γ levels are increased after acute naloxon administration in Balb/CJ mice (Figure 1) and after chronic administration in C57/Bl6 mice (Figure 2).

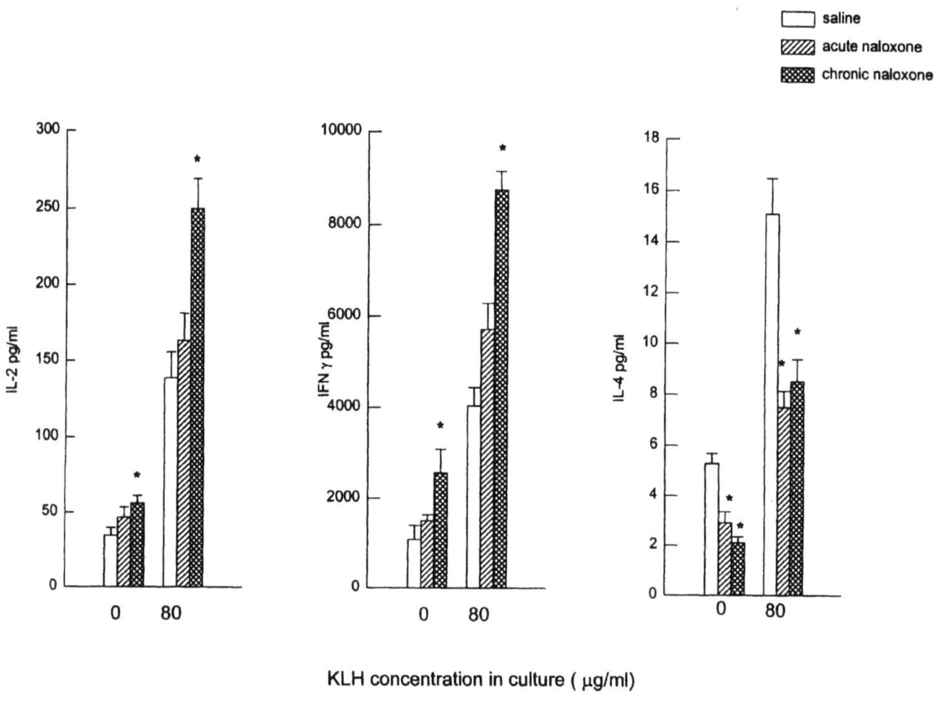

Figure 2. Effect of acute and chronic naloxone treatment on KLH-stimulated cytokine production in vitro by splenocytes from C57Bl/6 mice. Animals were killed 6 days after immunization. Spleen cells were cultured with 0 or 80 μg/ml KLH for 48 hr for IL-2 and IFN-γ and 72 h for IL-4. See legend of Figure 1 for methodological details.
* $p < 0.01$ vs saline

The administration of naloxone seems therefore to stimulate Th1 cytokines, while decreasing Th2 cytokines. Given the fact that naloxone is an almost pure antagonist at the μ-opioid receptor, devoid of any intrinsic activity, the effects of the drug are likely to be due to the removal of a regulatory tone exerted by endogenous opioid peptides.

Since an imbalance of Th1/Th2 cytokines is often at the basis of immune diseases, the effect of naloxone and/or BE on immune responses can be relevant.

For example, the Th1/Th2 paradigm has been involved in immune responses to organ transplantation (12). Under certain experimental conditions in fact, graft rejection has been associated with the presence of Th1 cytokines, while Th2 cytokines have been linked to graft survival. Therefore we also investigated the effect of the opioid receptor antagonist naloxone on the onset of allograft skin rejection in mice, and on the production of IL-2, IFN-γ and IL-4 during the development of the rejection.

As reported in Table 1 the continuous administration of BE significantly prolonged the time of rejection while naloxone significantly shortened it. This effect could be due to the effects of BE or naloxone on the Th1/Th2 balance. In fact, as shown in Table 2, the naloxone treatment affected cytokine production by splenocytes of transplanted mice evaluated at the moment of rejection. In naloxone treated animals a significant increase of IFN-γ levels was present. Interestingly the cytokine IL-2 was reduced at the moment of rejection. These low IL-2 concentrations could indicate a very rapid and massive utilisation of the cytokine by T lymphocytes. It is possible therefore that naloxone, by stimulating IL-2 production, leads to a shortening of the time necessary in order to reach the amount of IL-2 that is critical for the onset of rejection.

Table 1 Effect of BE or naloxone infusion on the survival time of skin graft

TREATMENT (continuous infusion with osmotic minipumps)	GRAFT SURVIVAL TIME (days)
Saline	10.4 ± 0.15
BE (100μg/Kg/day)	15.2 ± 0.35*
Naloxone (5mg/Kg/daily)	7.7 ± 0.25*

Protective bandages were removed on day 5, and grafts were scored daily until more than 95% of the surface was necrosed, and this was scored as complete rejection.
Saline, BE, or naloxone were administered by continuous infusion, starting the day of grafting. Results are expressed as the mean (± SEM) of the survival days of grafts in groups of 8 animals.
*= $p<0.01$ vs saline

On the whole, these data seem to confirm that the modulation of graft rejection by BE and naloxone could be due to their effects on cytokine production (13).

BETA-ENDORPHIN AND AUTOIMMUNITY

Considering the many regulatory inputs on beta-endorphin in cells of the immune system, it is conceivable that different physiological and pathological conditions characterized by alteration of neurotransmitters and cytokines could affect the concentrations of BE in immune cells, and, as a consequence, some immune functions.

BE concentrations are in fact modified in pathologies or experimental models where an activation or an inhibition of the immune system is present. In human, PBMC BE concentrations are decreased in comparison to normal subjects in rheumatoid arthritis,

Freund adjuvant induced arthritis, experimental autoimmune encephalitis and in MRL mice that are prone to spontaneous autoimmune disease (5, 14). In MLR mice, low levels of BE were present before the clinical appearance of the disease, suggesting that the BE concentrations are not a consequence of the disease.

Table 2. Splenocyte production of cytokines in control and transplanted animals

	IL-2 (pg/ml)		IFN-γ (pg/ml)	
	Medium	Con-A	Medium	Con-A
Control	15.3 ± 2.9	3369± 692	21±12.0	81260 ±15319
Graft+saline at rejection	2.6 ± 2.5	943± 89.7*	100±15.6	84202 ±21859
Graft+ naloxone at rejection	22.1 ± 3.9	2965±417	579±116*^	336186±19420*^

Controls were intact, not grafted mice. When more than 95% of graft surface was necrosed, spleens were removed and splenocytes cultured with or without Con-A (20 μg/ml), for cytokine production (13). Results are expressed as the mean (± SEM) of groups of 8 animals.
*=$p<0.05$ vs. Control
^= $p<0.01$ vs. Graft+saline

In most autoimmune diseases where BE is found decreased, the pathologies are organ specific. These diseases and other experimental models of autoimmune diseases, are characterized by the prevalence of Th1 cytokine pattern (15). In such animal models, administration of Th2 cytokines has been shown to be mostly protective against development of the disease.

Since, as previously discussed, BE seems to be involved in the modulation of Th1/Th2 balance, skewing it towards Th2, it can be speculated that low BE concentrations could be permissive for the development of the Th1 responses leading to autoimmunity. Most interestingly, low BE concentrations have been detected not only in splenocytes, but also in the hypothalamus of the autoimmune disease prone mice MRL, indicating that the impairment of the endorphinergic system is generalized (14). The opioid can therefore exert its regulation by acting at both central and peripheral sites of action.

FUTURE PERSPECTIVES

These data on effects of BE on Th1/Th2 balance suggest that modulation of the opioid peptide synthesis and/or blockade of its effects could be a new and interesting approach at immune modulation. Besides the reported effects of BE and naloxone on skin graft rejection, in consideration of the altered opiatergic system in clinical and experimental autoimmune diseases, the possibility to modulate the opioids as therapeutical approach is worthwhile of further development.

REFERENCES

1.Olson GA, Olson RD, Vaccarino AL, Kastin AJ, Endogenous opiates, 1997. Peptides 19:1791 (1998).

2 Heijnen CJ, Kavelaars A, Ballieux RE, ß-endorphin, cytokine and neuropeptide. Immunol Rev 119:41 (1991).

3. Sacerdote P, Bianchi M, Manfredi B, Panerai AE, Intracerebroventricular interleukin-1 alpha increases immunocyte beta-endorphin concentrations in the rat: involvement of corticotropin releasing hormone and neurotransmitters. Endocrinol 135:1346 (1994).

4. Sacerdote P, Manfredi B, Bianchi M, Panerai AE, Intermittent but not continuous inescapable footshock stress affects immune responses and immunocyte beta-endorphin concentrations in the rat. Brain Behav Immun 8: 251 (1994).

5. Panerai AE, Sacerdote P, Beta-endorphin in the immune system:a role at last?. ImmunolToday 18:317 (1997).

6. Sacerdote P, Rubboli F, Locatelli L, Ciciliato I, Mantegazza P, Panerai AE, Pharmacological modulation of neuropeptides in peripheral mononuclear cells. J Neuroimmunol 32: 35 (1991)

7. Manfredi B, Sacerdote P, Bianchi M, Veljic-Radulovic J, Panerai AE, Evidence for an opioid inhibitory effect on T cell proliferation. J Neuroimmunol. 44:43 (1993).

8. Panerai AE, Manfredi B, Granucci F, Sacerdote P, The beta-endorphin inhibition of mitogen-induced splenocytes proliferation is mediated by central and peripheral paracrine/autocrine effects of the opioid. J Neuroimmunol 58:71 (1995).

9. Sacerdote P, Bianchi M, Panerai AE: Involvement of Beta-endorphin in the modulation of paw inflammatory edema in the rat. Reg Peptides 63:79 (1996).

10. Carter LL, Dutton RW, Type 1 and Type 2: a fundamental dichotomy for all T-cell subsets. Curr Op Immunol 8:336 (1996).

11. Kruszewska B, Felten SJ, Moynihan JA, Alterations in cytokine and antibody production following chemical sympathectomy in two strains of mice. J Immunol 155:4613 (1995).

12. Nickerson P, Steurer W, Steiger J, Zheng X, Steele AW, Strom TB, Cytokines and the Th1/Th2 paradigm in transplantation. Curr Op Immunol 6:757 (1994).

13. Sacerdote P, Rosso di San Secondo VEM, Sirchia G, Manfredi B, Panerai AE, Endogenous opioids modulate allograft rejection time in mice :possible relation with Th1/Th2 cytokines. Clin Exp Immunol 113:465 (1998).

14. Sacerdote P, Lechner O, Sidman C , Wick G, PaneraiAE, Hypothalamic beta-endorphin concentrations are decreased in animals models of autoimmune disease J Neuroimmunol 97:129 (1999).

15. Charlton B, Lafferty KJ, The Th1/Th2 balance in autoimmunity. Curr Op Immunol 7:793 (1995).

MODULATION OF FAS/FASL IN A MURINE RETROVIRAL INFECTION BY AZT AND METHIONINE ENKEPHALIN

Rebecca Bowden, Sandi Soto, and Steven Specter

University of South Florida College of Medicine
Department of Medical Microbiology & Immunology
12901 Bruce B. Downs Blvd, MDC 10
Tampa, FL 33612

INTRODUCTION

Inappropriate induction of apoptosis has broad ranging pathologic implications and is associated with Alzheimer's, Hodgkin's and graft-versus-host diseases, transplant rejection, autoimmune disorders and diseases such as cancer and the acquired immunodeficiency syndrome (AIDS) (1). Induction of apoptosis of immune cells leading to a pathogenesis characterized by immunosuppression has been described in simian, bird, cat, mouse, and human retroviral diseases (2-4). There is substantial evidence that the significant increase of apoptosis in immune cells seen during retroviral infection may be regulated, at least in part, by Fas:FasL expression.

Studies on the mechanisms by which CD4+ cells progress to AICD reveal that T helper type 1 (Th1) cells expressing FasL can undergo AICD, but T helper type 2 (Th2) cells that do not express high levels of FasL do not, allowing for a more sophisticated regulation of CD4+ T cells (6, 7). Noble *et al.*(8) demonstrated that although activated CD8+ T cells that express FasL and Fas eliminate activated CD4+ T cell in a Fas-dependent mechanism, CD8+ T cells are resistant to Fas-dependent apoptosis. Differential susceptibility of Th subsets to Fas-mediated AICD by CD8+ T cells allows for a preferential removal of Th1 cells and an enhanced development of the Th2 immune response.

The murine retrovirus, Friend leukemia virus (FLV) was first described by Charlotte Friend in 1957 (9) and since has been studied as a model for several retroviral infections including human immunodeficiency virus (HIV) (10-12). FLV disease manifests several similarities to HIV infection, including immunosuppression and development of fatal disease. FLV provides a cost effective first screen for anti-retroviral therapy (13). Profound

Neuroimmune Circuits, Drugs of Abuse, and Infectious Diseases
Edited by Herman Friedman *et al.*, Kluwer Academic/Plenum Publishers, 2001

dysregulation of cytokine production and other immune parameters such as natural killer (NK) cell activity develops as infection progresses (14,15).

The endogenous opioid methionine enkephalin (MENK) was first reported to have an effect on human immune cells by Wybran *et al.* (16). Since that time, several studies have reported the immunostimulatory effects of MENK. *In vitro* studies done by Faith *et al.*(17) showed that peripheral blood mononuclear cells incubated with MENK demonstrated increased NK activity by a magnitude of 20 to 30%. Heagy and colleagues (18) demonstrated that MENK is a potent stimulator of T cell chemotaxis, suggesting that a local release of MENK may stimulate T cell movement from the peripheral blood to sites of inflammation. MENK enhances the formation of active T cell rosettes and this enhancement is blocked by naloxone treatment, indicating the presence of an opioid receptor-like structure for MENK on T cells (16,19). When high dose (5 mg/ml) MENK is injected intraperitoneally into rats, a decrease in CD4+ T cells but not in CD8+ T cells is observed, demonstrating that T cell subpopulations are also influenced by MENK (20). This selective role of MENK on T cell subpopulations is further supported by the finding that MENK treatment *in vivo* enhances IL-2 receptor expression and results in higher CD4+ cell populations, CD3+ cell percentages and T cell proliferative responses to mitogens (21).

Trials with MENK have been done in a small number of patients infected with HIV. HIV+ patients receiving high-dose treatment of MENK (125 μg/kg/week) exhibited significant increases in IL-2 receptors, NK, CD3+, CD4+, and CD8+ T cells and blastogenic responses to mitogens. A significant reduction in total lymph node size was noted after 8 and/or 12 weeks of treatment, with no significant toxicity experienced by any of the patients (22). The MENK therapy resulted in several appreciable increases in immune responses but the researchers did not follow the patients for a significant period of time to determine if there was any long-term clinical benefit. Specter and colleagues found that combination therapy using MENK and azidothymidine (AZT) enhances survival and results in decreased splenomegaly in mice infected with FLV (23). In contrast to combination treatments, no enhanced survival was observed for the MENK treatment group alone. Sin *et al.*(24) recently reported that the combination of AZT and MENK in FLV-infected *Mus dunni* cells and mouse spleen cells reduced viral replication compared to either of the treatments alone, an effect that was reversible by naloxone. MENK had little or no beneficial effect in the absence of spleen cells. Further expansion of these *in vitro* studies demonstrated that MENK works in an immunostimulatory manner to reduce infectivity when combined with AZT by stimulating IFNγ production, as this effect could be abrogated by the addition of anti-IFNγ antibodies (25). Thus, MENK used in combination with the antiretroviral drug AZT demonstrated antiviral activity mediated through the induction of cytokine production. However, the fact that MENK treatment alone provided no antiviral activity suggests that low levels of immune activation are insufficient to control retrovirus replication, demonstrating the need for MENK to be used in combination with specific antiretroviral therapy. Therefore, it is important to determine how the interaction of MENK with AZT influences immune responses in the host. This will lead to a better understanding of MENK for possible subsequent clinical application for AIDS patients. To this end, we undertook the present studies to determine how MENK and AZT in combination affect immunological and cellular parameters such as lymphocyte subpopulations, Fas:FasL expression and apoptosis of lymphocytes during FLV infection.

METHODS AND MATERIALS

Mice

Pathogen free female BALB/c mice were purchased from Harlan Sprague-Dawley (Indianapolis, IN) at 6-8 weeks of age. The mice were housed in groups of 10 per plastic cage and fed water and mouse food pellets *ad libitum* in accordance with the National Institutes of Health (NIH) Guidelines for the Care and Use of Laboratory Animals.

Drugs

Azidothymidine was supplied as a water soluble powder as a kind gift from Burroughs Wellcome Co. (Glaxo), Research Triangle Park, NC. MENK, a peptide consisting of Tyr-Gly-Gly-Phe-Met (MW 573.7 daltons), was purchased from Sigma Chemical Co., St. Louis, MO. It was suspended in phosphate buffered saline (PBS) solution for use *in vivo*.

Cytokines and Antibodies

Purified hamster phycoerythrin -conjugated IgG antibodies against mouse IgG_1, purified rat fluorescein isothiocyanate (FITC)-conjugated antibodies against mouse $IgG_{2a/2b}$, purified rat FITC-conjugated antibodies against mouse IgG_{2a}, and purified rat FITC-conjugated antibodies against mouse IgG_{2b} were purchased from PharMingen (San Diego, CA) and used as isotype controls for the flow cytometric studies. Purified mouse biotin-conjugated antibodies against mouse FasL (CD95L), purified hamster phycoerythrin conjugated antibodies against mouse Fas (CD95), purified rat phycoerythrin-conjugated antibodies against mouse CD8, purified hamster Cy-Chrome-conjugated antibodies against mouse CD3, and purified rat FITC-conjugated antibodies against mouse CD4 all were purchased from PharMingen. All antibodies for flow cytometric studies were stored at 4°C until use.

Virus

FLV used in these studies was the N/B tropic polycythemia inducing strain that contains the entire FLV complex, composed of a helper lymphatic leukemia virus and a defective spleen focus forming virus. Virus was prepared as a 10% suspension (w/v) of spleen homogenate in PBS from FLV infected mice. Stock preparations contained approximately 5,000 focus forming units (FFU) per ml and were stored at -70°C until use.

Infection and Treatment of Mice

Mice were divided into two groups: healthy and infected. Each mouse in the infected group received 100 FFU of FLV in 0.1 ml PBS, intraperitoneally (i.p.). Infected mice were divided into 4 treatment groups of 3 mice each: infected control group, MENK treatment group, AZT treatment group, and AZT plus MENK treatment group. The AZT treatment groups received AZT *ad libitum* in the drinking water at a final concentration of 0.0125 mg/ml beginning day 3 post infection (p.i.) MENK was administered i.p. at 3 mg/kg beginning day 3 p.i. Three days was selected to allow establishment of infection prior to initiation of therapy (23). Following the first administration of MENK, the mice were then treated 3 times per week until the end of the study. Treatment concentrations and schedules were established by previous studies to have the greatest reduction in morbidity and mortality in FLV infected mice (23).

Blood Cell Collection

Blood was obtained by cardiac puncture from healthy, infected, or infected treated mice. Mice were killed by inhalation of CO_2 gas and blood from the mice was collected by cardiac puncture using a Monoject heparinized syringe fitted with a 22 gauge needle. Approximately 0.5 ml of blood was collected per mouse. The blood was held on ice until use.

Spleen Cell Preparation

Spleens were aseptically removed from healthy, infected, or infected, treated mice on days 1, 3, 7, 14, 21, and 28 post treatment (p.t.). Spleens were weighed and then separated (Stomacher 80 Lab-Blender, Tekmar Co., Cincinnati, OH) individually in a sterile stomacher bag containing Hanks' balanced salt solution (HBSS) to make a single cell suspension. The spleen capsule and other debris were separated by transferring the cell suspension into a 15 ml conical tube (Fisher, Pittsburgh, PA). The cells were washed 3 times in HBSS. Cells were examined for viability using Trypan Blue exclusion and adjusted to 1 x 10^6 cells/ml PBS for fluorescent activated cell sorter (FACS) analysis.

Lymphocyte Cell Surface Marker Analysis

Purified splenocytes were layered onto a discontinuous Percoll (Fisher Scientific) density gradient consisting of 5 ml 50% Percoll, 2 ml 45% Percoll, 2 ml 35% Percoll, and 2 ml 30% Percoll layered into a 15 ml conical tube. Cells were pelleted in a centrifuge for 1½ hr at 4°C. The top layer of cells were removed from the gradient and washed twice with PBS. This layer of cells was ≥ 90% lymphocytes as determined by FACS analysis. The cells were then resuspended at 10^6 cells/100 μl PBS. Four μl of the antibodies for surface markers CD3, CD4, CD8, Fas, FasL, or the isotype controls was added to a Falcon 12 x 75 mm polystyrene tube (Becton Dickinson Labware, Lincoln Park, NJ). Cells were added to the tubes (100 μl) and incubated for 30 min on ice. Stained cells were then washed with 2 ml PBS and centrifuged for 10 min at 500 x g. Four μl of streptavidin-FITC was added to the cells stained with Fas and FasL and then incubated for an additional 30 min on ice. Cells were washed with 2 ml PBS. All cells were resuspended in 1 ml 1.0% paraformaldehyde, wrapped in aluminum foil to prevent exposure to light, and stored at 4°C until analyzed.

White blood cells were also analyzed for cell surface markers. Whole blood was added to tubes containing 4 μl of antibody as above. After labeling the samples with anti-Fas or anti-FasL and staining with strepavidin-FITC, the red blood cells were lysed in all samples by adding 1.0 ml of Immuno-Lyse (Coulter Corporation, Miami, FL) to each tube. Tubes were incubated for 1.0 min, followed by the addition of 0.25 ml of Immuno-Lyse fixative to each tube. Samples were washed with 2 ml PBS, followed by the addition of 1 ml 1% paraformaldehyde. Samples were wrapped in aluminum foil and stored at 4°C until analyzed. Samples were analyzed using a Becton Dickinson FACScaliber cytometer (San Jose, CA). Cells were gated for analysis using forward versus side scatter. A minimum of 10,000 cells were examined for each sample. Data were collected and analyzed using Cellquest. Positive marker thresholds were set based on the use of the isotype control antibodies. n=3 for all samples.

Apoptosis of Blood and Splenocytes

Lymphocytes purified by Percoll gradient centrifugation as described above and blood from healthy, infected, or treated infected mice were analyzed for apoptosis by dual color flow cytometry. Red blood cells were lysed from the peripheral blood samples using Immuno-Lyse as previously described. The enriched lymphocyte fraction and blood mononuclear cells (BMC) samples were added to 5 ml of 1% paraformaldehyde at a concentration of 10^6 cells/ 500 μl and incubated for 15 min on ice. Samples were then washed 2 times in 5 ml PBS and resuspended in 0.5 ml PBS. Samples were stored at -20°C in 5 ml 70% ethanol until analyzed.

Samples were analyzed for apoptosis using an APO-BRDU dual color flow cytometry kit (PharMingen). The ethanol was removed from the samples and cells were resuspended in 50 μl of DNA labeling solution (TdT reaction buffer, TdT enzyme, and Br-dUTP) and then incubated for 60 min at 37°C. At the end of the incubation time, samples were washed twice with rinse buffer and incubated in the dark for 30 min at rt in 0.1 ml FITC-labeled anti-BrdU antibody solution. Propidium iodide/RNase A solution (0.5 ml) was added and the samples were incubated another 30 min in the dark at rt. Samples were analyzed using a Becton Dickinson FACScaliber cytometer. Cells were gated for analysis using forward versus side scatter. A minimum of 5,000 cells was examined for each experimental set. Data were collected and analyzed using Cellquest. Gating thresholds were based on the use of negative and positive controls for each sample provided by and prepared according to the directions of the manufacturer.

RESULTS

Fas:FasL FACS Analysis

The expression of Fas:FasL on lymphocytes from both the spleen and the peripheral blood was affected during the course of FLV infection. On day 1 post treatment (p.t.), lymphocytes from the spleens demonstrated a slight increase in the expression of FasL in all of the infected animals regardless of treatment; however, expression of Fas was similarly decreased (Table 1). There was a substantial increase in the amount of double positive cells in all the infected animals as well. The amount of double positive cells in the peripheral blood were not significantly affected by FLV infection on day 1 p.t. (Table 1). However, FasL and Fas expression are significantly enhanced in infected controls. The MENK and AZT treatment groups demonstrated slight increases in the amount of Fas expression compared to noninfected controls. A substantial increase in FasL expression was also noted in AZT treatment groups. Combination therapy downregulated FasL expression and maintained normal levels of Fas.

Table 1. Fas:FasL expression on purified lymphocytes from spleen (A.) And whole blood (B.) Of BALB/c mice. BALB/c mice were infected with FLV and then treated 3 d p.i. with 3 mg/kg MENK, 0.0125 mg/ml AZT or a combination of both MENK and AZT. On days 1, 14, and 21 lymphocytes were purified from spleen and whole blood of 1 mouse per group. Lymphocytes were analyzed using FACS analysis. n=3. ±S.D.

		Noninfected	Infected	MENK	AZT	AZT/MENK
A. FAS						
Day	**1**	5.32 ± 2.31	2.13 ± 1.34	2.78 ± 1.69	2.03 ± 1.28	4.30 ± 1.79
	14	3.47 ± 1.97	1.65 ± 1.44	2.85 ± 1.08	2.67 ± 1.36	4.47 ± 1.53
	21	2.43 ± 1.05	2.27 ± 0.23	1.80 ± 0.35	1.33 ± 0.12	2.13 ± 2.14
FASL						
Day	**1**	0.14 ± 0.15	2.98 ± 1.70	2.89 ± 1.43	3.12 ± 1.54	2.87 ± 2.19
	14	0.66 ± 0.59	2.51 ± 2.36	1.39 ± 1.05	1.89 ± 2.70	3.94 ± 5.28
	21	0.82 ± 0.20	8.53 ± 0.35	12.47 ± 0.76	2.03 ± 0.55	1.51 ± 0.79
B. FAS						
Day	**1**	0.92 ± 0.99	2.92 ± 2.58	3.05 ± 2.70	3.03 ± 1.97	1.98 ± 1.20
	14	0.80 ± 0.26	0.77 ± 0.50	1.78 ± 1.29	1.11 ± 1.12	1.58 ± 0.85
	21	1.10 ± 0.66	40.17 ± 4.12	30.43 ± 6.0	35.27 ± 4.69	18.73 ± 3.1
FASL						
Day	**1**	1.56 ± 0.42	5.86 ± 3.54	1.94 ± 0.97	3.03 ± 1.95	0.82 ± 0.04
	14	1.15 ± 0.17	2.83 ± 0.29	0.80 ± 0.39	0.91 ± 0.15	0.62 ± 0.51
	21	1.40 ± 0.46	2.07 ± 0.80	0.73 ± 0.24	0.51 ± 0.34	0.65 ± 0.65

Day 21 p.t., lymphocytes from the spleens demonstrate significantly increased FasL expression in the infected controls, an effect that was further increased in the MENK treatment group (Figure 1, Table 1). Again, the expression of Fas was slightly decreased in all infected animals except in the combination treatment group. Interestingly, the amount of double positive cells was decreased in infected animals except for those in the MENK treatment group. The lymphocytes from the peripheral blood demonstrated a significant increase in the expression of Fas in all infected animals, regardless of treatment, on day 21 p.t. (Table 1). This effect was significantly decreased by combination therapy, although not to noninfected control levels.

Lymphocyte Apoptosis

The ability of the therapy groups to influence apoptosis of lymphocytes was Determined by using the TUNEL assay. Purified lymphocytes from the spleen and peripheral blood were gated by side and forward scatter (data not shown). Positive and negative controls were run for each sample (data not shown). On day 1 p.t., there was a significant increase in the amount of apoptotic lymphocytes in the spleen in the infected controls and MENK treatment group compared to noninfected controls (Table 2). This effect was even greater in the combination treatment group. AZT treatment maintained normal levels of apoptosis. FLV infection demonstrated no appreciable effect on the apoptosis of lymphocytes from the peripheral blood, regardless of treatment group (Table 2).

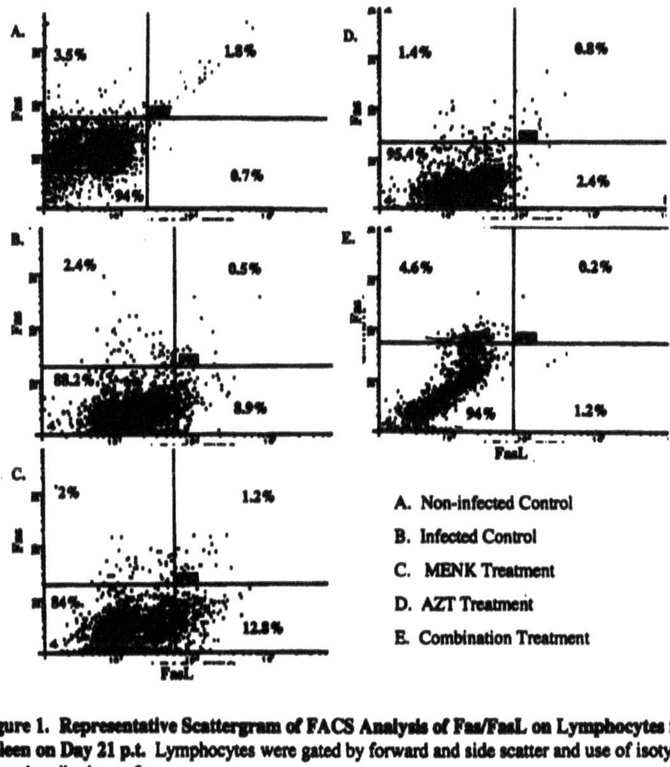

Figure 1. Representative Scattergram of FACS Analysis of Fas/FasL on Lymphocytes from Spleen on Day 21 p.t. Lymphocytes were gated by forward and side scatter and use of isotype control antibody. n=3.

Table 2. Apoptosis of purified lymphocytes from spleen and whole blood of BALB/c mice. BALB/c mice infected with FLV and then treated 3 d p.i. with 3 mg/kg MENK, 0.0125 mg/ml AZT or a combination of both MENK and AZT. On days 1, 14, and 21 lymphocytes were purified from spleen and whole blood of 1 mouse per group. Lymphocytes were analyzed using modified TUNEL methodology by FACS. n=3. ±S.D.

	Noninfected	Infected	MENK	AZT	AZT/MENK
Spleen					
Day 1	0.26 ± 0.11	3.52 ± 1.02	2.78 ± 0.20	0.23 ± 0.04	4.58 ± 1.29
14	1.72 ± 2.49	0.49 ± 0.13	1.33 ± 0.25	0.36 ± 0.14	0.20 ± 0.17
21	0.20 ± 0.07	89.47 ± 9.27	96.77 ± 3.28	95.80 ± 4.23	91.63 ± 5.12
Blood					
Day 1	0.14 ± 0.07	0.15 ± 0.13	0.12 ± 0.07	0.29 ± 0.18	0.25 ± 0.28
14	0.42 ± 0.24	0.40 ± 0.28	0.43 ± 0.06	0.54 ± 0.35	1.25 ± 0.25
21	2.23 ± 0.40	3.67 ± 0.85	22.23 ± 1.78	2.35 ± 0.16	2.70 ± 0.44

The majority of the lymphocytes from the spleens of infected mice were apoptotic by day 21 p.t. (Figure 2, Table 2). Interestingly, the lymphocytes from the peripheral blood, which had demonstrated no appreciable differences in amount of apoptosis through day 14 p.t., still only showed changes in one treatment group (Table 2). Treatment with MENK significantly increased apoptosis in lymphocytes from the peripheral blood on day 21 p.t. (Table 2).

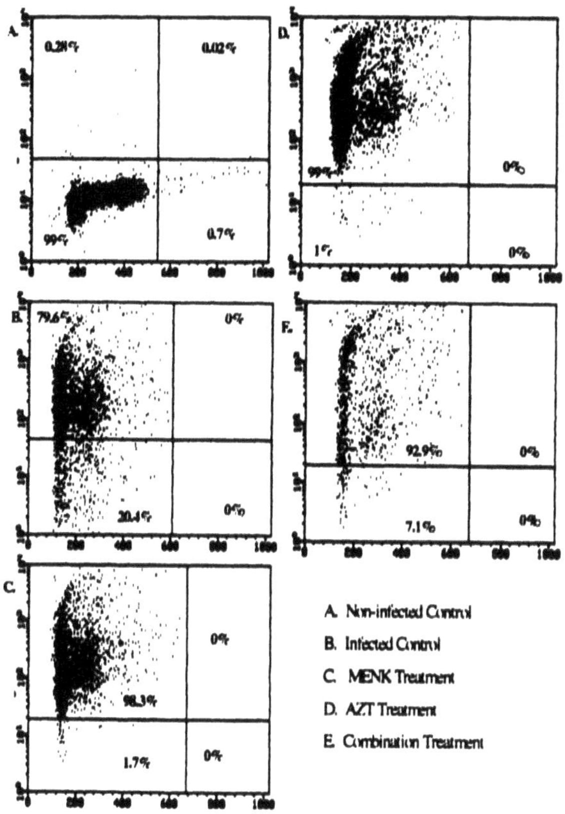

Figure 2. Representative Scattergrams of Apoptosis of Lymphocytes from the Spleen on Day 21 p.t. Lymphocytes were gated by side and forward scatter. DNA was labeled with Br-dUTP, followed by FITC labeled anti-BrdU antibodies. Cells were counter- stained with propidium iodide. n=3.

DISCUSSION

We investigated the role of Fas:FasL in FLV disease pathogenesis. Several studies have demonstrated the link between Fas:FasL expression and retroviral pathogenesis [26-29]. In agreement with these other studies, we found that infection with FLV significantly enhanced expression of FasL early in infection (Figure 1, Table 1). This effect continued throughout the course of the disease progression with even greater Fas and FasL expression seen by day 21 p.t. (Table 1). These studies also demonstrate that MENK treatment alone has the ability to upregulate the amount of Fas:FasL double positive T cells (data not shown), although double positive staining cells were evident in all groups regardless of treatment at this time point of disease. This upregulation of Fas:FasL expression was partially abrogated by combination treatment. Further studies dissecting the differences between treatment responders versus nonresponders could reveal several interesting aspects of the mechanism of FLV infection not currently known, such as the potential for co-receptors, soluble suppressive factors, etc.

Recent studies by Johnson *et al.(*30) demonstrated a long-term antiapoptotic effect of AZT therapy in HIV+ individuals. This treatment was shown to reverse HIV associated apoptosis in lymphocytes. We did not observe the same long-term protective effect in the spleens of FLV infected animals treated with AZT in these studies (Figure2, Table 2). None of the treatment groups appeared to have a significant effect on reducing the amount of T cell apoptosis over the 21 day period p.t. (Figure2, Table 2). Again, this may be due to lack of induction of a protective response in approximately half the mice. However, our results do corroborate reports by Cohen *et al. (*31) who reported that 60% of the CD4+ T cells in MAIDS mice were apoptotic during the late stages of infection.

This apoptotic effect may be due, in part, to the effect of cytokines. Aberrant cytokine production leading to an upregulation of Fas contributes to the depletion of T cells in HIV+ individuals (32). IL-12 has been shown to protect T cells from apoptosis in FIV infection [33] and Fas-mediated AICD in HIV+ individuals (34). Studies by Noble *et al.* (8) demonstrating that although activated CD8+ T cells that express FasL and Fas eliminate activated CD4+ T cells in a Fas-dependent mechanism, the CD8+ T cells themselves are resistant to Fas-dependent apoptosis. We see the same type of response in which the ratio of CD4+ to CD8+ T cells in the peripheral blood shifts dramatically during the course of FLV infection (data not shown). An interesting study will be to determine if the regulation of Fas:FasL by FLV is mediated by a soluble viral protein(s) as seen in other retroviral infections (27, 28, 35)]. If so, this would increase the probability that Fas:FasL expression regulation is a general mechanism of retroviral induced immunosuppression.

These studies demonstrated that combination therapy using AZT and MENK was able to modulate Fas:FasL expression early in FLV disease progression. The ability of MENK treatment alone to upregulate FasL expression underscores the need for this type of immunomodulatory substance to be used in combination with an antiviral chemotherapy. Thus, the identification of a treatment protocol using immunomodulators, such as MENK, that can stimulate an antiviral immune response in the host without causing potentially detrimental side effects along with the development of resistance remains a critical component in retroviral disease treatment. The present study extends our previous results that suggest this antiviral and immunostimulant approach to antiretroviral therapeutics has merit and should be expanded.

ACKNOWLEDGMENTS

The authors would like to thank Dr. Ray Widen and Rodney Arcenas for their invaluable help with the FACS analysis.

REFERENCES

1. Thompson, C.B., Apoptosis in the pathogenesis and treatment of disease. *Science*, 267: 1456-1462. 1995.
2. Guiot, A.-L., D. Rigal, and G. Chappuis, Spontanteous programmed cell death (PCD) process of lymphocytes of FIV-infected cats: pharmacological modulation in vitro. *Int. J. Immunopharmac.*, 19(3):167-179, 1997.
3. Hiromatsu, K., *et al.*, Increased Fas antigen expression in murine retrovirus-induced immunodeficiency syndrome, MAIDS. *Eur. J. Immunol.*, 24:2446-2451, 1994.
4. Lackner, A.A., *et al.*, Pathogenic (SIV mac-239) and nonpathogenic (SIV mac-1A11) molecular clones of SIV have distinct tissue distributions that vary with length of infection, in Retrovirus of human AIDS and Related Diseases. 1994. p. 27-34.
5. Alderson, M.R., *et al.*, Fas ligand mediates activation-induced cell death in human T lymphocytes. J. Exp. Med., 1995. 181`: p. 71-77.
6. Ramsdell, F., *et al.*, Differential ability of Th1 and Th2 T cells to express Fas ligand and to undergo activation-induced cell death. *Int. Immunol.*, 6(10):1545-1553. 1994.
7. Hanh, S., *et al.*, Down-modulation of CD4+ T helper 2 and type 0 cells by T helper type 1 cells via Fas/Fas-ligand interaction. *Eur. J. Immunol.*, 25(9):2679-2685, 1995.
8. Noble, A., G.A. Pestano, and H. Cantor, Suppression of immune responses by CD8 cells. I. Superantigen-actived CD8 cells induce unidirectional Fas-mediated apoptosis of antigen-activated CD4 cells. *J. Immunol.*, 159: 559-565. 1998.
9. Friend, C., Cell-free transmission in adult Swiss mice of a disease having the character of a leukemia. *J. Exp. Med.*, 105:307-318, 1957.
10. Johnson, C.S., *et al.*, Immunotherapeutic approaches to leukemia: the use of the Friend virus-induced erythroleukemia model system. *Cancer Research*, 50(Suppliment): 5682s-5686s, 1990.
11. Morrey, J.D., *et al.*, Effects of zidovudine on Friend virus complex infection in Rfv-$3^{r/s}$ genotype-containing mice used as a model for HIV. *J. AIDS*,. 3:500-510, 1990.
12. Portnoi, D., *et al.*, Zidovudine (azidodideoxythymidine) inhibits characteristic early alterations of lymphoid cell populations in retrovirus induced murine AIDS. *J. Immunol.*, 144:1705-1710, 1990.
13. Soldaini, E., *et al.*, Friend lekemia complex infection of mice as an experimental model for AIDS studies. *Veterinary Immunology and Immunopathology*, 21:97-110, 1989.
14. Lopez-Cepero, M., *et al.*, Altered interleukin production during Friend leukemia virus infection. *P.S.E.M.B.*, 188:353-363, 1988.
15. Moody, D.J., *et al.*, Suppression of natural killer cell activity by Friend leukemia virus. *JNCL*, 72(6):1349-1356, 1984.
16. Wybran, J., et al., Suggestive evidence for receptors for morphine and methionine enkephalin on normal human blood T lymphocytes. *J. Immunol.*, 123(3):1068-1070, 1979.
17. Faith, R.E., *et al.*, Neuro-immunomodulators with enkephalins: enhancement of human natural killer (NK) cell activity in vitro. *Clin. Immunol. Immunopathol.*,31: 412-418, 1984.
18. Heagy, W., *et al.*, Neurohormones regulate T cell function. *J. Exp. Med.*,May:1625-1633, 1990.
19. Miller, G.C., A.T. Murgo, and N.P. Plotnikoff, Enkephalin-enhancement of active T-cell rosettes from lymphoma patients. *Clin. Immun. Immunopharm.*, 26:446-451, 1983.
20. Jankovic, B.D. and D. Maric, In vivo modualtion of the immune system by enkephalins. *Intl. J. Neurosci.*, 51:167-169, 1990.
21. Wybran, J., *et al.*, Immunologic properties of methionine-enkephalin, and therapeutic implications in AIDS, ARC, and cancer. *Ann. N. Y. Acad. Sci*, 496:108-114, 1987.

22. Bihari, B., *et al.* Methionine enkephalin in the treatment of AIDS related complex. in *7th International Conference on AIDS*. Florence, Italy. 1991.
23. Specter, S., *et al.*, Methionine enkephalin combined with AZT therapy reduce murine retrovirus-induced disease. *Int. J. Immunopharmac.*, 16(11):911-917, 1994.
24. Sin, J.-I. and S. Specter, The role of interferon-gamma in antiretroviral ativity of methionine enkephalin and AZT in a murine cell culture. *The Journal of pharmacology and experimental therapeutics*, 279(3):1268-1273, 1996.
25. Sin, J.-I., N. Plotnikoff, and S. Specter, Anti-retroviral activity of methionine enkephalin and AZT in a murine cell culture. *Int. J. Immunopharmac.*,18(5):305-309, 1996.
26. Boirivant, M., *et al.*, HIV-1 gp120 accelerates Fas-mediated activation-induced human lamina propria T cell apoptosis. *J. Clin. Immunol.*, 18(1):39-47, 1998.
27. Bartz, S.R. and M. Emerman, Human immunodeficiency virus type 1 Tat induces apoptosis and increases sensitivity to apoptotic signals by up-regulating FLICE/caspase 8. *J. Virol.*,73(3):1956-1963, 1999.
28. Chen, X., et al., Role of the Fas/Fas ligand pathway in apoptotic cell death induced by the human T cell lymphotropic virus type 1 Tax transactivator. *Journal of General Virology*, 78(12):3277-3285, 1997.
29. Kanagawa, O., et al., Apoptotic death of lymphocytes in murine acquired immunodeficiency syndrome: involvement of Fas-FasL interaction. *Eur. J. Immunol.*, 25:2421, 1995.
30. Johnson, N. and J.M. Parkin, Anti-retroviral therapy reverses HIV-associated abnormalities in lymphocyte apoptosis. *Clin Exp Immunol*,113:229-234, 1998.
31. Cohen, D.A., *et al.*, Activation-dependent apoptosis in CD4+ T cells during murine AIDS. *Cell. Immunol.*,. 151:392-398, 1993.
32. Gehri, R., S. Hanh, and M. Rother, The Fas receptor in HIV infection: expression on peripheral blood lymphocytes and role in the depletion of T cells. *AIDS*, 10:9-16, 1996.
33. Mortola, E., *et al.*, Effect of interleukin-12 and interleukin-10 on the virus replication and apoptosis in T-cells infected with feline immunodeficiency virus. *J. Vet. Med. Sci.*, 60(11):1181-1185, 1996.
34. Estaquier, J., *et al.*, Fas-mediated apoptosis of CD4+ and CD8+ T cells from human immunodeficiency virus-infected persons: differential in vitro preventive effect of cytokines and protease antagonists. *Blood*, 87(12):4959-4966, 1996.
35. Mizuno, T., *et al.*, Apoptosis enhanced by soluble factor produced in feline immunodeficiency virus infection. *J. Vet. Med. Sci.*, 59(11):1049-1051, 1997.

ACUTE EFFECTS OF HEROIN ON THE CELLULARITY OF THE SPLEEN AND THE APOPTOSIS OF SPLENIC LEUKOCYTES

Karamarie Fecho[1] and Donald T. Lysle[1,2]

[1]Department of Psychology
[2]Curriculum in Neurobiology
University of North Carolina at Chapel Hill
Chapel Hill, North Carolina 27599-3270

INTRODUCTION

Although heroin use has long been associated with an increased incidence of several infections, including HIV (e.g., Hussey & Katz, 1950; Mathias, 1999), uncertainty exists as to whether this association results solely from factors secondary to the use of heroin, like increased exposure to pathogens, or whether heroin produces pharmacologically specific alterations of immune status that might influence disease susceptibility. Numerous studies have established that morphine produces pharmacological effects on many measures of immune status in animals (see Eisenstein and Hilburger, 1998 for review), but few studies have investigated whether heroin produces pharmacological alterations of immune status. Because heroin differs from morphine in both pharmacokinetic and pharmacodynamic properties, immunomodulatory differences between heroin and morphine might be expected. Heroin (also known as diacetylmorphine) is more lipid soluble than morphine and more potent than morphine in producing behavioral effects (Way et al., 1960; Oldendorf et al., 1972). Metabolism of heroin yields both 6-monoacetylmorphine and morphine. The biological properties of 6-monoacetylmorphine appear to differ from those of morphine and might contribute to heroin's unique effects (e.g., Way et al., 1960; Rady et al., 1994). Interestingly, heroin's antinociceptive effects appear to be mediated by a different subtype(s) of opiate receptor than that mediating morphine's antinociceptive effects (e.g., Brown et al., 1997).

Given the above considerations, recent studies by our laboratory have focused on defining the basic immunomodulatory properties of heroin in rats. Initial studies assessed the effects of a single injection of heroin on *ex vivo* measures of immune status. The results showed that acute heroin treatment produces dose-dependent decreases in the proliferative responses of mitogen-stimulated T and B cells, the production of interferon-γ by stimulated splenocytes and the cytotoxicity of natural killer (NK) cells in the spleen (Fecho et al., in press). During the course of our early investigations, an interesting observation was made that a one hour exposure to heroin produces a significant decrease in the number of leukocytes in the spleen. The purpose of the present study was to assess whether the heroin-induced decrease in splenic leukocytes is dose-dependent and naltrexone-reversible, and whether heroin affects a particular subset(s) of splenic leukocyte. Further studies were conducted to determine whether heroin alters leukocyte death in the spleen, by assessing the effects of heroin on necrosis and apoptosis.

METHODS AND RESULTS

The first set of experiments assessed whether acute exposure to heroin produces dose-dependent alterations in the total number of leukocytes in the rat spleen. The experiments involved treating rats with a sc injection of 0 (0.9% sterile saline), 0.01, 0.1, 1.0 or 10.0 mg/kg heroin sulfate (National Institute on Drug Abuse, Rockville, MD), isolating spleens one hour after the injections, and enumerating total numbers of leukocytes with a NOVA Celltrack II cell analyzer. The data were analyzed using an Analysis of Variance (ANOVA), with treatment and experimental replication included as main factors. A general contrast analysis was used to compare differences between the saline control group and each experimental group. Table 1 shows the results of those experiments. Heroin produced a significant, dose-dependent decrease in the number of splenic leukocytes ($F(4,45)=4.35$, $p<.01$ for main effect of heroin), with the most pronounced effect occurring after a dose of 10.0 mg/kg heroin ($t(45)=3.68$, $p<.001$ for comparison between 0 and 10.0 mg/kg heroin group). The decrease in splenic leukocytes produced by the 10.0 mg/kg dose of heroin represents a decrease of over 10% of the total number of leukocytes in the spleens of saline-treated rats.

Table 2 shows the results of experiments examining the naltrexone-reversibility of the heroin-induced decrease in splenic leukocyte numbers. In those experiments, animals received a sc injection of saline or 1.0 mg/kg naltrexone hydrochloride (Research Biochemicals International, Natick, MA) 15 min before a second sc injection of saline or 10.0 mg/kg heroin. Spleens were isolated 1 h after the heroin injection and total numbers of leukocytes then were counted. A one-way ANOVA was used to assess the overall treatment effect and a general contrast analysis was used to assess differences between select treatment groups. In agreement with the results presented in Table 1, the 10.0 mg/kg dose of heroin produced a significant decrease in splenic leukocytes ($F(3,12)=5.59$, $p<.05$ for main treatment effect; $t(12)=3.70, p<.01$ for comparison between saline/saline and saline/heroin groups). The results showed further that naltrexone, when administered at a dose that did not produce a significant effect on its own, completely antagonized the heroin-induced decrease in splenic leukocytes.

A study then was conducted to determine whether the heroin-induced decrease in total splenic leukocytes is selective for a particular leukocyte subset. In these experiments, animals received a sc injection of saline or different doses of heroin (0.01, 0.1, 1.0 or 10.0 mg/kg), and splenocytes were isolated one hour after the injections. Red blood cells were lysed using ACK lysing buffer and splenocytes were immunofluorescently labeled using several monoclonal antibody combinations (Pharmingen, Torreyana, CA; Harlan, Indianapolis, IN). Flow cytometry then was used to enumerate the relative number of cells in each major leukocyte subpopulation of the spleen. An ANOVA was used to assess main effects of treatment and experimental replication, and a general contrast analysis was used to assess differences between the saline control group and each experimental group. Table 1 shows the effect of increasing doses of heroin on the percent of $CD4^{+}CD3^{+}$ T cells, $CD8^{+}CD3^{+}$ T cells, $CD45RA/B^{+}$ B cells, $NKR\text{-}P1A^{lo}CD3^{+}$ T cells, $NKR\text{-}P1A^{hi}CD3^{-}$ NK cells,

CD11b/c$^+$ED1$^+$ monocytes/macrophages and CD11b/c$^+$ED1$^-$ granulocytes in the spleen. Heroin produced significant alterations only in the percent of NKR-P1A^{hi}CD3$^-$ NK cells in the spleen ($F(4,45)=18.81$, $p<.0001$ for main treatment effect). A general contrast comparison showed that the percent of NKR-P1A^{hi}CD3$^-$ NK cells was increased significantly by the 0.1 mg/kg dose of heroin ($t(45)=2.77$, $p<.01$) and decreased significantly by the 10.0 mg/kg dose of heroin ($t(45)=4.93$, $p<.0001$), when compared to values for the saline-injected animals. In terms of absolute numbers of cells, the 10.0 mg/kg dose of heroin produced a decrease of approximately 5.15×10^6 NKR-P1A^{hi}CD3$^-$ NK cells, which represents nearly 10% of the decrease in total splenic leukocytes produced by the 10.0 mg/kg dose of heroin. These results indicate that the heroin-induced decrease in splenic leukocytes is only partially selective for NKR-P1A^{hi}CD3$^-$ NK cells; heroin appears to produce a general decrease in the absolute size of each leukocyte subpopulation of the spleen.

Table 1. Effects of acute heroin treatment on total leukocyte numbers and relative numbers of leukocyte subsets in the spleen.[1]

	Heroin (mg/kg)				
	0	**0.01**	**0.1**	**1.0**	**10.0**
Change in Total Leukocyte Counts[2]	0.00±9.78	-8.50±17.1	-6.30±7.66	-22.9±10.4	-56.2±10.7
CD4$^+$CD3$^+$ T cells	37.7±0.97	37.0±0.67	37.5±0.82	37.5±1.03	36.8±0.84
CD8$^+$CD3$^+$ T cells	15.8±0.36	16.0±0.37	15.9±0.31	15.6±0.32	16.2±0.28
CD45RA/B$^+$ B cells	22.3±1.53	19.9±1.90	21.6±1.22	20.1±0.92	20.9±1.37
NKR-P1A^{lo}CD3$^+$ T cells	8.09±0.83	7.45±0.67	7.44±0.63	7.64±0.87	7.37±0.70
NKR-P1A^{hi}CD3$^-$ NK cells[3]	2.92±0.15	3.16±0.15	3.35±0.12	3.21±0.12	2.14±0.25
CD11b/c$^+$ED1$^+$ monocytes/macrophages	9.74±0.18	9.11±0.32	8.56±0.41	10.0±0.81	9.62±0.82
CD11b/c$^+$ED1$^-$ granulocytes	6.94±0.18	7.11±0.24	7.09±0.31	7.27±0.26	6.70±0.13

[1]Rats received a sc injection of saline or different doses of heroin (0.01, 0.1, 1.0 or 10.0 mg/kg), and spleens were isolated one hour after the injections. A NOVA Celltrack II Cell Analyzer was used to determine total numbers of splenic leukocytes, which are expressed for each treatment group as the change in the mean ± SEM x 10^{-6} number of leukocytes in the spleen, with respect to the saline control group (0 mg/kg heroin). Flow cytometry was used to calculate the relative size of different leukocyte subpopulations in the spleen, which are expressed for each treatment group as the mean ± SEM percent of different leukocyte subsets in the spleen. (n=12/group)

[2]Heroin produced a significant decrease in the number of leukocytes in the spleen ($F(4,45)=4.35$, $p<.01$).

[3]Heroin produced significant alterations in the relative number of NKR-P1A^{hi}CD3$^-$ NK cells in the spleen ($F(4,45)=18.81$, $p<.0001$).

Table 2. Effects of naltrexone on the heroin-induced decrease in splenic leukocyte numbers.[1]

Treatment Group	Change in Total Leukocyte Counts
saline/saline	$0.0 \pm 8.66 \times 10^6$
naltrexone/saline	$-16.5 \pm 4.98 \times 10^6$
saline/heroin	$-54.6 \pm 13.3 \times 10^6$
naltrexone/heroin	$-5.10 \pm 7.07 \times 10^6$

[1]Rats received a sc injection of saline or 1.0 mg/kg naltrexone 15 min before a sc injection of saline or 10.0 mg/kg heroin. One hour after the second injection, spleens were isolated and a NOVA Celltrack II cell analyzer was used to calculate total numbers of leukocytes in the spleen. The table shows for each treatment group the change in the mean ± SEM number of leukocytes in the spleen, with respect to the saline/saline control group. Heroin significantly decreased the number of splenic leukocytes ($F(3,12)=5.59$, $p<.05$) and naltrexone pretreatment antagonized the heroin-induced decrease. (n=4/group)

The observation that the 10.0 mg/kg dose of heroin decreased the number of leukocytes in the spleen prompted additional studies on whether heroin increases the rate of leukocyte death in the spleen. The first study investigated the effects of heroin on leukocyte necrosis in the spleen. Rats were administered a sc injection of 0 or 10.0 mg/kg heroin, and splenic mononuclear cells were isolated (*via* Histopaque-1083; Sigma Chemical Co., St. Louis, MO) one hour after the injection. Splenic mononuclear cells were cultured for either 45 min or 24 h, and levels of lactate dehydrogenase (LDH) were measured in supernatants. LDH is a cytosolic enzyme that is released from the intracellular fluid only after a cell loses its membrane integrity, the defining characteristic of necrosis (Cohen et al., 1992). A colorimetric, coupled enzymatic assay (CytoTox 96; Promega, Madison, WI) was used to measure LDH activity, which was reflected in the conversion of a tetrazolium salt to a red formazan product, whose absorbance was measured at 490 nm on a Bio-Tek EL312 microtiter plate reader. A standard curve generated with purified bovine heart LDH (Sigma) was used to convert absorbances into concentrations of LDH. An ANOVA was used to assess main effects of treatment, culture time and experimental replication. The results, shown in Table 3, indicated that supernatant concentrations of LDH increased significantly over time in culture ($F(1,24)=82.25$, $p<.0001$ for main effect of time), which reflects a general increase in necrotic cell death over time in culture. However, samples from heroin-treated animals did not differ from samples from saline-treated animals in terms of the amount of LDH released from splenic mononuclear cells after 45 min or 24 h of culture. These results suggest that heroin does not increase the rate of necrotic cell death in the spleen.

An investigation then was undertaken to assess whether heroin increases apoptosis in the spleen. In these experiments, animals received a sc injection of saline or 10.0 mg/kg heroin, splenic mononuclear cells were isolated (*via* Histopaque-1083; Sigma) one hour after the injections, and cells were analyzed for apoptosis either immediately after isolation or after

Table 3. Effects of acute heroin treatment on LDH levels in supernatants of splenic mononuclear cells cultured for 45 min or 24 h.[1]

Culture Time	Treatment Group	
	saline	**heroin**
45 min	0.228 ± 0.014	0.209 ± 0.032
24 h	0.477 ± 0.025	0.485 ± 0.044

[1]Splenic mononuclear cells were isolated one hour after rats received a sc injection of saline or 10.0 mg/kg heroin. The mononuclear cells were cultured for 45 min or 24 h, and LDH activity was measured in supernatants and converted into LDH concentrations on the basis of a standard curve generated with known concentrations of bovine heart LDH. The table shows the mean ± SEM U/μl of LDH in supernatants from 45 min or 24 h cultures of splenic mononuclear cells from saline- or heroin-treated rats. Heroin treatment did not alter the release of LDH from splenic mononuclear cells. (n=8/group)

24 h of culture. The number of apoptotic splenocytes was quantitated using flow cytometry and PE-conjugated Annexin V (Pharmingen). One of the earliest events in the apoptotic process is that cells translocate phosphatidylserine (PS) from the inner to the outer leaflet of the cell membrane. The translocation of PS serves as a molecular "tag" used by phagocytes to recognize apoptotic cells and engulf them. Annexin V is a highly specific, PS binding protein that can be used to identify cells in both early and later stages of apoptosis (Martin et al., 1995; Castedo et al., 1996). As shown in Figure 1, the percent of Annexin V^+cells was determined for two gated regions of the forward angle light scatter (fsc) versus side angle light scatter (ssc) histograms. The "live" cell gate encompassed cells showing a normal fsc versus ssc profile, whereas the "apoptotic" cell gate encompassed cells showing a decrease in fsc (a measure of cell size) and a small increase in ssc (a measure of cell granularity). The "apoptotic" cell gate therefore enriches cells showing morphological characteristics of apoptosis (Belloc et al., 1994). An ANOVA was used to assess main effects of treatment, culture time and experimental replication on the percent of Annexin V^+ cells within the "live" and "apoptotic" cell gates. General contrast comparisons then were used to assess differences between saline- and heroin-treated rats in the percent of Annexin V^+ cells within the "live" or "apoptotic" populations after 0 or 24 h of culture.

Table 4 shows the results of the Annexin V experiments. The number of Annexin V^+ cells within the "live" cell gate increased significantly over time in culture ($F(1,32)=17.50$, $p<.001$ for main effect of time). More importantly, heroin produced a significant increase in the percent of Annexin V^+ cells within the "live" cell gate ($F(1,32)=44.77$, $p<.0001$ for main effect of heroin), and the effect of heroin was apparent in both fresh and cultured mononuclear cells ($t(32)\geq 4.11$, $p<.001$). The number of Annexin V^+ cells within the "apoptotic" cell gate also increased significantly over time in culture ($F(1,32)=114.91$, $p<.0001$ for main effect of time). Heroin produced a significant increase in the number of Annexin V^+ cells within the "apoptotic" cell gate ($F(1,32)=41.26$, $p<.0001$ for main effect of heroin), but a general contrast comparison showed that the effect of heroin was apparent only at time 0 h ($t(32)=7.10$, $p<.0001$). After 24 h of culture, the majority of the cells within the "apoptotic" gate was Annexin V^+, and no differences were detected between heroin- and saline-treated rats.

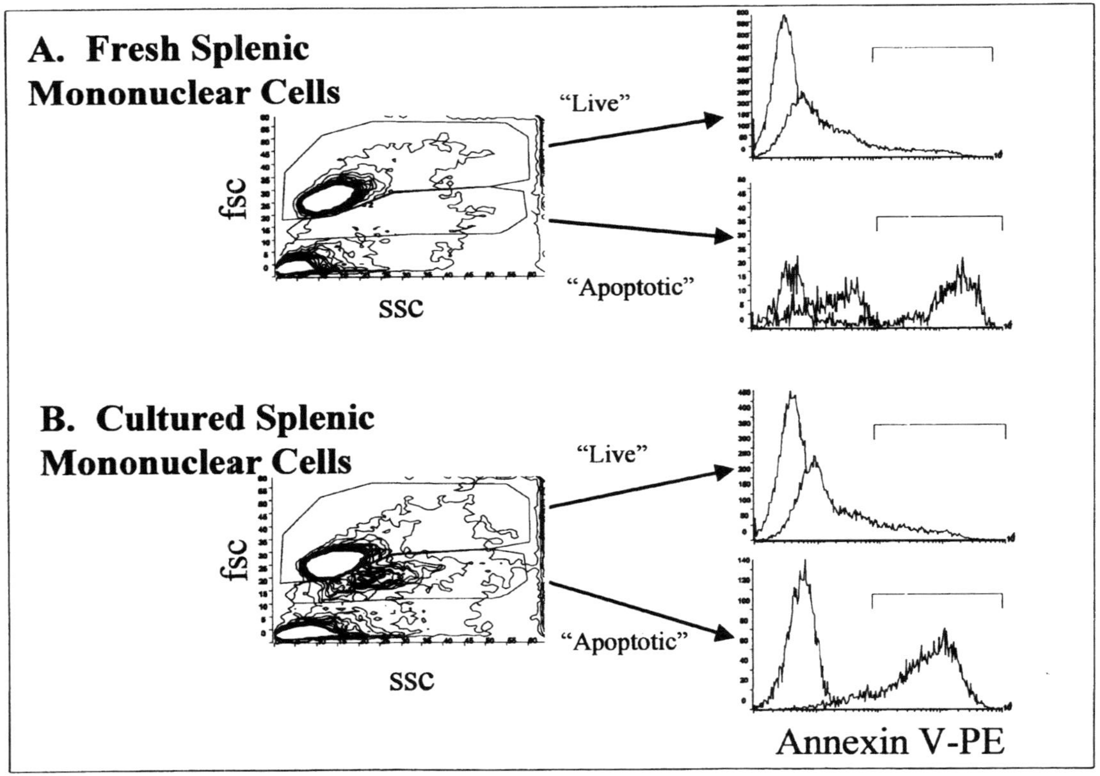

Figure 1. Methodology involved in analyzing the flow cytometry data from the Annexin V assay. Fresh splenic mononuclear cells (A) and splenic mononuclear cells that were cultured for 24 h (B) were stained with PE-conjugated Annexin V and analyzed by flow cytometry. For both fresh and cultured mononuclear cells, fsc versus ssc histograms were used to determine the percent of Annexin V^+ cells in the "live" gate (encompassing cells that showed a normal fsc versus ssc profile; these cells represented cells in early stages of apoptosis) and in the "apoptotic" gate (encompassing cells that showed a decrease in fsc and a small increase in ssc; these cells represented cells in later stages of apoptosis).

Table 4. Effects of acute heroin treatment on relative numbers of Annexin V^{+} cells in 0 or 24 h cultures of splenic mononuclear cells.[1]

	"Live" Gate		"Apoptotic" Gate	
	Treatment Group		Treatment Group	
Culture Time	saline	heroin	saline	heroin
0 h	15.40 ± 0.31	19.98 ± 0.74	60.61 ± 2.24	76.42 ± 1.83
24 h	18.00 ± 0.74	23.98 ± 1.17	83.18 ± 0.89	87.61 ± 0.64

[1]Splenic mononuclear cells were isolated one hour after rats received a sc injection of saline or 10.0 mg/kg heroin. Immediately after isolation (0h) or after a 24 h culture, the mononuclear cells were stained with Annexin V-PE and flow cytometry was used to enumerate the percent of Annexin V^{+} cells within the "live" gate (i.e., cells showing a normal fsc versus ssc profile) and within the "apoptotic" gate (i.e., cells showing a decrease in fsc and a small increase in ssc). The table shows the mean ± SEM percent of Annexin V^{+} cells within the "live" and "apoptotic" gates in fresh and cultured splenic mononuclear cells from saline- or heroin-treated rats. Heroin treatment produced a significant increase in the percent of Annexin V+ cells within both the "live" gate ($F(1,32)=44.77$, $p<.0001$) and the "apoptotic" gate ($F(1,32)=41.26$, $p<.0001$). (n=12/group)

DISCUSSION

The current results show that a one hour exposure of rats to heroin produces a decrease in the total number of leukocytes in the spleen. The effect of heroin is reversible by prior administration of naltrexone and dose-dependent, with a 10 mg/kg dose producing the most profound effect. These results indicate that heroin's effect is the result of a specific interaction with endogenous opiate receptors. Receptor binding studies show that heroin has little affinity for opiate receptors (Inturissi et al., 1983), and so heroin itself is unlikely to interact directly with opiate receptors to produce the decrease in splenic leukocytes. A more likely explanation is that heroin's metabolites bind to opiate receptors and thereby elicit the effector mechanism(s) responsible for the decrease in splenic leukocytes. The present study did not investigate the location of the opiate receptors involved in the heroin-induced decrease in splenic leukocytes. Substantial evidence exists for the importance of central nervous system opiate receptors in the immunomodulatory effects of *in vivo* opiate treatment (see Mellon and Bayer, 1998 for review). Nonetheless, the existence of functional opiate receptors on leukocytes (see Eisenstein and Hilburger, 1998 for review) suggests an important immunomodulatory role for these receptors.

Studies by our laboratory demonstrate that a one hour exposure of rats to heroin produces dose-dependent, naltrexone-reversible decreases in the proliferative responses of T and B cells, the production of interferon-γ, and the cytotoxicity of NK cells in the spleen (Fecho et al., in press). Insight into the mechanism(s) underlying these effects of heroin is provided by the analysis of heroin's effects on the distribution of leukocyte subpopulations within the spleen (Table 1). Indeed, a significant correlation exists between the effects of heroin on NK cell cytotoxicity and relative numbers of NK cells (Fecho et al., in press). This finding suggests that the effect of heroin in an *ex vivo* assay of NK cell cytotoxicity reflects, at least in part, a decrease in effector NK cells in the spleen. In contrast, the effects of heroin on *ex vivo* measures of T and B cell proliferation and interferon-γ production do not appear to reflect changes in the distribution of T cells, B cells or accessory cells in the spleen; instead, these effects likely result from specific alterations in leukocyte activities.

The present study showed that a single injection of heroin increases the relative number of apoptotic cells in the spleen, but does not change the number of necrotic cells in the spleen. The effect of heroin on apoptosis is stable over time in culture, since the heroin-induced increase in apoptotic cells was present in both fresh and cultured splenic mononuclear cells. Additional studies indicate that the effect of heroin on apoptosis also is stable over time *in vivo*. In particular, an increase in the relative number of apoptotic cells was present when splenic mononuclear cells were isolated one hour or three hours after the injection of heroin (Fecho, How and Lysle, unpublished observations). These results suggest that heroin is producing a very specific and long-lasting change in the rate of leukocyte apoptosis in the spleen.

One explanation for the observed increase in apoptotic cells after heroin treatment is that a one hour exposure to heroin induces splenic leukocytes to die by apoptosis. Many hours typically elapse between the induction of apoptosis by death receptor ligation, for instance, and the appearance of morphological and biochemical signs of apoptosis (Cohen et al., 1992). However, recent *in vivo* and *in vitro* studies demonstrate that certain external factors rapidly (i.e., within 2-3 h) induce apoptosis, as assessed by DNA fragmentation assays or morphological changes (Iwasa et al., 1996; Andréau et al., 1998; Collier et al., 1998; Matsue et al., 1999). Unlike those studies, the current study used a very early marker of apoptosis, translocation of PS, to identify apoptotic cells, thereby allowing for the detection of heroin-induced changes in apoptosis at even earlier time points. The observed increase in the percent of Annexin V^{+} cells after heroin might reflect a heroin-related induction of apoptosis in the spleen. Another possibility is that heroin is accelerating the death of cells already induced to undergo apoptosis. Because low levels of apoptosis normally take place within the spleen (and other organs of the body), heroin might simply increase this "background" level of apoptosis. Perhaps heroin facilitates the biochemical cascade of events that are initiated *after* a cell receives a specific external or internal signal to undergo apoptosis.

An interesting, though unresolved, question concerns the relationship between the decrease in total leukocytes and the increase in apoptotic cells in the spleen after heroin treatment. In the present study, the effect of heroin on both leukocyte numbers and apoptosis was observed when spleens were harvested one hour after the injection of heroin. Heroin might decrease leukocyte numbers in the spleen by increasing their apoptotic death and clearance. Another possibility is that the heroin-induced decrease in splenic leukocytes results from a heroin-induced alteration in the migration of leukocytes to/from the spleen. Additional studies by our laboratory indicate that heroin does not increase the relative number of circulating leukocytes or NK cells (Fecho and Lysle, unpublished observations). However, heroin might trigger the migration of leukocytes away from the spleen to other lymphoid organs, like the bone marrow, or nonlymphoid organs, like the skin. For example, studies have shown that NK cells migrate selectively to the bone marrow after periods of stress or hormonal treatments (Ottaway & Husband, 1992). In a similar manner, opiates produce well-known vasodilatory effects in the periphery, and so the shunting of blood away from the spleen and other involuntary organs toward the periphery might reduce the number of leukocytes in the spleen after heroin treatment.

The present study complements and extends recent studies that suggest an increase in splenocyte apoptosis after acute or chronic morphine treatment (Singhal et al., 1997; Yin et al., 1999). Singhal et al. (1997) showed that repeated injections of morphine produce a time-dependent increase in the number of apoptotic cells in the rat spleen, such that 10 days of twice daily injections produce more apoptotic cells than 3 days of twice daily injections. Yin et al. (1999) showed that the cellularity of the mouse spleen is reduced 24 h after a single injection of morphine. Because that same study showed that the morphine-induced decrease in splenic leukocytes is blocked by administering morphine in combination with Fas-Ig, the suggestion was made that morphine decreases splenocyte numbers by increasing Fas-mediated apoptosis. Yin and colleagues (1999) also showed that *in vitro* exposure to morphine primes a T cell hybridoma, human peripheral blood lymphocytes and mouse splenocytes to undergo apoptosis by inducing Fas expression on those cells. Thus, it is possible that opiates interact directly with leukocytes to induce changes in their rate of apoptosis.

In summary, the present study shows that acute exposure to heroin produces a dose-dependent, naltrexone-reversible decrease in the total number of leukocytes in the spleen. The heroin-induced decrease in splenic leukocytes appears to reflect a general decrease in all major leukocyte subpopulations of the spleen, with NK cells being particularly sensitive to the effects of heroin. Heroin does not increase the necrosis of splenic leukocytes, but heroin does produce a significant increase in the apoptosis of splenic leukocytes, and the effect of heroin on leukocyte apoptosis is evident in both fresh and cultured splenocytes.

ACKNOWLEDGMENTS

The authors wish to acknowledge Tam How for expert technical assistance, and Larry Arnold and Bill Nostrom for maintenance of the Flow Cytometry Facility at the University of North Carolina at Chapel Hill.

This work was supported by U.S. PHS grants to D.T.L. (DA10167; DA07481). K.F. is the recipient of an award from the University Research Council at the University of North Carolina at Chapel Hill, and D.T.L. is the recipient of a Research Scientist Award (DA00334).

REFERENCES

Andréau, K., Lemaire, C., Souvannavong, V., and Adam, A., 1998, Induction of apoptosis by dexamethasone in the B cell lineage. *Immunopharmacology* 40:67.

Brown, G.P., Yang, K., King, M.A., Rossi, G.C., Leventhal, L., Chang, A., and Pasternak, G.W., 1997, 3-methoxynaltrexone, a selective heroin/morphine-6β-glucuronide antagonist. *FEBS Letters* 412:35.

Castedo, M., Hirsch, T., Susin, S.A., Zamzami, N., Marchetti, P., Macho, A., and Kroemer, G., 1996, Sequential acquisition of mitochondrial and plasma membrane alterations during early lymphocyte apoptosis. *J. Immunol.* 157:512.

Cohen, J.J., Duke, R.C., Fadok, V.A., and Sellins, K.S., 1992, Apoptosis and programmed cell death in immunity. *Annu. Rev. Immunol.* 10:267.

Collier, S.D., Wu, W.-J., and Pruett, S.B., 1998, Endogenous glucocorticoids induced by a chemical stressor (ethanol) cause apoptosis in the spleen in B6C3F1 female mice. *Toxicol. Appl. Pharmacol.* 148:176.

Eisenstein, T.K., and Hilburger, M.E., 1998, Opioid modulation of immune responses: Effects on phagocyte and lymphoid cell populations. *J. Neuroimmunol.* 83:36.

Fecho, K., Nelson, C.J., and Lysle, D.T., in press, Phenotypic and functional assessments of immune status in the rat spleen following acute heroin treatment. *Immunopharmacology*.

Hussey, H.H., and Katz, S., 1950, Infections resulting from narcotic addiction. *Am. J. Med.* 9:186.

Inturrisi, C.E., Schultz, M., Shin, S., Umans, J.G., Angel, L., and Simon, E.J., 1983, Evidence from opiate binding sites that heroin acts through its metabolites. *Life Sci.* 3:773.

Iwasa, M., Maeno, Y., Inoue, H., Koyama, H., and Matoba, R., 1996, Induction of apoptotic cell death in rat thymus and spleen after a bolus injection of methamphetamine. *Int. J. Legal Med.* 109:23.

Martin, S.J., Reutelingsperger, C.P.M., McGahon, A.J., Rader, J.A., van Schie, R.C.A.A., LaFace, D.M., and Green, D.R., 1995, Early redistribution of plasma membrane phosphatidylserine is a general feature of apoptosis regardless of the initiating stimulus: Inhibition by overexpression of bcl-2 and abl. *J. Exp. Med.* 182:1545.

Mathias, R., 1999, Heroin snorts risk transition to injection drug use and infectious disease. *NIDA Notes* 14(1):1.

Matsue, H., Edelbaum, D., Hartmann, A.C., Morita, A., Bergstresser, P.R., Yagita, H., Okumura, K., and Takashima, A., 1999, Dendritic cells undergo rapid apoptosis in vitro during antigen-specific interaction with $CD4^+$ T cells. *J. Immunol.* 162:5287.

Mellon, R.D., and Bayer, B.M., 1998, Evidence for central opioid receptors in the immunomodulatory effects of morphine: Review of potential mechanism(s) of action. *J. Neuroimmunol.* 83:19.

Oldendorf, W.H., Hyman, S., Braun, L., and Oldendorf, S.Z., 1972, Blood-brain barrier: Penetration of morphine, codeine, heroin, and methadone after carotid injection. *Science* 178:984.

Ottaway, C.A., and Husband, A.J., 1992, Central nervous system influences on lymphocyte migration. *Brain Behav. Immun.* 6:97.

Rady, J.J., Aksu, F., and Fujimoto, J.M., 1994, The heroin metabolite, 6-monoacetylmorphine, activates *delta* opioid receptors to produce antinociception in Swiss-Webster mice. *J. Pharmacol. Exp. Ther.* 268:1222.

Singhal, P.C., Reddy, K., Frankl, N., Sanwal, V., and Gibbons, N., 1997, Morphine induces splenocyte apoptosis and enhanced mRNA expression of cathepsin-B. *Inflammation* 21:609.

Way, E.L., Kemp, J.W., Young, J.M., and Grassetti, D.R., 1960, The pharmacologic effects of heroin in relationship to its rate of biotransformation. *J. Pharmacol.* 129:144.

Yin, D., Mufson, R.A., Wang, R., and Shi, Y., 1999, Fas-mediated cell death promoted by opioids. *Nature* 397:218.

ALTERATION OF EARLY T CELL DEVELOPMENT BY OPIOID AND SUPERANTIGEN STIMULATION

Lois E. McCarthy[1] and Thomas J. Rogers[1,2,3]

[1]Department of Microbiology and Immunology
[2]Center for Substance Abuse Research
and [3]the Fels Institute for Cancer Research and Molecular Biology
Temple University School of Medicine
Philadelphia, PA 19140

INTRODUCTION

It is well established that a wide variety of immune responses can be modulated by opioid alkaloids and peptides, through the activation of opioid receptors present on immune cells.[1-5] This is supported by evidence from radioligand binding studies and analysis by RT-PCR which demonstrate the presence of μ-, δ-, and κ-opioid receptors by various cells of the immune system.[6-9] Evidence suggests that opioids may be capable of influencing the development of T cells. Administration of morphine (a relatively μ receptor-selective agonist) to experimental animals results in thymic atrophy with reduced cellularity, reflecting a profound deficit in the number and percentage of CD4-, CD8-double positive (DP) thymocytes.[10,11] Morphine has been shown to induce apoptosis in macrophages, splenocytes, peripheral blood T lymphocytes, and the Jurkat T cell line.[12-14] The κ receptor agonist U50,488H also appears to be capable of modulating T cell development. Administration of this agent in culture to a DP thymocyte cell line prevented the differentiation of these cells to a more mature CD4-single positive (SP) phenotype in response to stimulation by superantigen.[15] There is evidence that the endogenous δ-opioid receptor ligands, met- and leu-enkephalins, may play a role in T cell maturation and development. Murine CD4 cells can express met-enkephalin upon activation with mitogens or cytokines.[16] Immature CD4-, CD8-double negative (DN) T cells from the murine fetal thymus at gestational day 15 have been shown to express proenkephalin A, the precursor of the enkephalins.[17]

The superantigen staphylococcal enterotoxin B (SEB) exerts a profound impact on T cell development. When added to cultures of fetal thymus glands, SEB induces a deletion of thymocytes expressing the Vβ7, 8.1, 8.2, and 8.3 T cell receptor (TCR) alleles.[18] In contrast, thymocytes expressing other Vβ alleles are not deleted by SEB. The present study examined the influence of DPDPE, a δ receptor-selective agonist, together

Neuroimmune Circuits, Drugs of Abuse, and Infectious Diseases
Edited by Herman Friedman *et al.*, Kluwer Academic/Plenum Publishers, 2001

with SEB on the distribution of thymocyte subpopulations generated in fetal thymic organ culture (FTOC).

MATERIALS AND METHODS

Fetal Thymic Organ Culture

Thymus glands were removed from fetal BALB/c mice on gestational day 14, and were cultured at the air-liquid interface on polycarbonate membranes in inserts of Transwell plates (Costar, Cambridge, MA) at 37°C with 5% CO_2, according to the method of Ceredig.[19] The medium for FTOC consisted of DMEM supplemented with FCS (10%) and 2-ME (5×10^{-5} M). SEB (final concentration 50 μg/ml) and/or DPDPE (final concentration 1 μM) were added at the initiation of culture and again on days 3 and 6, and the FTOC was terminated at 9-10 days.

Flow Cytometry

Following FTOC, single-cell suspensions were prepared from the thymuses, and thymocytes were washed and resuspended in Hank's balanced salt solution supplemented with 2% FCS. Thymocytes were incubated for 30 min at 4°C with the following fluorescence-conjugated monoclonal antibodies: PE-anti-CD4, PE-anti-Vβ8, PE-anti-Vβ6, Cy-Chrome-anti-CD3, and Cy-Chrome-anti-CD8 (PharMingen, San Diego, CA). Following incubation, cells were washed and fixed with 1% paraformaldehyde. Two-color flow cytometric analysis of 5,000 gated events was performed.

RESULTS

The impact of stimulation by SEB and DPDPE on the development of T cells expressing CD4 and CD8 was examined using FTOC. Treatment of gestational day-14 murine FTOCs for 9-10 days with SEB, DPDPE, or the combination of SEB and DPDPE, resulted in reductions in the percent of CD4-SP T cells. The three treatments yielded equivalent decrements (approximately 35-45%), compared to the proportion of this subset generated in untreated FTOCs (approximately 7-8% of the total thymocyte population). The decreases in mature CD4-SP cells induced by these agents appear to be selective, because no alterations in mature CD8-SP cells or in the immature DN or DP populations were observed (Table 1).

Table 1. SEB and DPDPE treatment alter the development of CD4-SP T cells in FTOC.[1]

	SEB	DPDPE	SEB + DPDPE
$CD4^-CD8^-$	No change	No change	No change
$CD4^+CD8^+$	No change	No change	No change
$CD4^+CD8^-$	Decreased	Decreased	Decreased
$CD4^-CD8^+$	No change	No change	No change

[1] Gestational day-14 murine thymus glands were cultured for 9-10 days with SEB (50 μg/ml) and/or DPDPE (1 μM). Thymocyte subpopulations were quantitated by flow cytometry.

Development of the mature T cell repertoire within the thymus depends upon the processes of positive selection and negative selection (deletion) of thymocytes expressing particular variable-region alleles of the TCR. Administration of SEB to FTOC induces negative selection, due to deletion of thymocytes bearing the $V\beta8^+$ TCR. Consistent with this observation, the present study found that SEB-treated FTOCs exhibited a 35% reduction in $CD3^+V\beta8^+$ cells compared to control cultures. DPDPE-treated and SEB + DPDPE-treated FTOCs yielded reductions in $V\beta8^+$ T cells similar to that seen after SEB treatment (Table 2). In contrast, the percent of Vβ6-TCR-expressing $CD3^+$ thymocytes, which are not deleted by SEB, was similar in all groups.

Table 2. SEB and DPDPE treatment alter the development of Vβ8-expressing T cells in FTOC.[1]

	SEB	DPDPE	SEB + DPDPE
$CD3^+V\beta8^+$	Decreased	Decreased	Decreased
$CD3^+V\beta6^+$	No change	No change	No change

[1] Gestational day-14 murine thymus glands were cultured for 9-10 days with SEB (50 μg/ml) and/or DPDPE (1 μM). Thymocyte subpopulations were quantitated by flow cytometry.

DISCUSSION

The present study shows that the development of T lymphocytes in the thymus early in ontogeny can be altered by exogenous agents. Our results indicate that CD4-SP thymocytes may be particularly susceptible to deletion by both superantigen and opioids. Exposure to either the bacterial superantigen SEB or the δ-opioid peptide agonist DPDPE resulted in substantial and equivalent losses in this population. Treatment with SEB together with DPDPE did not synergize in this deletion. The superantigen-induced loss in T cells is considered to resemble the process of clonal deletion of self-reactive T cells which normally occurs in the thymus as a result of negative selection.[18] Currently, the mechanism whereby DPDPE exposure to FTOC resulted in the loss in CD4 cells is unknown. However, it is possible that apoptosis is responsible for the reduction in this subset, because other studies have demonstrated that morphine can induce apoptosis in macrophages, splenocytes, peripheral blood lymphocytes, and the Jurkat T cell line.[12-14] Nevertheless, the thymocyte alteration mediated by either SEB or DPDPE must be selective, because both mature CD8-SP T cells and immature thymocytes (both DN and DP) were unaffected by these treatments. Further, the data suggest that the stage of T cell maturation vulnerable to loss by opioid exposure is at or subsequent to the DP stage (in which positive and negative selection occurs). This suggestion is consistent with the observation of Guan *et al.*[15] that the superantigen-driven differentiation of the DPK-DP cell line to a CD4-SP phenotype was inhibited by the κ-opioid receptor-selective agonist, U50,488H.

A role for the endogenous δ-opioid receptor-selective peptides, the enkephalins, in the natural development of T cells in the thymus has been proposed by Linner *et al.*[17] They found that mRNA for the enkephalin precursor proenkephalin A is constitutively expressed by DN cells in the murine fetal thymus on day 15 of gestation. Subsequently, the expression of this δ-opioid peptide precursor is lost. They further observed that treatment of day-15 fetal thymocytes with the δ-opioid receptor antagonist naltrindole enhanced thymocyte proliferation, while the δ-receptor agonist deltorphin inhibited proliferation of day-15 fetal thymocytes. Thus, it was hypothesized that during a discrete period in fetal development, proenkephalin A is expressed in the thymus to restrict spontaneous proliferation of thymocytes, in favor of allowing their differentiation into mature T lymphocytes. The observation in the present study that treatment of FTOC with either SEB or DPDPE resulted in a loss of thymocytes expressing the $V\beta8^+$ TCR suggests that these agents may eliminate cells in a manner analogous to the process of negative selection, rather than a generalized loss of cells in the thymus. The specificity of the reduction in thymocytes generated in FTOC following exposure to SEB or DPDPE is supported by the finding that the proportion of TCR Vβ6-bearing thymocytes was unaffected by either treatment. Further study is necessary to determine whether the reduction in CD4-SP T cells induced by SEB or DPDPE is restricted to $V\beta8^+$ thymocytes.

The present study demonstrates that a δ-opioid receptor agonist is capable of modulating T cell development in the thymus. Moreover, the influence of the opioid peptide appears to be selective in that it inhibited CD4-SP cell development, while sparing CD8-SP T cells. The effect of the opioid on thymocyte development also appears to be stage specific, i.e., affecting thymocytes during or following the DP stage, in which positive and negative selection occurs. The finding that DPDPE also resulted in a reduction in the proportion of $CD3^+V\beta8^+$ T cells suggests that opioids can further modulate the development of the mature T lymphocyte repertoire by inhibiting T cell subpopulations expressing particular TCR alleles.

ACKNOWLEDGMENTS

This study was funded by NIDA grants DA06650, DA11130, and T32DA07237.

REFERENCES

1. C. Alicea, S. Belkowski, T.K. Eisenstein, M.W. Adler and T.J. Rogers, Inhibition of primary macrophage cytokine production in vitro following treatment with the κ-opioid agonist U50,488H, *J. Neuroimmunol.* 64:83-90 (1996).
2. L. Guan, R. Townsend, T.K. Eisenstein, M.W. Adler and T.J. Rogers, Both T cells and macrophages are targets of κ-opioid-induced immunosuppression, *Brain Behav. Immun.* 8:229-240 (1994).
3. N.A. Shahabi and B.M. Sharp, Anti-proliferative effects of δ-opioids on highly purified CD4+ and CD8+ murine T cells, *J. Pharm. Exp. Ther.* 273:1105-1113 (1995).
4. D.D. Taub, T.K. Eisenstein, E.B. Geller, M.W. Adler and T.J. Rogers, Immunomodulatory activity of μ- and κ-selective opioid agonists, *Proc. Natl. Acad. Sci. USA* 88:360-364 (1991).
5. P. Van Den Bergh, R. Dobber, S. Ramlal, J. Rozing and L. Nagelkerken, Role of opioid peptides in the regulation of cytokine production by murine $CD4^+$ T cells, *Cell. Immunol.* 154:109-122 (1994).
6. C. Alicea, S.T. Belkowski, J.K. Sliker, J. Zhu, L.-Y. Liu-Chen, T.K. Eisenstein, M.W. Adler and T.J. Rogers, Characterization of κ-opioid receptor transcripts expressed by T cells and macrophages, *J. Neuroimmunol.* 91:55-62 (1998).
7. L.F. Chuang, T.K. Chuang, K.F. Killam, Jr., A.J. Chuang, H.-F. Kung, L. Yu and R.Y. Chuang, Delta opioid receptor gene expression in lymphocytes, *Biochem. Biophys. Res. Commun.* 202:1291-1299 (1994).

8. M. Sedqi, S. Roy, S. Ramakrishnan, R. Elde and H.H. Loh, Complementary DNA cloning of a μ-opioid receptor from rat peritoneal macrophages, *Biochem. Biophys. Res. Commun.* 209:563-574 (1995).
9. B.M. Sharp and T. Eisenstein, Expression of opioid receptor by immune cells, *J. Neuroimmunol.* 69:3-13 (1996).
10. Y. Sei, K. Yoshimoto, T. McIntyre, P. Skolnick and P. Arora, Morphine-induced thymic hypoplasia is glucocorticoid-dependent, *J. Immunol.* 146:194-198 (1991).
11. D.O. Freier and B.A. Fuchs, Morphine-induced alterations in thymocyte subpopulations of B6C3F1 mice, *J. Pharm. Exp. Ther.* 265:81-88 (1993).
12. P.C. Singhal, K. Reddy, N. Franki, V. Sanwal and N. Gibbons, Morphine induces splenocyte apoptosis and enhanced mRNA expression of cathepsin-B, *Inflammation* 21:609-615 (1997).
13. P.C. Singhal, P. Sharma, A.A. Kapasi, K. Reddy, N. Franki and N. Gibbons, Morphine enhances macrophage apoptosis, *J. Immunol.* 160:1886-1893 (1998).
14. P.C. Singhal, A.A. Kapasi, K. Reddy, N. Franki, N. Gibbons and G. Ding, Morphine promotes apoptosis in Jurkat cells, *J. Leuk. Biol.* 66:650-658 (1999).
15. L. Guan, T. Eisenstein, M. Adler and T. Rogers, Modulation of DPK cell function by the kappa opioid agonist U50,488H, in: *Drugs of Abuse, Immunomodulation, and AIDS,* H. Friedman *et al.*, ed., Plenum Press, New York (1998).
16. K. Linner, H. Quist and B. Sharp, Met-enkephalin-containing peptides encoded by proenkephalin A mRNA expressed in activated murine thymocytes inhibit thymocyte proliferation, *J. Immunol.* 154:5049-5060 (1995).
17. K. Linner, H. Quist and B. Sharp, Expression and function of proenkephalin A messenger ribonucleic acid in murine fetal thymocytes, *Endocrinology* 137:857-863 (1996).
18. Y. Takeuchi, T. Horiuchi, K. Hamamura, T. Sugimoto, H. Yagita and K. Okumura, Role of CD4 molecule in intrathymic T-cell development, *Immunology* 74:183-190 (1991).
19. R. Ceredig, Differentiation potential of 14-day fetal mouse thymocytes in organ culture: Analysis of CD4/CD8-defined single-positive and double-negative cells, *J. Immunol.* 141:355-362 (1988).

EFFECT OF OPIOIDS ON ORAL SALMONELLA INFECTION AND IMMUNE FUNCTION

Toby K. Eisenstein,[1,2] Amanda Shearer MacFarland,[2*] Xiaohui Peng,[2] Mary E. Hilburger,[2†] Rahil T. Rahim,[2] Joseph J. Meissler, Jr.,[2] Thomas J. Rogers,[1,2] Alan Cowan,[1,3] and Martin W. Adler[1,3]

[1]Center for Substance Abuse Research
[2]Department of Microbiology and Immunology
[3]Department of Pharmacology
Temple University School of Medicine
Philadelphia, PA 19140

INTRODUCTION

It is clearly established that morphine and other opioids are immunomodulatory.[1] Many of the studies demonstrating effects of the opioids on the immune system using animal models have used either single acute injections of drug[2,3] or for more prolonged drug administration, implantation of slow-release morphine pellets.[4,5] Our laboratories have shown that 75-mg slow-release morphine pellets result in marked depression of the capacity of the spleen cells taken 48 hr later to mount an in vitro antibody response to sheep red blood cells as a test antigen.[6] It has been difficult to carry out extensive dose response studies with morphine using slow-release pellets because they are only available in 75 mg and a few lower doses. Also, testing of opioids selective for other receptors has been hampered by a lack of pellets to deliver these compounds. In this paper we present data showing that osmotic mini-pumps can be used to deliver opioids that are selective for mu, kappa, or delta receptors, and that compounds in all three classes are immunosuppressive.

*Current address: Department of Microbiology, University of Pennsylvania School of Medicine, 209 Johnson Pavilion, Philadelphia, PA 19104.
†Current address: Laboratory of Tumor Immunology & Biology, National Cancer Institute, NIH, Bldg. 10, Rm. 8B07, 10 Center Dr., MSC-1750, Bethesda, MD 20892.

It is of interest to relate the immunomodulatory properties of opioids to effects on host resistance to infection. It is well known that intravenous drug users (IVDUs) have increased rates of infection,[7-11] but it is less clear if this is due to direct inoculation with pathogens during the course of heroin injection or to opioid-induced immunosuppression. In this regard, there is a significant literature showing that HIV infected patients have increased incidence and severity of infections with Salmonella.[12-18] As one-third of HIV patients are also IVDUs, it is not unreasonable that there may be an intersection between drug abuse and sensitization to Salmonella infection in the HIV-infected population. However, the epidemiologic analysis has not yet been carried out to establish a correlation in human heroin addicts between drug use and Salmonella incidence. In the present studies, the question of the effect of morphine and other opioids was directly tested by implanting mice with slow-release morphine pellets or osmotic mini-pumps and giving Salmonella orally. The results show that morphine dramatically potentiated oral Salmonella infection in this laboratory model.

METHODS

Animals

C3HeB/FeJ female mice, 6 to 8 weeks old purchased from Jackson Laboratories were used in all experiments. Animals were rested for one week before use.

Drug Administration

Mice were anesthetized with methoxyflurane and areas of the back were shaved. An incision was made in the skin and a 75-mg slow-release morphine pellet (NIDA) inserted. The wound was closed with a surgical clip. Alternatively, an Alzet osmotic mini-pump model 1003D, filled with either saline, U50,488H (kappa) or deltorphin II ($delta_2$) was implanted subcutaneously to deliver drug or placebo. Controls received a placebo pellet, a 30-mg naltrexone pellet, or a morphine plus a naltrexone pellet.

Salmonella Infection

Salmonella typhimurium, strains SL3235 (avirulent)[19] and W118-2 (virulent)[20] were used. Inocula were prepared from log phase BHI cultures, and suspended to indicated doses in physiologic saline. Mice were fasted for 4 hr prior to oral inoculation, anesthetized with methoxyflurane, and Salmonella were administered at indicated doses via an oral feeding needle in a volume of 0.2 ml. Following inoculation, pellets were implanted subcutaneously. On the designated days post-infection, mice were sacrificed and spleen, liver, Peyer's patches (PP), or mesenteric lymph nodes (MLN) were aseptically removed. The organs were homogenized in sterile water using a Tekmar tissuemizer, and bacteria were enumerated as the number of colony-forming units (CFU) by plating dilutions of the homogenates on Levine eosin-methylene blue (EMB) agar plates. Lactose non-fermenting colonies were counted after incubation overnight at 37°C.

Plaque-Forming Cell Assay

Mice were sacrificed 48 hr after pellet or pump implantation and their spleens removed. Spleen cells from two to three mice in each group were pooled. Immune function was assessed using an in vitro plaque-forming cell (PFC) assay described previously,[4,21] which measures the capacity of spleen cells to mount an IgM response to sheep red blood cells (SRBCs) added to the spleen cell cultures. Cells were harvested after 5 days of incubation, washed in RPMI, and the number of PFCs quantitated using the Cunningham modification of the Jerne hemolytic plaque assay.[22] Data from the PFC assay were calculated as the mean number of PFC/10^7 cells ± SD. Percent suppression was calculated as the ratio of PFCs in opioid-treated cells to PFCs in cells of placebo-pelleted mice. Values given for maximal suppression in Table 1 were calculated from assays on two pooled cohorts with each of these cohorts tested in at least triplicate.

RESULTS

Morphine Sensitizes to Oral Salmonella Infection

In order to ascertain if morphine could sensitize to oral Salmonella infection, mice were implanted with slow-release morphine pellets and inoculated at the time of pellet implantation with either an avirulent strain of Salmonella typhimurium, SL3235, or a virulent strain, W118-2. As shown in Fig. 1, morphine enhanced the replication of the avirulent strain by about 100-fold in the mesenteric lymph nodes by day 3 after inoculation, and naltrexone blocked the effect. When a similar experimental design was used with inoculation of virulent Salmonella, the sensitizing effect of morphine was even more dramatic (Fig 2). Morphine markedly potentiated growth of virulent Salmonella, with differences in microbial burden between placebo and opioid-treated mice of approximately a million-fold. Naltrexone blocked the increase in microbial burden, but not completely. Morphine treated mice that were challenged with various doses of virulent Salmonella all showed dramatic decreases in mortality and in mean survival time (Fig. 3). Other studies to ascertain the mechanism by which morphine potentiated oral Salmonella infection showed that the bacteria found in the PP were predominantly daughter cells of the organisms in the inoculum. Therefore, opioid treatment did not just allow more of the cells in the original inoculum to enter the gut-associated lymphoid tissue, but potentiated their replication.[23]

Effect of Kappa and Delta Agonists on Alteration of Immune Function and Resistance to Oral Infection

In order to continuously administer opioid agonists other than morphine, osmotic mini-pumps were used to deliver drugs in solution at a controlled rate. Dose response curves were constructed for the capacity of the various agonists to induce immunosuppression 48 hr after pump implantation using an assay for antibody formation by mouse spleen cells routinely employed by our laboratory. Using pumps it was found that morphine (primarily mu selective), U50,488H (kappa selective), and deltorphin II ($delta_2$ selective) were all immunosuppressive when spleens were tested 48 hr after pump implantation. Table 1 shows the doses that gave maximal suppression for each agonist. U50,488H and deltorphin were used at these doses to test for capacity to increase sensitization to Salmonella infection. The kappa agonist had no effect on potentiation of oral infection with virulent Salmonella at a dose of 1.8 x 10^4 CFU/ml. When the effect of deltorphin II was tested it was found that 1/5 mice had a low level (approximately 10^2 organisms) in the PP and MLN, but 4/5 animals had 10^2 organisms in the spleen.

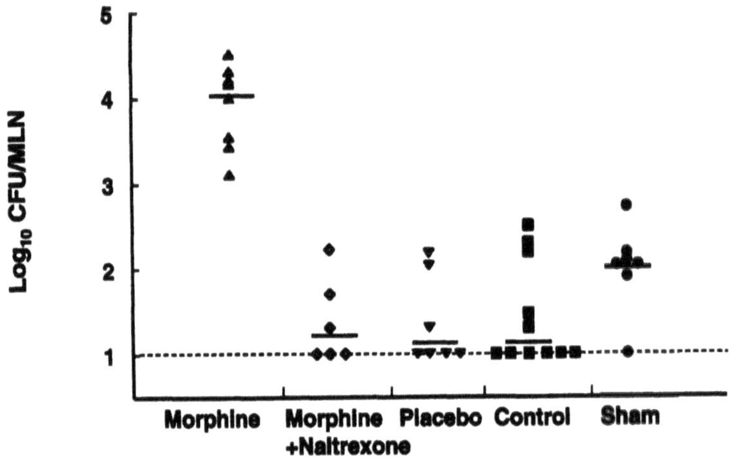

Figure 1. Effect of morphine on growth of attenuated *S. typhimurium*, strain SL3235, in mesenteric lymph nodes (MLN). Mice were subcutaneously implanted with a morphine pellet, a morphine plus a naltrexone pellet, a placebo pellet, or given a sham operation (no pellet implanted) and immediately afterward orally inoculated with 1.7 x 10^8 CFU of SL3235. Seventy-two hours post-inoculation mice were sacrificed, mesenteric lymph nodes removed and homogenized, and CFU determined by plate counts. Each point represents the number of CFU of an individual mouse. Horizontal lines represent median CFU for each group. The dotted line represents the limit of detection of Salmonella.

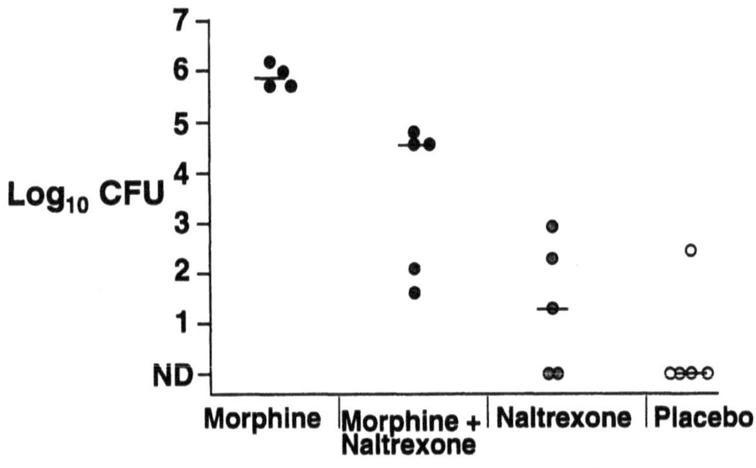

Figure 2. Effect of morphine on growth of virulent *S. typhimurium*, strain W118-2, in the MLN. Mice received drug or placebo pellets as described above, with the addition of a group which was implanted with a naltrexone pellet only. Mice were inoculated at the time of pellet implantation with 3 x 10^4 CFU of W118-2. Seventy-two hours post-implantation and inoculation, the number of CFU were determined as described above. Each point represents the number of CFU of an individual mouse. Horizontal lines represent median CFU for each group. ND = None detected. (Reprinted from MacFarlane et al, Journal of Infectious Diseases, volume 181(4), (c) 2000 by the Infectious Diseases Society of America. All rights reserved.)

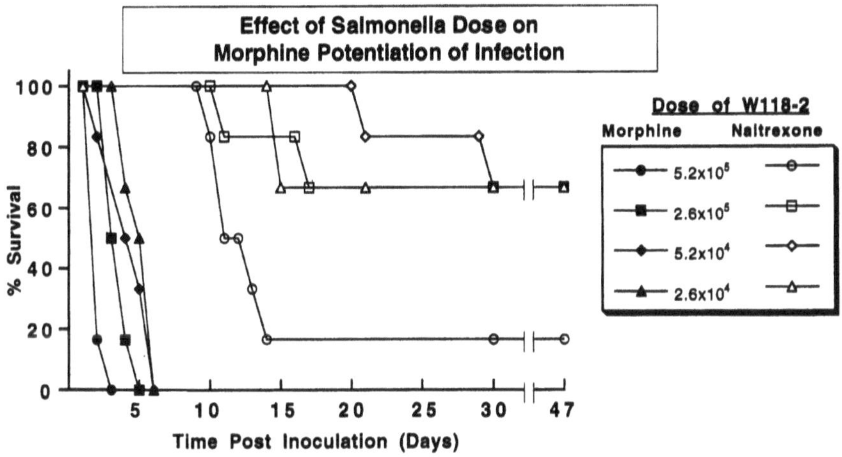

Figure 3. Effect of Salmonella dose on survival of mice implanted with morphine. Cohorts of mice (N = 6) were implanted with either a morphine pellet or a morphine plus a naltrexone pellet. Mice were then orally inoculated with the indicated doses of *S. typhimurium*, W118-2. Closed symbols represent groups implanted with morphine only; open symbols represent groups implanted with morphine plus naltrexone pellets.

Table 1. Optimal dose for induction of immunosuppression by opioids administered in osmotic minipumps.[1]

Compound	Dose (mg/kg/day)	Maximal immunosuppression
Morphine	0.5 to 1.0	61%
U50,488H	1.0	63%
Deltorphin II	0.5	62%

[1] PFC responses compared with responses of placebo-pelleted mice. All mice sacrificed 48 hr post-pump implantation.

DISCUSSION

These studies show that morphine administered by a 75 mg slow-release pellet markedly potentiates oral Salmonella infection with both an avirulent and a virulent strain of Salmonella. The increase in microbial burden was most dramatic when infection was with the virulent strain, where over a million-fold difference in bacterial burden in the mesenteric lymph nodes was observed in morphine-treated compared with control or placebo-pelleted mice. Mice given morphine pellets had dramatic differences in mean survival time compared with those receiving placebo pellets over a 25-fold range of challenge doses. In other studies, we have observed that morphine resulted in rapid death even with a challenge dose as low as 231 cells.[23] Thus, morphine seems to have a major potentiating effect on infection with this oral pathogen. The results raise the question as to whether the increased incidence and severity of Salmonella infections in patients infected with HIV correlates with use of intravenous heroin. The answer to this question awaits further epidemiologic investigation.

In regard to the mechanism by which morphine enhances oral Salmonella infection, it is well established that the drug decreases peristalsis and increases muscle tone and spasms.[24] The effect of morphine in the intestinal tract is via opioid receptors present in the circular muscle, submucosal plexus and the myenteric plexus of the small intestine.[24-26] It was of interest to see if opioid agonists with selectivity for the kappa and $delta_2$ receptors induced similar sensitivity to oral Salmonella infection. A kappa agonist has been shown to delay gastrointestinal transit time in the guinea pig;[27] in mice classical pharmacologic analysis suggests that mu and delta opioid receptors mediate intestinal stasis.[28] As slow-release pellets were not available for opioids other than morphine, we developed a procedure for administering kappa and $delta_2$ agonists by mini-pumps. Our results show that the kappa agonist, U50,488H, used at an immunosuppressive concentration, did not potentiate oral Salmonella infection. The $delta_2$ agonist, also administered at the immunosuppressive dose, resulted in a very modest effect on Salmonella infection, with retrieval of some organisms from the lymph nodes and PPs of one animal, and from the spleen in 4/5 animals. However, the numbers of bacteria were low and clearly not in the same order of magnitude as observed with morphine administered by pellet. Experiments are in progress to test the effect of morphine when it is administered by pump in order to permit a parallel comparison with the other agonists. Recently, using mu opioid receptor knock-out mice,[29] mu-receptor specific antibodies,[30] or mu-receptor antisense,[30] it was concluded that inhibition of gastrointestinal transit by morphine is essentially a mu-opioid receptor mediated function. While the studies reported in this paper using kappa- and delta-receptor selective opioids were confined to a single dose of each drug, and a single challenge dose of Salmonella, the results are consistent with the mu opioid receptor as the major mediator of enhanced Salmonella infectivity.

We have reported that morphine results in sepsis in mice,[31] which represents translocation of endogenous flora out of the gastrointestinal tract into the liver, spleen, and peritoneal cavity. It remains to be determined whether morphine potentiates endogenous infection or exogenous Salmonella infection because it creates conditions which allow the organisms to grow to higher numbers as a consequence of bowel stasis, or whether the immunosuppressive and anti-inflammatory effects of the drug[1] inhibit the antibacterial response which should occur in the gut-associated lymphoid tissues. It is also possible that morphine alters gut permeability allowing greater numbers of organisms to escape from the intestinal lumen. Analysis of microbial burdens in PPs, MLNs, and spleen at various times after virulent Salmonella infection suggests that organisms are mainly leaving the gastrointestinal tract via the PPs and then migrating to the MLN and finally the spleen.[23] The pattern of Salmonella growth in the organs over time would argue against major breaches in intestinal integrity along the length of the small intestine. This pattern fits well with the route of pathogenesis of Salmonella, which is invasion via the PPs, movement to the MLN, and finally seeding via lymph and blood to the spleen.[32] It appears more likely that morphine simply potentiates infection by increasing the dose of organisms or by retarding their killing.

ACKNOWLEDGMENTS

This work was supported by NIH grants DA-11134 and DA-06650.

REFERENCES

1. Eisenstein, T.K., and M.E. Hilburger, Opioid modulation of immune responses: effects on phagocyte and lymphoid cell populations, *J. Neuroimmunol.* 83:36 (1998).
2. Flores, L.R., M.C. Hernandez, and B.M. Bayer, Acute immunosuppressive effects of morphine: lack of involvement of pituitary and adrenal factors, *J. Pharmacol. Exp. Ther.* 268:1129 (1994).
3. Weber, R.J., B. Ikejiri, K.C. Rice, A. Pert, and A.A. Hagan, Opiate receptor mediated regulation of the immune response in vivo, *NIDA Res. Mono. Ser.* 48:341 (1987).
4. Bussiere, J.L., M.W. Adler, T.J. Rogers, and T.K. Eisenstein, Differential effects of morphine and naltrexone on the antibody response in various mouse strains, *Immunopharmacol. Immunotoxicol.* 14:657 (1992).
5. Bryant, H.U., E.W. Bernton, and J.W. Holaday, Immunosuppressive effects of chronic morphine treatment in mice, *Life Sci.* 41:1731 (1987).
6. Bussiere, J.L., M.W. Adler, T.J. Rogers, and T.K. Eisenstein, Cytokine reversal of morphine-induced suppression of the antibody response, *J. Pharmacol. Exp. Ther.* 264:591 (1993).
7. Louria, D.B, T. Hensle, and J. Rose, The major medical complications of heroin addiction, *Ann. Int. Med.* 67:1 (1967).
8. Hussey, H.H., and S. Katz, Infections resulting from narcotic addiction, *Am. J. Med.* 9:186 (1950).
9. Scheidegger, C., and W. Zimmerli, Infectious complications in drug addicts: seven year review of 269 hospitalized narcotic abusers in Switzerland, *Rev. Infect. Dis.* 11:486 (1989).
10. Haverkos, H.W., and R.W. Lange, Serious infections other than human immunodeficiency virus among intravenous drug users, *J. Infect. Dis.* 161:894 (1990).
11. Horsburgh, C.R., R.A. Anderson, and E.J. Boyko, Increased incidence of infections in intravenous drug use, *Infect. Control Hosp. Epidemiol.* 10:211 (1989).
12. Jacobs, J.L., J.W.M. Gold, H.W. Murray, R.B. Roberts, and D. Armstrong, Salmonella infections in patients with the acquired immunodeficiency syndrome, *Ann. Int. Med.* 102:186 (1985).
13. Glaser, J.B., L. Morton-Kute, and S.R. Berger, Recurrent *Salmonella typhimurium* bacteremia associated with the acquired immunodeficiency syndrome, *Ann. Int. Med.* 102:189 (1985).
14. Levine, W.C., J.W. Buehler, N.H. Bean, and R.V. Tauxe, Epidemiology of nontyphoidal Salmonella bacteremia during the human immunodeficiency virus epidemic, *J. Infect. Dis.* 164:81 (1991).
15. Sperber, S.J., and C.J. Schleupner, Salmonellosis during infection with human immunodeficiency virus, *Rev. Infect. Dis.* 9:925 (1987).
16. Gruenewald, R., S. Blum, and J. Chan, Relationship between human immunodeficiency virus infection and Salmonellosis in 20- to 59-year-old residents of New York City, *Clin. Infect. Dis.* 18:358 (1994).
17. Celum, C.L., R.E. Chaisson, G.W. Rutherford, J.L. Barnhart, and D.F. Echenberg, Incidence of Salmonellosis in patients with AIDS, *J. Infect. Dis.* 156:998 (1987).
18. Angulo, F.J., and D.L. Swerdlow, Bacterial enteric infections in persons infected with human immunodeficiency virus, *Clin. Infect. Dis.* 21:S84 (1995).
19. Hoiseth, S.K., and B.A.D. Stocker, Aromatic-dependent *Salmonella typhimurium* are non-virulent and effective as live vaccines, *Nature* 291:238 (1981).
20. Killar, L.M., and T.K. Eisenstein, Immunity to Salmonella infection in C3H/HeJ and C3H/HeNCrlBR mice: studies using an aromatic-dependent live Salmonella strain as a vaccine, *Infect. Immun.* 47:605 (1985).
21. Mishell, R.I., and R.W. Dutton, Immunization of dissociated spleen cell cultures from normal mice, *J. Exp. Med.* 126:423 (1967).
22. Cunningham, A., and A. Szenberg, Further improvements in the plaque technique for detecting single antibody producing cells, *Immunology* 14:599 (1968).

23. MacFarlane, A.S., X. Peng, J.J. Meissler, Jr., T.J. Rogers, E.B. Geller, M.W. Adler, and T.K. Eisenstein, Morphine increases susceptibility to oral *Salmonella typhimurium* infection, *J. Infect. Dis.* 181:in press (April issue) (1999).
24. Kromer, W., Endogenous and exogenous opioids in the control of gastrointestinal motility and secretion, *Pharmacol. Rev.* 40:121 (1988).
25. Manara, L., and A. Bianchetti, The central and peripheral influences of opioids on gastrointestinal propulsion, *Ann. Rev. Pharmacol. Toxicol.* 25:249 (1985).
26. Ruoff, H.-J., B. Fladung, P. Demol, and T.R. Weihrauch, Gastrointestinal receptors and drugs in motility disorders, *Digestion* 48:1 (1991).
27. Culpepper-Morgan, J.A., M.J. Kreek, P.R. Holt, D. LaRoche, J. Zhang, and L. O'Bryan, Orally administered kappa as well as mu opiate agonists delay gastrointestinal transit time in the guinea pig, *Life Sci.* 42:2073 (1988).
28. Pol, O., I. Ferrer, and M.M. Puig, Diarrhoea associated with intestinal inflammation increases the potency of mu and delta opioids on the inhibition of gastrointestinal transit in mice, *J. Pharmacol. Exp. Ther.* 270:386 (1994).
29. Roy, S., H. Liu, and H.H. Loh, mu-opioid receptor-knockout mice: the role of mu-opioid receptor in gastrointestinal transit, *Mol. Brain Res.* 56:281 (1988).
30. Pol, O., L. Valle, P. Sánchez-Bláquez, J. Garzón, and M.M. Puig, Antibodies and antisense oligodeoxynucleotides to mu-opioid receptors selectively block the effects of mu-opioid agonists on intestinal transit and permeability in mice, *Brit. J. Pharmacol.* 127:397 (1999).
31. Hilburger, M.E., M.W. Adler, A.L. Truant, J.J. Meissler, Jr., V. Satishchandran, T.J. Rogers, and T.K. Eisenstein, Morphine induces sepsis in mice, *J. Infect. Dis.* 176:183 (1997).
32. Carter, P.B., and F.M. Collins, The route of enteric infection in normal mice, *J. Exp. Med.* 139:1189 (1974).

ALTERED T-CELL RESPONSIVENESS IN MORPHINE "TOLERANT" RATS: EVIDENCE FOR A POTENTIAL ROLE OF THE PARAVENTRICULAR NUCLEUS OF THE HYPOTHALAMUS

R. Daniel Mellon, Nassim E. Noori, Monica C. Hernandez, and Barbara M. Bayer

Departments of Pharmacology and Neuroscience
Georgetown University Medical Center
3900 Reservoir Road, N.W.
Washington, DC 20007

INTRODUCTION

The acute administration of morphine produces significant alterations in several parameters of the immune system including lymphocyte proliferation responses, natural killer cell cytolytic activity, antibody production and macrophage function[1,2]. Our laboratory has examined the effects of acute morphine on blood lymphocyte proliferation to the polyclonal mitogen concanavalin A (ConA) in the male Sprague-Dawley rat. These studies demonstrated that the acute suppressive effect of morphine on this *in vitro* measure of lymphocyte function is mediated by central opioid receptors[3], predominantly via the μ-subtype of opioid receptors[4].

Mechanistic studies suggested that the effect of morphine on the blood lymphocyte proliferation response is largely independent of either pituitary or adrenal-derived factors[5]. However, additional experiments implicated a role for the autonomic ganglia in the effects of acute morphine since the blockade of peripheral nicotinic receptors was found to completely antagonize the acute effect of morphine on this parameter[6]. Further, the stimulation of peripheral nicotinic receptors with nicotine and epibatidine, a selective nicotinic receptor agonist, mimicked the effects of morphine on immune cells[7]. It was concluded from these experiments that the acute effects of morphine on the blood lymphocyte proliferation response were mediated primarily through indirect stimulation of peripheral autonomic neurons.

Studies examining the effects of chronic morphine administration have also demonstrated significant alterations in the same parameters of the immune system, including, lymphocyte proliferation[8], antibody production[9-11], natural killer cell cytolytic activity[12,13] and macrophage function[14,15]. However, in contrast to the primary mechanism mediating the effect of acute administration of morphine, the effect of chronic morphine on many of these parameters appears to be largely attributed to the prolonged elevations in glucocorticoids associated with chronic morphine treatment[16].

Our laboratory has demonstrated that an apparent tolerance to the suppressive effect of morphine treatment on blood lymphocyte proliferation can be induced via an injection protocol employing escalating doses of morphine over 5-7 days[17,18]. Although tolerance does appear to develop under these conditions, further studies have suggested that the chronic morphine treatment induced an enhanced sensitivity to the immunological impact of restraint, considered to be a psychological stressor[18]. Further studies were therefore conducted to determine if the delayed-type hypersensitivity response also demonstrated morphine tolerance in these animals.

Although acute morphine appears to be acting via central opioid receptors, the exact site of action for morphine-induced suppression of blood lymphocyte proliferation is not clear. Our laboratory has provided evidence that the anterior hypothalamus is one region where microinjection of morphine can alter this parameter[3]. Given the clear suggestion that the autonomic nervous system is the primary mediator of this response, the potential role of the paraventricular nucleus of the hypothalamus (PVN) was examined in both acute and chronic morphine treated animals via c-fos immunohistochemistry.

METHODS

Animals

Pathogen-free male Sprague-Dawley rats were obtained from Taconic Farms (Germantown, NY) and housed 3 per cage with microisolator tops. Animals were allowed to acclimate to this new environment for 1 week prior to experimentation.

Drug Treatment Protocol

Morphine sulfate was generously provided by the National Institute on Drug Abuse (Research Triangle Park, NC). Morphine was dissolved in sterile saline and injected in a volume of 1 ml/kg. Animals were injected with either saline (1 ml/kg, s.c.) or increasing doses of morphine according to the following schedule (Table 1):

Table 1. Morphine Treatment Protocol

	Morphine (mg/kg, s.c.)	
Day	8:00 AM	4:00 PM
1	10	10
2	20	20
3	30	30
4	40	40
5	40	40
6	40	40
7	10	

Antinociception

Antinociception was determined via a modified tail-flick assay[18]. All animals were acclimated to both handling and the tail-flick apparatus for 2-3 days prior to testing and tested every other day for the development of tolerance to a 10 mg/kg dose of morphine. Baseline latency to tail-flick was between two and three seconds. A cutoff of eight seconds was established in order to prevent tissue damage. The mean of three consecutive tests was used to calculate the percent maximum possible effect (%MPE) as follows:

$$\%MPE = \frac{\text{Postdrug latency (sec) - predrug latency (sec)}}{\text{Cutoff (8 sec) - predrug latency (sec)}} \times 100$$

Lymphocyte Proliferation

Blood proliferation assays were completed as previously described[18]. Briefly, 2 hrs after drug treatment, animals were killed via decapitation, trunk blood was collected in sterile heparinized polypropylene tubes and kept on ice until processed. Blood was diluted 1:5 with cold RPMI-1640 medium containing 1% fetal bovine serum and gentamicin. Diluted blood was then plated into a 96 well microtiter plate with increasing concentrations of the T-cell mitogen concanavalin A (ConA: 0, 2, 4 and 6 μg/well) and incubated for 72 hrs. Cells were then pulsed with [^{3}H]thymidine for an additional 24 hrs and harvested onto glass fiber filters (Brandel, Gaithersburg, MD). The amount of labeled DNA was determined by liquid scintillation spectrophotometry (Betaplate model 1205, Wallac, Gaithersburg, MD). Data were presented as the maximal response (typically 4 μg/well) using these doses of ConA.

Delayed-Type Hypersensitivity Response

Animals were sensitized via a subcutaneous injection of bovine serum albumin (BSA) with Freund's Complete Adjuvant at the base of the tail. Seven days later, animals received an injection (100 μl) of heat aggregated BSA in the right hind paw and saline (100 μl) in the left hind paw as a challenge. Seventy-two hrs after the challenge injection paw width was measured with calipers. Data were expressed as a mean % the of saline-injected paw.

Immunohistochemistry for c-fos

Animals were acclimated to handling and tested for nociception responses to establish the presence of tolerance for the entire week prior to experimentation. In addition, to minimize the effects of injection stress on c-fos expression in the PVN, animals were also acclimated for 3 days via handling and administration of a sterile saline injection (1 ml/kg, i.p.). On the day of the assay, animals were administered a lethal injection of pentobarbital (25 mg, i.p.) and perfused with saline containing 2% sodium nitrate solution followed by fixative (acrolein in 4% paraformaldehyde). Brains were removed and stored in 25% sucrose. Forty micron slices were cut on a freezing microtome and c-fos in brain tissues was detected via immunohistochemistry. Primary antibody to c-fos (Santa Cruz Biotechnology, Inc., Santa Cruz, CA) was detected via biotin conjugated goat anti-rabbit IgG followed by 1 hour exposure to ABC (Vectastain ELITE ABC Kit, Vector Laboratories, Inc., Burlingame, CA). Matched brain sections were drawn via camera lucida and the number of c-fos positive cells were counted bilaterally and divided by 2 to obtain the number of c-fos positive cells per region.

Statistical Analysis

Data were analyzed by one-way ANOVA with post-hoc comparisons employing a Newman-Keuls test. A value of $p < 0.05$ was considered to be statistically significant.

RESULTS

Development of Tolerance to Effects of Morphine

Our laboratory has previously demonstrated that acute administration of morphine (10 mg/kg, s.c.) to the Sprague-Dawley rat produces both antinociception and suppression of blood lymphocyte proliferation response to the polyclonal mitogen concanavalin A (ConA). Tolerance appears to develop to both of these parameters. Figure 1A demonstrated that tolerance to the effect of morphine (acute challenge of 10 mg/kg) on antinociception can be induced over a period of 6 days as measured by the tail-flick latency to a radiant heat source. Panel B in Figure 1 indicates that via this same injection protocol, tolerance also appears to develop to the effect of morphine on blood lymphocyte proliferation.

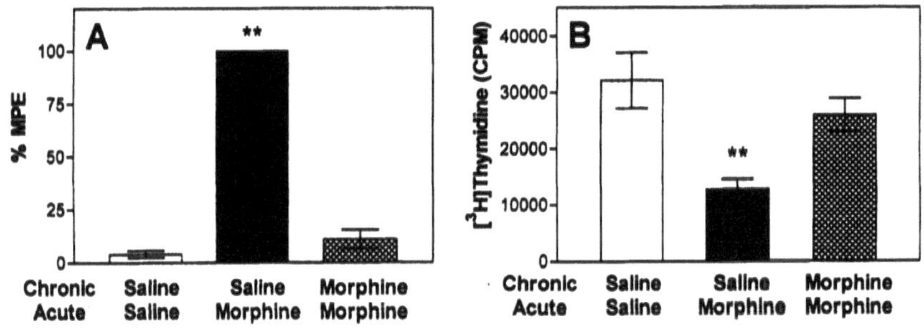

Figure 1. Development of tolerance to morphine: (A) nociception and (B) suppression of blood lymphocyte proliferation. Animals (7 per group) were injected twice a day with morphine according to the tolerance protocol described in Methods. Nociceptive responses were measured 30 min after acute saline or morphine (10 mg/kg, s.c.) and blood proliferation was measured 2 hrs after drug treatment. Unstimulated cell proliferation was less than 1000 cpms for all groups. ** $P < 0.01$ compared to both saline/saline and morphine/morphine group (ANOVA, Newman-Keuls).

Delayed-Type Hypersensitivity Response

To determine if the "tolerant" state also extended to a more complex immune parameter, the delayed-type hypersensitivity response was examined in morphine "tolerant" animals. Table 2 demonstrates that the paw swelling in the animals that received an acute dose of morphine administered 1 hour prior to primary antigen challenge was slightly lower than the saline treated controls; however, this effect was not statistically different from either the saline or chronic morphine treated animals. In contrast, the paw swelling in animals chronically treated with morphine was significantly lower than the swelling detected in saline treated control animals.

Table 2. Paw diameter 72 hrs after antigen challenge.

Treatment Group (Chronic/Acute)	% Saline-injected paw (mm)
Saline/Saline	29.5 ± 5.4
Saline/Morphine	20.3 ± 4.2
Morphine/Morphine	12.9 ± 1.8 *

* $P < 0.05$ compared to saline saline treated animals

Immunohistochemistry for c-fos

To determine if central autonomic nuclei in the brain also demonstrated tolerance to morphine administration, immunohistochemistry for c-fos, a generalized marker of neuronal activation was examined. As the effects of acute morphine on blood lymphocyte proliferation appear to be mediated primarily via the autonomic nervous system, we hypothesized that the PVN, one of the key autonomic centers of the brain, was activated following acute morphine treatment. Immunohistochemical analysis for c-fos demonstrated that acute morphine treatment significantly increased the number of c-fos positive cells within the PVN (Figure 2). Interestingly, like the effects noted for both nociception and blood lymphocyte proliferation, animals made tolerant to an acute dose of morphine also failed to demonstrated increased c-fos expression in the PVN under these conditions (Figure 2).

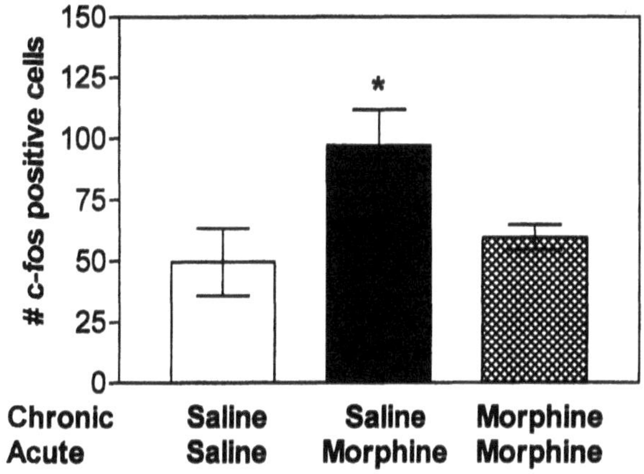

Figure 2. The effect of acute morphine administration on c-fos expression in the paraventricular nucleus (PVN) of the hypothalamus in naive and morphine "tolerant" animals. Animals (n=7-8 per group) were treated with increasing doses of morphine as described in Methods. On Day 7, animals were challenged with either saline or morphine (10 mg/kg, s.c.). Two hrs after treatment, animals were anesthetized with a lethal dose of sodium pentobarbital, perfused, and brain c-fos expression in the PVN was examined as described in Methods. * $P < 0.05$ compared to both chronic saline/acute saline and chronic morphine/acute morphine treatment groups (ANOVA, Newman-Keuls).

DISCUSSION

Development of Tolerance to the Acute Effect of Morphine on Lymphocyte Proliferation

Chronic administration of opioids have been reported to produce a wide variety of immune alterations. Many of these studies have employed the use of implanted morphine pellets. Although the use of morphine pellets has been demonstrated to rapidly induce a high degree of tolerance, implantation of a single 75 mg/kg pellet into mice produces dramatic alterations in immune tissues, including atrophy and decreased cellularity of the spleen and thymus, decreased body weight, adrenal hypertrophy as well as decreased lymphocyte proliferation responses[8,19]. The studies described here employed twice daily injections of morphine to induce a tolerant state without significantly altering body mass. Under these conditions, tolerance developed to both the antinociceptive effect of morphine as well as morphine-induced suppression of blood lymphocyte proliferation. These studies initially suggested that the controlled exposure to morphine may not necessarily be immunosuppressive, as such, further characterization of the "morphine tolerant" animal was completed.

Delayed-type Hypersensitivity Response

Although apparent tolerance had developed to the *in vitro* measures of cell-mediated immunity, we examined whether this was also accompanied by tolerance to an *in vivo* measure of cellular immunity, namely delayed-type hypersensitivity (DTH) response. The DTH response is the primary mechanism by which the host defends itself against intracellular bacteria, such as *Listeria monocytogenes* and *Mycobacterium tuberculosis*. As shown in Table 2, animals receiving chronic injections of morphine over 7 days demonstrated significantly reduced DTH responses when examined 72 hrs after secondary antigen exposure. These results are consistent with other published studies reporting that chronic morphine administration produced a decreased DTH response in mice implanted with morphine pellets[20] and in both the rat[21] and pig model[22] following multiple injections of morphine. The decreased DTH response in the studies reported here demonstrating tolerance to the effects of morphine on lymphocyte proliferation supports the hypothesis that despite the apparent tolerance to the nociceptive effects of morphine these animals remain immunocompromised.

Potential Role of the PVN in Acute Morphine-induced Immunomodulation

In addition to the well characterized role of the PVN in neuroendocrine regulation, particularly the HPA axis, the PVN is also a key regulator of the autonomic nervous system[23]. Retrograde tracing studies demonstrate that the PVN directly innervates the intermediolateral cell column as well as other primary autonomic motor nuclei in the CNS[24]. As such, it is neuroanatomically positioned to potentially play a role in the centrally-mediated effects of morphine on the immune system. Acute morphine administration has been reported to either increase[25,26] or produce no change in c-fos expression in the PVN[27]. In contrast to the effect of acute morphine administration, chronic treatment with morphine via the implantation of morphine pellets (two 75 mg morphine pellets on day 1 and four 75 mg pellets on day two) induces tolerance to the effects of acute central morphine as well as interleukin-1β on c-fos expression in the PVN suggesting that chronic morphine treatment can produce tolerance to PVN neuronal activation[28] by either stimulus.

The studies described here support the conclusion that acute administration of morphine increases c-fos expression in the PVN and that this effect appears to show tolerance upon repeated injection of morphine. In addition, these studies demonstrate that the same conditions that induce tolerance to the lymphocyte proliferation assay also induce tolerance to c-fos induction in the PVN. Although this correlation does not directly implicate the PVN in the acute immunosuppressive effect of morphine on blood lymphocyte proliferation, the data support the hypothesis that central opioid receptors regulate this known central autonomic nucleus and establish the PVN as a likely effector nucleus in the neuronal circuitry whereby the CNS communicates with the immune system.

Although tolerance appears to readily develop to morphine-induced nociception, blood lymphocyte proliferation and even c-fos expression in the PVN, the immunological status of chronic morphine treated animals remains uncertain. Evidence is mounting that opioids alter physiological responses that may impact the organisms ability to respond to other stimuli. Our laboratory has previously demonstrated that following the same injection protocol described in these studies to induce an apparent tolerant state, morphine "tolerant" rats demonstrated an increased sensitivity to restraint stress[18]. Since the form of restraint employed in this laboratory does not lead to significant alterations in lymphocyte proliferation[29], this enhanced sensitivity suggests that chronic morphine treatment has altered the ability for the animal to respond normally to this psychological stress paradigm. Since the HPA axis is essential for the stress response, it is tempting to speculate that the lack of neuronal activation within the PVN of morphine tolerant animals represents a mal-adaptive response induced by chronic morphine administration. This conclusion is in agreement with the observations of Chang et al. that chronic morphine exposure resulted in a loss of the normal c-fos induction in the PVN following microinjection of interleukin-1 directly into the CNS[28].

Although the loss of c-fos expression in the PVN correlated with the lack of an acute morphine-induced suppression of blood lymphocyte proliferation response, the DTH response was significantly suppressed in animals made "tolerant" to morphine. This observation suggests that c-fos activation in the PVN does not correlate with morphine-induced suppression of the DTH response. Whether or not the lack of a response of the PVN plays a role in suppressed DTH response noted in these studies, however, remains to be determined. Regardless, these studies support the more global hypothesis that alterations in central neuronal structures, like the PVN, appear to be altered by both acute and chronic opioid administration and that these alterations may contribute to neuroimmunomodulation.

ACKNOWLEDGMENTS

This work was supported by NIH Grant DA04358 (BMB) and NIH Fellowship DA05779 (RDM).

REFERENCES

1. Bryant HU and Holaday JW: Opioids in immunological processes. In Opioids II, ed. by A.Herz, H.Akil and E.J.Simon, pp. 361-392, Springer-Verlag, New York, 1993.

2. Eisenstein TK and Hilburger ME. 1998. Opioid modulation of immune responses: Effects on phagocyte and lymphoid cell populations. *J. Neuroimmunol.* 83:36-44.

3. Hernandez MC, Flores LR and Bayer BM. 1993. Immunosuppression by morphine is mediated by central pathways. *J. Pharmacol. Exp. Ther.* 267:1336-1341.

4. Mellon RD and Bayer BM. 1998. Role of central opioid receptor subtypes in morphine-induced alterations in peripheral lymphocyte activity. *Brain Res.* 789:56-67.

5. Flores LR, Hernandez MC and Bayer BM. 1994. Acute immunosuppressive effects of morphine: Lack of involvement of pituitary and adrenal factors. *J. Pharmacol. Exp. Ther.* 268:1129-1134.

6. Flores LR, Dretchen KL and Bayer BM. 1996. Potential role of the autonomic nervous system in the immunosuppressive effects of acute morphine administration. *Eur J Pharm* 318:437-446.

7. Mellon RD and Bayer BM. 1999. The effects of morphine, nicotine and epibatidine on lymphocyte activity and hypothalamic-pituitary-adrenal axis responses. *J. Pharmacol. Exp. Ther.* 288:635-642.

8. Bryant HU, Bernton EW and Holaday JW. 1987. Immunosuppressive effects of chronic morphine treatment in mice. *Life Sci.* 41:1731-1738.

9. Güngör M, Genç E, Sagduyu H, Eroglu L and Koyuncuoglu H. 1980. Effect of chronic administration of morphine on primary immune response in mice. *Experientia* 36:1309-1310.

10. Eisenstein TK, Meissler JJ, Jr., Geller EB and Adler MW. 1990. Immunosuppression to tetanus toxoid induced by implanted morphine pellets. *Ann NY Acad Sci* 594:377-379.

11. Molitor TW, Morilla A, Risdahl JM, Murtaugh MP, Chao CC and Peterson PK. 1992. Chronic morphine administration impairs cell-mediated immune responses in swine. *J. Pharmacol. Exp. Ther.* 260:581-586.

12. Shavit Y, Lewis JW, Terman GW, Gale RP and Liebeskind JC. 1984. Opioid peptides mediate the suppressive effect of stress on natural killer cell cytotoxicity. *Science* 223:188-190.

13. Freier DO and Fuchs BA. 1994. A mechanism of action for morphine-induced immunosuppression: Corticosterone mediates morphine-induced suppression of natural killer cell activity. *J. Pharmacol. Exp. Ther.* 270:1127-1133.

14. Levier DG, Brown RD, McCay JA, Fuchs BA, Harris LS and Munson AE. 1993. Hepatic and splenic phagocytosis in female B6C3F1 mice implanted with morphine sulfate pellets. *J. Pharmacol. Exp. Ther.* 267:357-363.

15. Tomei EZ and Renaud FL. 1997. Effect of morphine on Fc-mediated phagocytosis by murine macrophages in vitro. *J. Neuroimmunol.* 74:111-116.

16. Levier DG, McCay JA, Stern ML, Harris LS, Page D, Brown RD, Musgrove DL, Butterworth LF, White KL, Jr. and Munson AE. 1994. Immunotoxicological profile of morphine sulfate in B6C3F1 female mice. *Fundamental and Applied Toxicology* 22:525-542.

17. Bayer BM, Hernandez MC and Ding XZ. 1996. Tolerance and crosstolerance to the suppressive effects of cocaine and morphine on lymphocyte proliferation. *Pharmacol. Biochem. Behav.* 53:227-234.

18. Bayer BM, Brehio RM, Ding XZ and Hernandez MC. 1994. Enhanced susceptibility of the immune system to stress in morphine-tolerant rats. *Brain Behav Immun* 8:173-184.

19. Bryant HU, Bernton EW, Kenner JR and Holaday JW. 1991. Role of the adrenal cortical activation in the immunosuppressive effects of chronic morphine treatment. *Endocrinology* 128:3253-3258.

20. Bryant HU and Roudebush RE. 1990. Suppressive effects of morphine pellet implants on in vivo parameters of immune function. *J. Pharmacol. Exp. Ther.* 255:410-414.

21. Pellis NR, Harper C and Dafny N. 1986. Suppression of the induction of delayed hypersensitivity in rats by repetitive morphine treatments. *Exper. Neurol.* 93:92-97.

22. Schoolov YN, Pampusch MS, Risdahl JM, Molitor TW and Murtaugh MP: Effect of chronic morphine treatment on immune responses to keyhole limpet hemocyanin in swine. In The Brain Immune Axis and Substance Abuse, ed. by B.M.Sharp, T.K.Eisenstein, J.J.Madden and H.Friedman, pp. 169-174, Plenum Press, New York, 1995.

23. Saper CB: Central autonomic system. In The Rat Nervous System, ed. by G.Paxinos, pp. 107-135, Academic Press, San Diego, 1995.

24. Strack AM, Sawyer WB, Hughes JH, Platt KB and Loewy AD. 1989. A general pattern of CNS innervation of the sympathetic outflow demonstrated by transneuronal pseudorabies viral infections. *Brain Res.* 491:156-162.

25. Bot G and Chahl LA. 1996. Induction of fos-like immunoreactivity by opioids in guinea-pig brain. *Brain Res.* 731:45-56.

26. Gutstein HB, Thome JL, Fine JL, Watson SJ and Akil H. 1998. Pattern of c-fos mRNA induction in rat brain by acute morphine. *Can. J. Physiol. Pharmacol.* 76:294-303.

27. Garcia MM, Brown HE and Harlan RE. 1995. Alterations in immediate-early gene proteins in the rat forebrain induced by acute morphine injection. *Brain Res.* 692:23-40.

28. Chang SL, Patel NA, Romero AA, Thompson J and Zadina JE. 1996. FOS expression induced by interleukin-1 or acute morphine treatment in the rat hypothalamus is attenuated by chronic exposure to morphine. *Brain Res.* 736:227-236.

29. Flores CM, Hernandez MC, Hargreaves KM and Bayer BM. 1990. Restraint stress-induced elevations in plasma corticosterone and β-endorphin are not accompanied by alterations in immune function. *J. Neuroimmunol.* 28:219-225.

ACTIONS OF ENDOTOXIN AND MORPHINE

Sulie L. Chang,[1] Bernardo Felix,[1] Yuhui Jiang,[1] and Milan Fiala[2]

[1]Department of Biology
Seton Hall University
South Orange, NJ 07079
[2]Departments of Neurology, Medicine, and Microbiology and Immunology
UCLA School of Medicine
Los Angeles, CA 90095

INTRODUCTION

Immune cells circulate within blood vessels that are lined by microvascular endothelial cells. Indeed, the initial interaction between immune cells and endothelial cells (leukocyte-endothelial interaction) in venules is an indicator of a functional immune response (House and Lipowsky, 1987). The endothelial cells are also critical components of microvascular physiological barriers, such as the blood-brain barrier (BBB) and the blood-thymus barrier (BTB) [Takata et al., 1990]. The integrity of these microvascular barriers is critical in protecting the host from a variety of pathogens, including HIV (Fiala et al., 1998). Therefore, both immune cells and endothelial cells play an important role in a host's defense mechanisms. In this article, the interaction between the actions of morphine and the bacterial endotoxin, lipopolysaccharide (LPS), on both endothelial and immune cells will be presented to illustrate possible mechanisms by which opioids impact immune function.

LIPOPOLYSACCHARIDE INDUCES THE EXPRESSION OF MU OPIOID RECEPTORS

Previously, Vidal et al. (1998) reported that co-treatment with interleukin-1α (IL-1α) and interleukin-1β (IL-1β) induces the expression of mu opioid receptors in brain microvascular endothelial cells. IL-1 is a pro-inflammatory cytokine. Its secretion increases in disease or stress conditions, and can be stimulated by endotoxins, such as lipopolysaccharide (LPS). IL-1 mediates a cascade of immunological responses by binding to specific receptors and inducing a variety of transcriptional factors, such as NF-IL6 and NFkB (O'Neill, 1995). These transcriptional factors, in turn, bind to certain cytokine response elements, such as NF-IL-6 and NF-GM, on target genes. The promoter region of the mu opioid receptor gene has been shown to contain several cytokine response elements (Min et al., 1994). We have hypothesized that conditions that stimulate the production of cytokines may also induce the expression of mu opioid receptors, even in tissues that do not express mu opioid receptors in the basal state.

In our recent study, eight adult male Harlan Sprague-Dawley rats were randomly assigned to receive an intraperitoneal (IP) injection of either 5 mg/kg or 30 mg/kg LPS, saline, or no treatment (n = 2 for each group) for 12 h. At the end of the treatment period, the mesentery and the caudate putamen (CPU) of the brain were collected from each rat for RNA isolation and RT-PCR analysis. The CPU was selected and served as a positive control because mu opioid receptors are abundantly expressed in this particular area of the brain (Mansour et al., 1987).

Total cellular RNA was isolated using a phenol-quanidine thiocyanate extraction method with minor modifications (Promega Technical Bulletin, TB087:12/89). Specific oligonucleotide primers for the mu opioid receptor and β-actin genes were synthesized from the published sequences (Wang et al., 1994; Nudel et al., 1983). RT-PCR analysis was conducted according to a previously published protocol (Vidal et al., 1998).

There was no detectable mu opioid receptor mRNA in the mesentery of either naive animals or saline-treated rats. Mu opioid receptor mRNA was readily detectable in the CPU, as expected. Following an injection of 30 mg/kg, but not 5 mg/kg LPS, the expression of mu opioid receptors was clearly increased in the mesentery. It appears that the mesentery does not have detectable mu opioid receptors under basal conditions, but the expression of mu opioid receptors can be induced in response to stress produced by endotoxin shock.

These data, together with those from our previous studies showing that co-treatment with IL-1α and IL-1β induces mu opioid receptor expression in endothelial cells (Vidal et al., 1998), suggests that opioid-dependent pathways in endothelial cells may be altered by endotoxins, possibly via an IL-1-mediated mechanism.

MORPHINE STIMULATES THE EXPRESSION OF ADHESION MOLECULES ON ENDOTHELIAL CELLS

As described previously, endothelial cells line the venules of the mesentery. It is also within these venules that white blood cells, such as neutrophils, first adhere to the endothelial lining of the vessel wall in order to exit the vascular system and enter the tissues to function in the body's immune defense mechanisms. This microvascular endothelial cell barrier is also a critical component in the systemic protection against invasion by various pathogens in the CNS, including HIV. There is no detectable expression of mu opioid receptor in these endothelial cells at the basal level; however, mu opioid receptor expression can be induced by cytokines, as shown in our studies. Previous *in vivo* studies have shown that the activation of mu opioid receptors by morphine in brain microvascular endothelial cells causes an increase in BBB permeability to molecules that would otherwise not be able to cross the barrier (Oishi et al., 1989). It is our hypothesis that mu opioid receptors may be induced in pathological conditions in these endothelial cells. We, therefore, subsequently examined the effects of morphine on the expression of adhesion molecules in a primary culture prepared from human brain microvasculature endothelial cells (HBMEC) isolated from individuals with pathological conditions.

Human brain microvascular endothelial cells (HBMEC) were isolated from surgical brain tissue, and grown to confluency in 96-well plates. They were then treated with either morphine (0.42, 4.2, or 42 μM) or control vehicle for 4 hr or 20 hr. The medium was then removed, and the monolayer cells were fixed with acetone/methanol (1:1) for 20 min, air dried, and stored at -20°C. The adhesion molecules, ELAM, VCAM and ICAM, were examined using ELISA analysis, in triplicate, according to previously published procedures (Stins et al., 1997).

Treatment with morphine (0.42 μM) increased the expression of ELAM, VCAM and ICAM-1 at 4 hr, with the greatest increase seen in ICAM-1 expression (Fig. 1). Induction

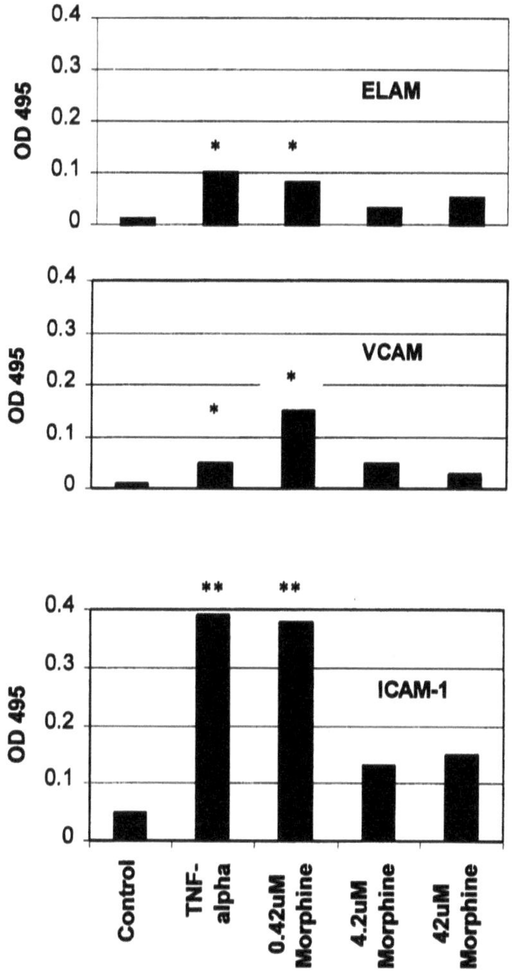

Figure 1. Morphine treatment increases the expression of ELAM, VCAM, and ICAM-1 at 4 h ($p<0.05$).

of expression of all three adhesion molecules by tumor necrosis factor alpha (TNF-α) was used as a positive control.

At 20 hr, morphine's effects on both ELAM and VCAM were greater than those seen at 4 hr, but less on ICAM-1 than at the 4 h time point. However, the expression of ICAM-1 was still the most affected.

Since expression of ICAM-1 was the most affected by morphine treatment, we then evaluated the reversibility of this effect by naloxone. Co-treatment with morphine and naloxone was found to reverse morphine's activation of ICAM-1, indicating that morphine's modulation of adhesion molecules in endothelial cells may be mediated through an opioid receptor-dependent pathway.

These results suggest that, in pathological conditions, expression of mu opioid receptors may be induced in endothelial cells, and that morphine exposure, under such conditions, may result in some unexpected effects due to the induced expression of the mu opioid receptor.

MORPHINE INCREASES THE PERMEABILITY OF MICROVASCULAR ENDOTHELIAL CELL BARRIERS

After determining that treatment with morphine can increase the expression of adhesion molecules on endothelial cells, we then established an *in vitro* microvascular endothelial cell barrier model system in order to measure the changes in permeability produced by morphine on an endothelial cell barrier, and to study the cytotoxic effects of morphine on microvascular endothelial cells.

Rat brain microvascular endothelial cells (BMVEC), kindly provided by Dr. Danica Stanimirovic of the National Research Council of Canada, were used to construct a microvascular endothelial cell barrier model on uncoated 8 μm micro-porous inserts (Zhang et al., 1998). The inserts were placed into 24-well plates with complete medium added to both the top and bottom chambers of the insert. The cells were incubated at 37°C until maximum confluence was reached. The medium was then replaced with medium containing 1 μM, 10 μM, or 100 μM morphine, or medium only, and incubated for 4 hrs. Medium containing [^{14}C]-inulin was then added to the upper chamber of the model constructs, and the cells were incubated for 2 hr. The permeability of the endothelial cell barrier was determined by scintillation counting of [^{14}C]-inulin that had permeated from the upper to the lower chamber.

Treatment of the microvascular endothelial cell BMVEC barrier with morphine (1 μM, 10 μM, or 100 μM) resulted in a significant dose-dependent increase in permeability to inulin (Fig. 2). However, naloxone did not reverse the effect; instead, there seemed to be a synergistic effect with morphine.

MORPHINE DECREASES CELL VIABILITY AND POTENTIATES THE EFFECTS OF LPS ON CELL VIABILITY OF ENDOTHELIAL CELLS

Rat BMVEC cells were grown to confluency in 12-well plates, and treated with 1, 10, or 100 μM morphine, or saline for 24 hr. At the end of treatment, the endothelial cells were trypsinized, and viability was assessed using a hemacytometer and the trypan blue dye exclusion assay.

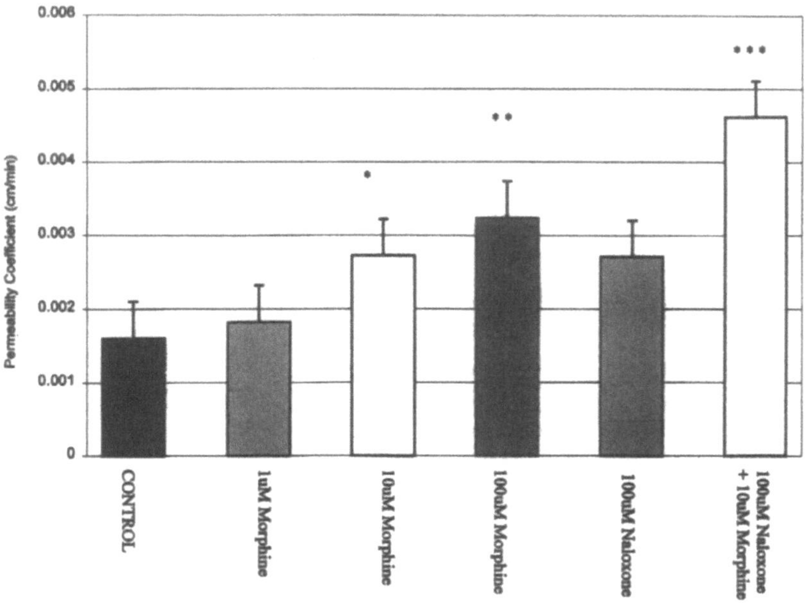

Figure 2. Morphine treatment increases inulin permeability across a rat BMVEC barrier ($p<0.05$).

As shown in Figure 3, in addition to an increase in the number of detached cells from the confluent endothelial cell monolayer, cell viability decreased with morphine treatment in a dose-dependent manner. The higher the concentration of morphine, the lower the percentage of viable cells in the endothelial cell monolayer.

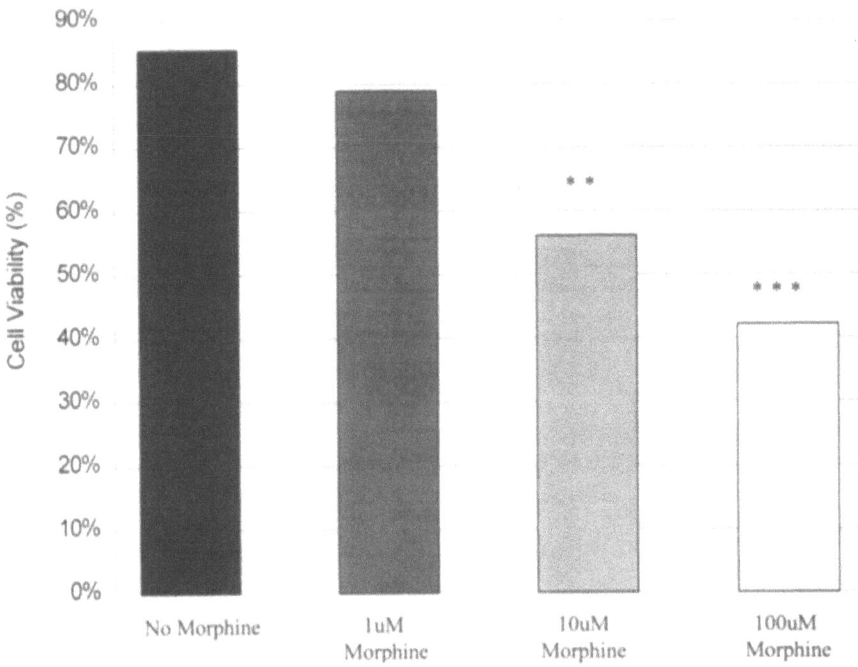

Figure 3. The effects of morphine on rat BMVEC viability after a 24-h treatment ($p<0.05$).

In a subsequent study, we investigated the effects of LPS on the viability of rat BMVEC, and of morphine pre-treatment on LPS' actions. Confluent BMVEC cells, pre-treated with and without 1 μM morphine for 24 hr, were then incubated for an additional 24 hr with fresh medium containing various concentrations of LPS. At the end of the LPS treatment, the monolayer cells were trypsinized and counted as above.

The first three bars in Fig. 4 demonstrate that treatment with LPS decreased the percentage of viable cells in the endothelial cell monolayer. Bar 4 shows the percentage of viable cells after treatment with 1 μM morphine alone. Not only did the cells which were pre-treated with 1 μM morphine for 24 h, followed by treatment with LPS at various concentrations (Fig. 4, bars 5 and 6), show an increase in the number of detached cells, but there was also a further decrease in the percentage of viable cells in the monolayer.

These data support morphine's potentiation of LPS' adverse effects on the viability of endothelial cells, which could lead to increased penetration of small pathogenic viruses, such as HIV-1, through a host's microvascular endothelial cell barriers.

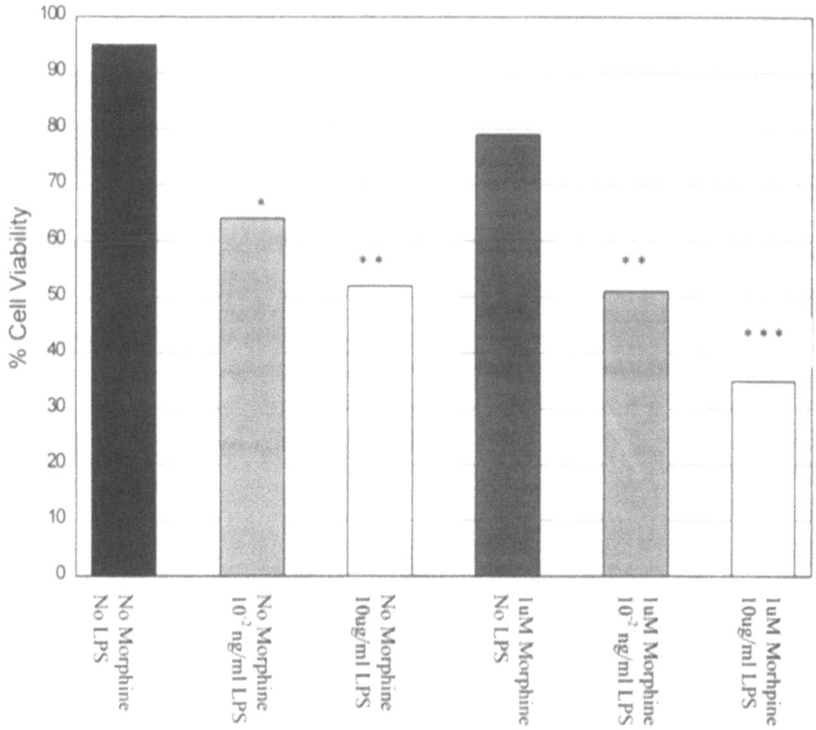

Figure 4. Morphine pre-treatment potentiates the effects of LPS on BMVEC viability ($p<0.05$).

MORPHINE POTENTIATES LPS-INDUCED CYTOKINE PRODUCTION FROM HL-60 PROMYELOCYTIC LEUKEMIA CELLS

In a parallel study, we examined morphine's ability to potentiate LPS' effects on immune cells using human promyelocytic leukemia HL-60 cells. HL-60 cells can be selectively induced to differentiate into monocyte/macrophage cells by 1α, 25-dihydroxy-vitamin D3, or to granulocytes by retinoic acid or DMSO. We have shown that HL-60 cells express mu opioid receptors using RT-PCR and primers derived from the sequence for the brain mu opioid receptor (Vidal et al., 1998). Treatment with either retinoic acid or 1α, 25-dihydroxy-vitamin D3 does not appear to change the levels of mu opiod receptor mRNA. However, treatment with DMSO decreases mu opioid receptor expression, as determined by semi-quantitative RT-PCR (Cheng et al., 1999). Although classical binding sites for mu opioid receptors were not demonstrated on either untreated or 1α, 25-

dihydroxy-vitamin D3-treated HL-60 cells using receptor binding assays (data not shown), morphine has been shown to potentiate LPS-induced secretion of IL-1β from 1α, 25-dihydroxy-vitamin D3-treated HL-60 cells.

HL-60 cells were treated with 10^{-7} M 1α, 25-dihydroxy-vitamin D3 for 5 days. At the end of treatment, the cells were harvested, re-suspended in fresh culture medium, and placed in 24-well plates. To each well, one of the following was added: (1) naloxone (100 μM) and phosphate buffered saline (PBS); (2) naloxone (100 μM) and morphine (10 μM); (3) PBS and PBS, or (4) PBS and morphine (10 μM). Fifteen minutes after treatment, LPS was added to each well to a final concentration of 10 μg/mL. The supernatant was collected at 5 min, 4.5 hr, and 24 hr after the addition of LPS for the ELISA assays of IL-1β.

Table 1 shows the effect of a 24-hr treatment with LPS on the secretion of IL-1β from 1α, 25-dihydroxy-vitamin D3-treated HL-60 cells pre-treated with PBS, morphine, naloxone, or naloxone and morphine. LPS-induced IL-1β secretion is not detectable in HL-60 cells without treatment with 1α, 25-dihydroxy-vitamin D3 (0.04 pg/mL). LPS treatment of 1α, 25-dihydroxy-vitamin D3-treated HL-60 cells, pre-treated with PBS only, results in secretion of 8.76 pg/mL of IL-1β. IL-1β secretion, induced by a 24-hr LPS treatment, is potentiated by pre-treatment of the cells with morphine (11.81 pg/mL). Pre-treatment with naloxone alone does not affect LPS induction of IL-1β secretion (6.75 pg/mL); however, co-treatment with naloxone and morphine appears to reverse the potentiation effect of morphine to 8.95 pg/mL.

Although classical binding sites for mu opioid receptors on HL-60 cells have not been demonstrated, the reversibility by naloxone of morphine's potentiation of LPS induction of IL-1β secretion from 1α, 25-dihydroxy-vitamin D3-treated HL-60 cells indicates the presence of functional classical mu opioid receptors on these cells.

Table 1. Morphine potentiates LPS-induced secretion of hIL-1β from 1α, 25-dihydroxy-vitamin D3-treated HL-60 cells.

Treatments	Concentration of hIL-1β, pg/mL
PBS + PBS	8.76 +/- 0.38
PBS + Morphine	11.81 +/- 0.01 **
Naloxone + PBS	6.75 +/- 0.48
Naloxone + Morphine	8.59 +/- 0.05

**p<0.001

SUMMARY

In summary, our current studies show that treatment with a bacterial endotoxin, lipopolysaccharide (LPS), induces the expression of mu opioid receptors in the rat mesentery. This induction may be mediated through IL-1's actions on mu opioid receptors. Morphine stimulates the expression of adhesion molecules in human brain microvascular endothelial cells (HBMEC) isolated from pathological tissues. Under pathological conditions, mu opioid receptor-dependent pathways may be modulated through the induction of mu opioid receptors, especially in endothelial cells. Treatment with morphine increases [^{14}C]-inulin permeability of an *in vitro* microvascular endothelial cell barrier, and decreases endothelial cell viability. Morphine pre-treatment potentiates the effects of LPS on endothelial cell viability, and on LPS induction of IL-1β secretion from 1α, 25-dihydroxy-vitamin D_3-treated HL-60 human leukemia cells.

Previously, it was suggested that an opioid-dependent pathway may be involved in the recovery from endotoxin shock (D'Amato and Holaday, 1984). Induction of mu opioid receptors by treatment with high doses of endotoxin suggests that mu opioid receptor-dependent pathways may be involved in mediating the response to endotoxins. Taken together, these data provide valid evidence for an association between endotoxins and opioid actions. These studies suggest that opioid-dependent pathways in disease or in endotoxin exposure may be modified by cytokine-induced expression of opioid receptors in endothelial cells. In a pathological condition, an alteration of the opioid-dependent pathway may be expected. When morphine is used for its therapeutic values, it may, indeed, potentiate LPS' effects in an adverse manner. From a clinical perspective, these data indicate that morphine and an endotoxin, such as LPS, may interact in a positive feedback type of reaction, and thereby modulate the body's immune responses with unexpected and detrimental results.

ACKNOWLEDGMENTS

This work is supported, in part, by grants DA 07058 (to SLC), DA 10442 (to MF), and HL 63065 (to MF).

NOTES

Requests may be forwarded to Sulie L. Chang. Fax: (1) (973) 275-2489; E-mail: changsul@shu.edu.

REFERENCES

Cheng, H., Rahimi, H., and Chang, S.L., 1999, Expression of mu opioid receptors in HL-60 promyelocytic leukemia cells, *30th International Narcotics Research Conference Abstract.*

D'Amato, R., and Holaday, J.W., 1984, Multiple opioid receptors in endotoxic shock: evidence for delta involvement and mu-delta interactions in vivo, *Proc. Natl. Acad. Sci.*, 81:2898-2901.

Fiala, M., Gan, M., Zhang, X-H., House, S.D., Newton, T., Graves, M.C., Shapshak, P., Stins, M., Kinm, K.S., Witte, M., and Chang, S.L., 1998, Cocaine enhances monocyte migration across the blood-brain barrier: cocaine's connection to AIDS dementia and vasculitis, In: *Drugs of Abuse, Immunomodulation, and AIDS*, H. Friedman, J. Madden, and T. Klein, eds., Plenum Publishing Corp., New York and London, pp. 199-206.

House, S.D., and Lipowsky, H.H., 1987, Leukocyte-endothelial adhesion: Microhemodynamics in mesentery of the cat, *Microvasc. Res.*, 34:363-379.

Mansour, A., Khachaturian, H., Lewis, M.E., Akil, H., and Watson, S.J., 1987, Autoradiographic differentiation of mu, delta, and kappa opioid receptors in the rat forebrain and midbrain, *J. Neurosci.*, 7:2445-2464.
Min, B.H., Augustin, L.B., Felsheim, R.F., Fuchs, J.A., and Loh, H.H., 1994, Genomic structure and analysis of promoter sequence of a mouse μ opioid receptor gene, *Proc. Natl. Acad.. Sci.*, 91: 9081-9085.
Nudel, U., Zakut, R., Shani, M., Neuman, S., Levy, Z., and Yaffe, D., 1983, The nucleotide sequence of the rat cytoplasmic β-actin gene, *Nucleic Acids Res.*, 11:1759-1771.
Oishi, R., Baba, M., Nishibori, M., Itoh, Y., and Saeki, K., 1989, Involvement of central histaminergic and cholinergic systems in the morphine-induced increase in blood-brain barrier permeability to sodium fluorescein in mice, *Naunyn Schmiedebergs Arch Pharmacol*, 339:159-165.
O'Neill, L.A.J., 1995, Towards an understanding of the signal transduction pathways for interleukin 1, *Biochimica et Biophysica Acta*, 1266:31-44.
Stins, M.F., Gilles, F., Kim, K-S., 1997, Selective expression of adhesion molecules on human brain microvascular endothelial cells, *J. Neuroimmunol.* 76:81-90.
Takata, K., Kasahara, T., Kasahara, N., Ezaki, O., and Hirano, H., 1990, Erythrocyte/HepG2-type glucose transport is concentrated in cells of blood-tissue barriers, *Biochem. Biophy. Res. Commun.* 173:67-73.
Vidal, E.L., Patel, N.A., Wu, G., Fiala, M., and Chang, S.L., 1998, Interleukin-1 induces the expression of opioid receptors in endothelial cells, *Immunopharmacol.*, 38:261-266.
Wang, J-B., Johnson, P.S., Persico, A.M., Hawkins, A.L., Griffin, C.A., and Uhl, G.R., 1994, Human μ opiate receptor: cDNA and genomic clones, pharmacologic characterization and chromosomal assignment, *FEBS Lett.* 338:217-222.
Zhang, L., Taub, D., Looney, D., Chang, S.L., Way, D., Witte, M., Graves, M.C., and Fiala, M., 1998, Cocaine opens the blood-brain barrier to HIV-1 invasion, *J. Neurovirol.*, 4:619-626.

PHARMACONEUROIMMUNOLOGY IN THE INTESTINAL TRACT: OPIOID AND CANNABINOID RECEPTORS, ENTERIC NEURONS AND MUCOSAL DEFENSE

David R. Brown, Benedict T. Green, Anjali Kulkarni-Narla, Sutthasinee Poonyachoti and DeWayne Townsend, IV

Department of Veterinary PathoBiology, College of Veterinary Medicine
University of Minnesota Academic Health Center
Minneapolis/St. Paul, Minnesota 55108

INTRODUCTION

The mucous membranes lining the gut, bronchi, endometrium and other organ systems represent the largest surface area of the body that is in constant contact with the external environment. They share common structural and functional features which allow them to selectively absorb beneficial substances and vigorously exclude noxious substances, such as microbial pathogens, which might produce systemic infection. Mucosa-associated lymphoid tissue represents a unique immunologic compartment with the capability of generating secretory immunoglobulin A (sIgA) in defense of mucosal surfaces.[1] Specific host defense in the intestine is mediated by the gut-associated lymphoid tissue (GALT) which comprises the largest mass of immune cells in the body.[2] In addition to immunocytes, over 100 million neurons are contained within the wall of the intestine which extensively innervate the mucosa and GALT. The enteric nervous system (ENS) regulates intestinal blood flow, motility, and mucosal ion transport.[3] It also mediates some aspects of intestinal host defense, such as epithelial barrier function and inflammatory reactions, but its role in GALT function is not clearly understood.

The intestine has traditionally been an important model system for the pharmacological characterization of opioid receptors, the study of tolerance and dependence phenomena, and the discovery of endogenous opioid peptides. The antipropulsive actions of opioids in the intestinal tract are well known and contribute to both the therapeutic antidiarrheal and deleterious constipating effects of this drug class. In addition to their antimotility effects, increasing evidence gathered over the past three decades indicates that opioids also modify mucosal function.[4] Cannabinoids, like the opioids, alter intestinal motility, but their effects on mucosal function are not defined.[5] With its complex and unique nervous and immune systems, the intestine is an excellent system for investigations of the effects of opioids, cannabinoids and other drugs of abuse on the neuroimmune axis. In this chapter, we will summarize our recent findings concerning the distribution and functions of opioid and cannabinoid systems in the gut as they relate to mucosal host defense.

OPIOID AND CANNABINOID RECEPTORS AND THEIR COGNATE LIGANDS ARE EXPRESSED IN ENTERIC NEURONS

The ganglionated plexuses of the intestine form a dense, interconnected neural network which can be organized into reflex arcs containing intrinsic primary afferent neurons, interneurons, and motor or secretomotor neurons.[3] Within a 10 mm ileal segment from the guinea pig, for example, there are some 6500 intrinsic primary afferent neurons, 4200 ascending or descending interneurons and 7000 motor neurons.[6] The myenteric plexus (MP) is located between the longitudinal and circular smooth muscle coats and coordinates intestinal propulsion and segmentation. The intestinal tract of large mammals, such as the pig, also contains an inner submucosal plexus (ISP), which lies beneath the villous epithelium, and an outer submucosal plexus (OSP), which lies above the circular muscle. The OSP is thought to relay information between the myenteric plexus and the ISP. Lying closest to the lumen, the ISP has many sensory and secretomotor fibers; it conveys sensory information from the mucosa and modulates epithelial function through secretomotor neurons. Interneurons in the OSP project to myenteric and ISP neurons to coordinate the secretory and mechanical functions of the intestine.[7]

The ENS contains neural pathways whose activity is physiologically modulated by endogenous opioids and presumably, endocannabinoids. Chemical and immunohistochemical analyses have indicated that enkephalins and dynorphin are expressed in enteric neurons.[8] Although they can interact with intestinal μ-opioid receptors, endomorphins have not yet been detected in the gut.[9] The endocannabinoid 2-arachidonylglycerol and fatty acid aminohydrolase, an enzyme that can degrade arachidonylethanolamide (anandamide), have been detected in the intestinal wall as well.[10, 11] Through interactions with their cognate receptors on enteric neurons, both classes of substances have been found to inhibit neuronal activity and decrease transmitter release from enteric neurons. Although their endogenous ligands are chemically dissimilar, opioid and cannabinoid receptors share some common features: they represent a heterogeneous class of proteins, are coupled to similar transducer and effector molecules, and are expressed in neurons and immune cells. It is important to note that, as in the central nervous system, the differential expression of opioid receptor types within subregions of the ENS manifests considerable species variability.

We have employed the porcine small intestine in our laboratory as an experimental model for the human bowel. The porcine ENS differs from that in the commonly-studied guinea pig ileum or other rodent intestinal preparations in its greater neural complexity and different neurochemical coding. Because of its mass, the porcine intestine has been the original source for many gut peptide hormones and neurotransmitters; it also expresses many drug receptor mRNAs and proteins which can be isolated in greater abundance than from rodents or other small animals. The porcine gut has been a valuable model for characterizing drug actions within functionally- and neurochemically-distinct gut segments and ENS subregions. In the context of infectious disease, the porcine and human intestines are susceptible to infection by a similar constellation of enteropathogens.

Opioid receptors and the CB_1-cannabinoid receptor within the ENS of porcine ileum have been characterized through the use of molecular biological, radioligand binding, immunocytochemical and pharmacological techniques. Unlike the myenteric plexus of the guinea pig ileum for example, the porcine ileal MP does not express μ-opioid receptors (Table 1). Furthermore, δ-opioid receptors appear to be the predominant opioid receptor type present in neurons and nerve fibers throughout the porcine ENS.[12, 13] Preliminary findings in our laboratory indicate that CB_1-cannabinoid receptors display a similarly wide pattern of ENS distribution (Table 1). Both receptors are linked to an inhibition of neurogenic, atropine-sensitive contractions in circular muscle strips from porcine ileum, but only δ-opioid receptors inhibit active anion secretion by the intestinal mucosa that is evoked by electrical stimulation of submucosal secretomotor nerves (Table 1) or serosal administration of inflammatory mediators (Table 2). These functional differences may related to the differential expression of the receptors on neurochemically-distinct subpopulations of cholinergic neurons (i.e. neurons and fibers manifesting immunoreactivity to the acetylcholine-synthesizing enzyme, choline acetyltransferase or ChAT) in the porcine ileum. Our immunohistochemical studies indicate that both δ- and κ-opioid receptors appear to be expressed on ChAT- and substance P-positive neurons, whereas CB_1-cannabinoid receptors are expressed on ChAT-immunoreactive neurons that only infrequently display substance P immunoreactivity.

Table 1. Immunohistochemical distribution of receptor-like immunoreactivities and their relationship to receptor-linked drug effects in the porcine small intestine[1]

Receptor type	Myenteric ganglia	Smooth muscle nerve fibers	Inhibit neurogenic circular muscle contraction?	Submucosal ganglia	Mucosal nerve fibers	Inhibit neurally-mediated mucosal secretion?
CB_1-Cannabinoid	Yes	Yes	Yes	Yes	Yes	No
delta-Opioid	Yes	Yes	Yes	Yes	Yes	Yes
kappa-Opioid	Yes	?	Yes	No	No	No
mu-Opioid[2]	?	?	No	?	?	Yes

[1]All antisera used in immunohistochemical studies were raised against epitopes in the extracellular N-termini of these receptors.

[2]The N-terminus of the porcine μ-opioid receptor has approximately 70% sequence identity to its human counterpart. The anti-μ-opioid receptor antisera that are currently available are directed against epitopes in the N-terminus of the human receptor and do not appear to recognize the porcine receptor.

? - unknown.

EFFECTS OF OPIOIDS AND CANNABINOIDS ON MUCOSAL SECRETORY DEFENSE

The columnar epithelial cells of the intestinal mucosa actively absorb and secrete ions, nutrients and water. The active secretion of water and ions by these cells [1] acts to dilute and purge microorganisms or toxins in the bowel, [2] promotes the transfer of secretory IgA, antimicrobial defensin peptides and mucin into intestinal mucus and the gut lumen and, [3] by affecting intraluminal pH, may alter the growth characteristics of enteric microflora. Mucosal secretion is modulated by many enteric neurotransmitters, as well as inflammatory mediators released by mucosal mast cells that may affect transport indirectly through their ability to stimulate enteric neurons.[14] Opioids inhibit mucosal fluid and electrolyte secretion in the small intestines of several mammalian species both *in vitro* and *in vivo*.[15] We have examined the actions of opioids in isolated sheets of mucosa with attached submucosa from porcine ileum. This preparation contains submucosal ganglia and nerve fibers, but is devoid of smooth muscle coats and the myenteric plexus. After mucosal sheets are mounted in Ussing-type flux chambers under conditions in which the electrochemical driving forces for passive ion movements between epithelial cells are eliminated, short-circuit current (Isc) can be monitored continuously. The Isc is an electrical measure of active, transcellular ion transport across each sheet. The apparatus permits the addition of drugs, pathogens or other agents to either the luminal or serosal aspect of each mucosal sheet and changes in Isc can be measured. Mucosal sheets can also be subjected to electrical transmural stimulation which transiently evokes active anion secretion through the depolarization of secretomotor neurons in the submucosa. We have found that stable enkephalin analogs, such as the selective *delta*-opioid agonist [D-Pen^2, D-Pen^5]enkephalin (DPDPE) rapidly decrease baseline- and electrically-evoked Isc in mucosal sheets in a naloxone-reversible manner. Measurements of the transepithelial isotopic fluxes of the major extracellular ions sodium and chloride indicate that these opioid effects are due respectively to an enhancement of salt absorption and a reduction in the net secretory flux of chloride ions. The actions of opioids are inhibited by neuronal conduction blockers such as saxitoxin (STX) or tetrodotoxin, a result indicating that these substances act *via* submucosal neurons.[16]

Table 2. Mechanisms underlying short-circuit current elevations produced by inflammatory mediators across porcine ileal mucosa-submucosa sheets and the effects of δ-opioid and cannabinoid receptor agonists

Inflammatory Mediator	Inhibited by 0.1 μM STX?	Inhibited by 0.1 μM atropine?	Inhibited by 0.1 μM DPDPE?	Inhibited by 0.1 μM HU-210?
Compound 48/80[1]	Yes	No	Yes	No
Histamine	Yes	No	Yes[2]	No
Kallidin	Yes	No	Yes	No
Prostaglandin E_2	No	N.T.	No	N.T.
Serotonin	Yes	No	Yes	No
Tryptase (trypsin)	Yes	No	Yes	No

[1]Prosecretory effect of the mast cell degranulator 48/80 was also inhibited by the H_1-histamine antagonist, diphenhydramine.

[2]Inhibitory action of DPDPE was not reversed by 0.1 μM naltrindole.

N.T., not tested.

Inflammation is an initial defensive reaction against intestinal pathogen invasion. For example, mice with a diminished capacity to mount acute inflammatory reactions are more susceptible to *Salmonella* infections.[17] We have discovered that DPDPE dramatically inhibits mucosal Isc responses to several different classes of inflammatory mediators, and the actions of the δ-opioid agonist could be prevented by the selective δ-opioid antagonist naltrindole.[18,19] These mediators include the biogenic amines histamine and serotonin; mast cell tryptase (using trypsin as an analogous protease to activate type 2 proteinase-activated receptors); the mast cell degranulator, compound 48/80; and the tissue kinin, kallidin (Table 2). Each of these substances produce elevations in Isc that are attributable to active anion secretion and the effects of each mediator are prevented by STX and the cyclooxygenase inhibitor, indomethacin. In mucosal sheets from milk-fed pigs, elevations in Isc associated with immediate hypersensitivity reactions to the milk protein, β-lactoglobulin, are also attenuated by DPDPE or STX (Poonyachoti and Brown, unpublished observations). Based on these results, it appears that their secretory effects are attributable to interactions with submucosal secretomotor neurons through a prostanoid-dependent mechanism. We hypothesize that histamine, tryptase, kallidin and serotonin act through a common neural circuit which is modulated by δ-opioid receptors. Although many enteric neurons contain acetylcholine, it is interesting that the secretory actions of these inflammatory mediators are not altered by atropine, which suggests that the submucosal neural circuit(s) mediating them do not contain a muscarinic cholinergic synapse (Table 2). It is important to note that some mediators of inflammation, such as prostaglandin E_2, produce large increases in Isc in the porcine ileal mucosa through direct actions on intestinal epithelial cells. Prostaglandin E_2-induced secretion is not altered by either STX or DPDPE (Table 2).

There are no published reports of cannabinoid actions on mucosal host defense. We have conducted studies in our laboratory with potent, non-selective cannabinoid agonists such as HU-210 to address this paucity of information. The results of our studies suggest that HU-210 does not affect electrogenic ion transport evoked by inflammatory mediators (Table 2). This difference in the actions of opioids and cannabinoids is probably due to the expression of their receptors on neurochemically-distinct neural pathways (see above). Alternatively, cannabinoid actions on the mucosa may be mediated by myenteric neurons which are removed in the experimental preparation or they may be expressed in mucosal preparations from other regions of the gut or in other species. Nevertheless, these results highlight an important difference in the functional roles of δ-opioid receptors and cannabinoid receptors in the mucosa of the small intestine.

NEUROREGULATION OF ENTEROPATHOGEN INVASION

The intestinal epithelium constitutes a cellular barrier to subepithelial invasion by luminal pathogens. Intestinal epithelial cells adhere to each other through a series of junctional complexes. The apical tight junctions (or zonulae occludens), which circumferentially belt these cells, have been the focus of intense study because they are disrupted by stress, soluble bacterial products, cytokines and other immune mediators, and various neurotransmitters.[14] The transmigration of subepithelial neutrophils into the intestinal lumen, a prominent feature of several disease states, occurs across tight junctions.[20] Opium preparations have been employed in the past to render experimental animals susceptible to enteric bacterial infections.[21] Subcutaneous infusions or pellet implants of morphine in mice and rats enhance the growth of Gram-negative enteric bacteria in the intestinal lumen and promote translocation of these bacteria to mesenteric lymph nodes and distant sites.[22, 23] These opiate actions have been attributed to their ability to impair gut motility and thereby increase the contact time of luminal bacteria with the mucosa. A role for the intestinal mucosa itself has not been examined.

The organized GALT functions in antigen presentation to lymphocytes and includes the Peyer's patches (PP) of the small intestine which, in humans, pigs and other large animals, occur as single lymphoid follicles in jejunum and aggregated follicles in ileum. The jejunal PP disseminate immunological information from the gut lumen to mucosal surfaces throughout the body. The specialized epithelium covering the dome of these follicles consists of cuboidal epithelial cells, referred to as microfold or "M" cells. They endocytose and process luminal antigens for subsequent delivery to underlying antigen-presenting cells.[24] M cells associated with jejunal PP are most active in taking up particles from the intestinal lumen.[25]

In addition to their role in long-term mucosal defense, M cells serve as portals for rapid intestinal invasion by a number of microorganisms, including Gram-negative bacteria such as *Shigella*, *Yersinia* and *Salmonella*, and viruses such as human immunodeficiency virus (HIV), reoviruses and poliovirus. Through this route, enteropathogens can invade the lamina propria and infect epithelial cells and mucosal macrophages.[26] Although the intestine has been viewed as a passive participant in the infection process, this may not correct. We and others have reported that enteric neurons appear to innervate PP in the porcine jejunum.[27-29] Kulkarni-Narla *et al.*[27] recently detected cholinergic and adrenergic neurons in enteric ganglia lying between jejunal PP follicles. Cholinergic fibers could also be observed on the border of PP follicles, as well as within PP (Table 3). Of particular interest, nerve fibers expressing immunoreactivity to vasoactive intestinal peptide or substance P were observed lying under the PP dome (Table 3). This finding implies that the functions of the PP, including antigen or pathogen transport, may be under ENS control. These neuropeptides have also been shown to alter gut lymphocyte function, proliferation and differentiation, and may affect leukocyte trafficking in the intestine.[30]

Table 3. Immunohistochemical localization of neurotransmitter/neuromodulator substances in porcine jejunal Peyer's patches

Transmitter substance	Interfollicular ganglia	Perifollicular nerve fibers	Intrafollicular nerve fibers
Acetylcholine (based on ChAT immunoreactivity)	Yes	Yes	Yes
Catecholamine (based on TH/VMAT-2 immunoreactivity)	Yes	No	No
Substance P	Yes	Yes	Yes
Vasoactive intestinal peptide	Yes	Yes	Yes

Abbreviations: ChAT, choline acetyltransferase; TH, tyrosine hydroxylase; VMAT-2, type 2 vesicular monoamine transporter. See Kulkarni-Narla et al.[27] for more details.

Most studies of pathogen invasion of PP cells have been performed in cultured cells. Many of these cell lines do not completely reproduce all M cell characteristics, and they do not possess an attached nervous system for investigating functional interactions between enteric neurons and M cells. We tested the hypothesis that bacterial uptake by PP is under ENS control. Single PP were isolated from the porcine jejunum, mounted in Ussing chambers (2 cm^2 flux area) and the effects of the neural conduction blocker STX were examined on PP invasion by *Salmonella cholerasuis* SC54. These bacteria represent a vaccine strain of *Salmonella* which invades PP and causes an extensive infection and inflammation of PP and mesenteric lymph nodes.[31] SC54 was added to the luminal bathing medium and was exposed to PP sheets for 60 min. Our bacterial quantification protocol was based on a standard gentamicin resistance invasion assay.[32] After tissues were removed from chambers and treated with gentamicin, serial dilutions of PP homogenates were spread-plated onto MacConkey's, LB, and XLD agars. LB agar was chosen because it is a non-selective agar that supports the growth of Gram-negative and -positive bacteria to provide an estimation of total bacteria present in each PP. MacConkey agar was used for the selective isolation of Gram-negative enteropathogens and for differentiating *Salmonella* from coliforms based on lactose fermentation. XLD agar was used to differentiate *Salmonella* from nonpathogenic bacteria. *S. cholerasuis* SC54 is identified by xylose fermentation, lysine decarboxylation and hydrogen sulfide production.

This bacterial strain effectively invaded the isolated porcine jejunal PP. The majority of bacteria isolated from tissue homogenates and cultured on MacConkey's and XLD agars had the colony characteristics of *S. cholerasuis.* STX pretreatment produced a 3.5-fold increase in the number of colony-forming units (CFU) per gram of tissue. This result suggests that bacterial translocation into jejunal PP is under inhibitory neural control by the host. Further investigations of this phenomenon may extend our understanding of the neural underpinnings of intestinal infection; further delineate the impact of opioids, cannabinoids and other drugs of abuse on intestinal disease resistance; and aid in identifying potential "pharmacological adjuvants" which would increase the efficacy of oral vaccines through their ability to reduce enteric neural activity and enhance antigen sampling by intestinal PP.

CONCLUDING REMARKS

In the intestinal tract, opiate abuse is directly associated with some severe complications, including toxic megacolon, necrotizing enterides, and necrotizing angiitis.[33] In gastroenterological practice, the use of opioids as antidiarrheal agents is contraindicated in inflammatory colitides (for example, *Shigella* dysentery, amebiasis, pseudomembranous colitis etc.) because they are associated with enhanced mucosal invasion of enteroinvasive pathogens.[34] Because opioid actions on the intestine are not widely viewed in the context of host defense processes in standard pharmacological or medical literature, it is perhaps not surprising there have been no clinical reports in opiate abusers which specifically correlate the frequency of opioid use with enteric infections. In addition, there is a paucity of information on the actions of opioids in GALT. Marijuana is the most commonly-abused illegal drug in the United States, but interest in its legitimate medicinal use and the development of nonpsychoactive, synthetic cannabinoids for use in a variety of medical disorders is rapidly growing. Although natural and synthetic cannabinoids can decrease disease resistance, there is no information as to whether they can act as specific co-factors for intestinal infection.[35] Because they may have either adverse or beneficial effects on the intestinal tract, pharmacological, microbiological, and immunological research on the gastrointestinal actions of cannabinoids is urgently needed.

Impairments in host defense processes of the intestine due to abuse of opioids, cannabinoids or other illicit, neuroactive drugs could set the stage for the initiation and systemic spread of bacterial and viral infections. Similarly, the legitimate administration of opioid (and potentially, cannabinoid) analgesic drugs to critically ill or immunocompromised patients may exacerbate preexisting intestinal infections or enhance seeding of enteric microorganisms to extra-intestinal sites in the body. As a major neuroimmune locus, the intestinal mucosa is an important and clinically-relevant system for investigating the actions of neuroactive drugs on specific and non-specific host defense processes and for the identification of novel therapeutic strategies to enhance mucosal immunity and combat microbial infections.

ACKNOWLEDGEMENTS

The studies described in this article were funded in part by NIH grants R01 DA-10200, T32 DA-07234, and T32 DA-07239.

REFERENCES

1. J.P. Kraehenbuhl and MR Neutra, Molecular and cellular basis of immune protection of mucosal surfaces, *Physiol. Rev.* 72:853 (1992).
2. A.M. Mowat, and J.L.Viney, The anatomical basis of intestinal immunity, *Immunol. Rev.* 156:145 (1997).
3. R.K. Goyal and I. Hirano, The enteric nervous system, *New Engl. J. Med.* 334:1106 (1996).
4. D.R. Brown, Antidiarrheal drugs: pharmacologic control of intestinal hypersecretion, in: *Principles of Pharmacology: Basic Concepts and Clinical Applications*, P.L. Munson, ed., Chapman & Hall, London, p. 1083 (1994).
5. R.G. Pertwee, Pharmacology of cannabinoid CB1 and CB2 receptors, *Pharmacol. Ther.* 74:129 (1997).
6. W.A. Kunze and J.B. Furness, The enteric nervous system and regulation of intestinal motility, *Ann. Rev. Physiol.*, 61:117 (1999).
7. J.P. Timmermans, D. Adriaensen, W. Cornelissen and D.W. Scheuermann, Structural organization and neuropeptide distribution in the mammalian enteric nervous system, with special attention to those components involved in mucosal reflexes, *Comp. Biochem. Physiol.* 118A:331 (1997).
8. W. Kromer, Endogenous opioids, the enteric nervous system and gut motility, *Dig. Dis.*, 8:361 (1990).
9. K. McConalogue, E.F. Grady, J. Minnis, B. Balestra, M. Tonini, N.C. Brecha, N.W. Bunnett and C. Sternini, Activation and internalization of the mu-opioid receptor by the newly discovered endogenous agonists, endomorphin-1 and endomorphin-2, *Neuroscience* 90:1051 (1999).
10. R. Mechoulam, S. Ben-Shabat, L. Hanus, M. Ligumsky, N.E. Kaminski, A.R. Schatz, A. Gopher, S. Almog, B.R. Martin, D.R. Compton, R.G. Pertwee, G. Griffin, M. Bayewitch, J. Barg and Z. Vogel, Identification of an endogenous 2-monoglyceride, present in canine gut, that binds to a cannabinoid receptor, *Biochem. Pharmacol.* 50:83 (1995).
11. N. Ueda, S.K. Goparaju, K. Katayama, Y. Kurahashi, H. Suzukiand S. Yamamoto, A hydrolase enzyme inactivating endogenous ligands for cannabinoid receptors, *J. Med Invest.*, 45:27 (1998).
12. D.R. Brown, S. Poonyachoti, T.R. Kowalski, M.A. Osinski, M.S. Pampusch, R.P.Elde and M.P. Murtaugh, *delta*-Opioid receptor mRNA expression and immunohistochemical localization in porcine ileum, *Digestive Dis. Sci.* 43: 1402 (1998).
13. D. Townsend, IV, K. Ham-Lammé and D.R. Brown, Opioid and $alpha_2$-adrenergic receptors: ileal binding characteristics and distribution in porcine intestinal tract, *FASEB J.*, 13:A732 (1999).
14. M.H. Perdue and D.M. McKay, Integrative immunophysiology in the intestinal mucosa, *Am. J. Physiol.*, 267:G151 (1994).
15. A. De Luca and I.M. Coupar, Insights into opioid action in the intestinal tract, *Pharmacol. Ther.*, 69:103 (1996).
16. F.L. Quito and D.R. Brown, Neurohormonal regulation of ion transport in the porcine distal jejunum. enhancement of sodium and chloride absorption by submucosal opiate receptors, *J. Pharmacol. Exp. Ther.*, 256:833 (1991).
17. L.M. Araujo, O.G. Ribeiro, M. Siqueira, M. De Franco, N. Starobinas, S. Massa, W.H. Cabrera, D. Mouton, M. Seman and O.M. Ibanez, Innate resistance to infection by intracellular bacterial pathogens differs in mice selected for maximal or minimal acute inflammatory response, *Eur J Immunol.* 28:2913 (1998).

18. S. Poonyachoti, M.S. Pampusch and D.R. Brown, Histamine-induced ion transport in ileal mucosa: neuroimmunomodulation by selective opioids, *FASEB J.*, 12: A764 (1998).
19. B.T. Green, N.W. Bunnett and D.R. Brown, Type 2 protease-activated receptor (PAR-2) in porcine ileal mucosa: neuroregulation of active ion transport and modulation by δ-opioid receptors, *FASEB J.*, 13:A733 (1999).
20. C.A. Parkos, S.P. Colgan and J.L. Madara, Interactions of neutrophils with epithelial cells: lessons from the intestine, *J. Am. Soc. Nephrol.* 5:138 (1994).
21. S.B. Formal, G.J. Dammin, E.H. LaBrec and H. Schneider, Experimental *Shigella* infections: characteristics of a fatal infection produced in guinea pigs, *J. Bacteriol.* 76:604 (1958).
22. N.S.F. Runkel, F.G. Moody, G.S. Smith, L.F. Rodriguez, Y. Chen, M.T. Larocco and T.A. Miller, Alterations in rat intestinal transit by morphine promote bacterial translocation, *Dig. Dis. Sci.*, 38:1530 (1993).
23. M.E. Hilburger, M.W. Adler, A.L. Truant, J.J. Meissler, Jr., V. Satishchandran, T.J. Rogers, and T.K. Eisenstein, Morphine induces sepsis in mice, *J. Infect. Dis.* 176:183 (1997).
24. M.R. Neutra, Role of M cells in transepithelial transport of antigens and pathogens to the mucosal immune system, *Am. J. Physiol.* 274:G785 (1998).
25. E.M. Liebler, C. Lemke and J.F. Pohlenz, Ultrastructural study of the uptake of ferritin by M cells in the follicle-associated epithelium in the small and large intestines of pigs, *Am. J. Vet. Res.* 56:725 (1995).
26. P.J. Sansonetti and A. Phalipon, M cells as ports of entry for enteroinvasive pathogens: mechanisms of interaction and consequences for the disease process, *Sem. Immunol.* 11:193 (1999).
27. A. Kulkarni-Narla, A.J. Beitz and D.R. Brown, Catecholaminergic, cholinergic and peptidergic innervation of gut-associated lymphoid tissue in porcine jejunum and ileum, *Cell Tiss. Res.* 298: 275 (1999).
28. H.J. Krammer and W. Kühnel, Topography of the enteric nervous system in Peyer's patches of the porcine small intestine, *Cell Tiss. Res.* 272:267 (1993).
29. O.B. Balemba, M.L. Grøndahl, G.K. Mbassa, W.D. Semuguruka, A. Hay-Smith, E. Skadhauge and V. Dantzer, The organisation of the enteric nervous system in the submucous and mucous layers of the small intestine of the pig studied by VIP and neurofilament protein immunohistochemistry, *J. Anat.* 192::257 (1998).
30. F. Chen and M.S. O'Dorisio, Peptidergic regulation of mucosal immune function, in: *Gastrointestinal Regulatory Peptides*, D.R. Brown, ed., Springer-Verlag, Berlin, p. 363 (1993).
31. M.B. Roof and D.D. Doitchinoff, Safety, efficacy, and duration of immunity induced in swine by use of an avirulent live *Salmonella choleraesuis*-containing vaccine, *Am. J. Vet. Res.* 56:39 (1995).
32. E.A. Elsinghorst, Measurement of invasion by gentamicin resistance, *Meth. Enzymol.* 236: 405 (1994).
33. M.H. Roszler, K.A. McCarroll, and I.J. Jacobs, Radiologic study of intravenous drug abuse complications, in: *Infections in Intravenous Drug Abusers*, D.P. Levine and J.D. Sobel, eds., Oxford Univ. Press, New York, p. 96 (1991).
34. G. Kandel and M. Donowitz, Antidiarrhoeal drugs for the treatment of infectious enteritis, in: *Enteric Infection*, M.J.G. Farthing and G.T. Keusch, eds., Raven Press, New York, p. 453 (1988).
35. T.W. Klein, C. Newton and H. Friedman, Cannabinoid receptors and immunity, *Immunol. Today*, 19:373 (1998).

CANNABINOID-MEDIATED INHIBITION OF INDUCIBLE NITRIC OXIDE PRODUCTION BY RAT MICROGLIAL CELLS: EVIDENCE FOR CB_1 RECEPTOR PARTICIPATION

Guy A. Cabral, Katharine Nowell Harmon, and Steven J. Carlisle

Department of Microbiology and Immunology
Medical College of Virginia, Virginia Commonwealth University
Richmond, Virginia 23298-0678

ABSTRACT

Activated brain microglial cells release inflammatory mediators such as nitric oxide (NO) that may play important roles in central nervous system antibacterial, antiviral, and antitumor activities. However, excessive release of these factors has been postulated to elicit immune-mediated neurodegenerative inflammatory processes and to cause brain injury. Recent studies using the rat animal model indicate that select cannabinoids may modulate production of these inflammatory factors. Treatment of neonatal rat brain cortical microglial cells with the cannabinoid paired enantiomers CP55940 and CP56667 resulted in a stereoselective differential effect on inducible NO production. The analog CP55940 exerted a dose-dependent inhibition of interferon gamma (IFNγ)/bacterial lipopolysaccharide (LPS)-inducible NO production which was significantly greater than that exerted by CP56667. Pretreatment of microglial cells with the CB_1 cannabinoid receptor-selective antagonist SR141716A reversed this CP55940-mediated inhibition. MRT-PCR demonstrated the presence of CB_1 receptor mRNA within microglial cells consistent with the presence of CB_1 receptors. Collectively, these results indicate that the cannabinoid analog CP55940 selectively inhibits inducible NO production by microglial cells and that this inhibition is effected, at least in part, through the CB_1 receptor.

1. INTRODUCTION

Cannabinoids have been reported to alter the functional activities of macrophages and macrophage-like cells. Δ^9-tetrahydrocannabinol (THC), the major psychoactive component of marijuana, has been shown to inhibit the proliferation and phagocytic activity of murine P388D_1 cells (Tang et al., 1992) and to reduce the amount of tumor necrosis factor-alpha (TNF-α) and nitric oxide (NO) produced by murine macrophage-like RAW264.7 cells in response to bacterial lipopolysaccharide (LPS) (Fischer-Stenger et al., 1993; Jeon et al., 1996). In addition, THC

Neuroimmune Circuits, Drugs of Abuse, and Infectious Diseases
Edited by Herman Friedman *et al.*, Kluwer Academic/Plenum Publishers, 2001

and anandamide, an endogenous cannabinoid receptor ligand, have been reported to inhibit peritoneal macrophage-mediated killing of TNF-α sensitive murine L929 fibroblasts (Devane et al., 1992; Cabral et al., 1995). Taken together, these studies indicate that endogenous and exogenous cannabinoids have the ability to modulate various macrophage activities. However, the manner in which cannabinoids exert their modulatory effects on macrophages remains to be defined although both receptor and non-receptor mediated mechanisms have been proposed (Felder et al., 1992). To date, two cannabinoid receptors have been identified. The first of these is designated CB_1 and is localized primarily in neuronal tissue (Matsuda et al., 1990). The second has been designated CB_2 and is found in immune cells including blood monocytes and splenic macrophages (Munro et al., 1993; Galiègue et al., 1995). Both cannabinoid receptors belong to the seven transmembrane domain super-family based on extrapolation of their cDNA coding sequences and are coupled to pertussis toxin-sensitive G_i proteins which inhibit adenylate cyclase activity and decrease the amount of forskolin-stimulated cAMP (Howlett et al., 1986; Kaminski et al., 1994).

Brain microglial cells are a resident population of macrophages capable of migration, differentiation, and proliferation. In the adult brain, these cells are relatively quiescent and ramified in appearance whereas, during early development and after brain injury, they become activated and ameboidal, phagocytose tissue debris, and produce cytokines such as interleukin-1β, interleukin-6, and TNF-α. Activated microglial cells may play a key role in brain injury and in pathophysiological neurodegenerative disorders such as AIDS-encephalitis (Gehrmann and Kleihues, 1994). Indeed, it has been proposed that these disorders may be due to the action of monokines and NO released from activated macrophages and microglial cells rather than the result of direct cytopathology induced by the human immunodeficiency virus [HIV] (Merrill and Martinez-Maza, 1993).

The release of NO from microglial cells under basal conditions is negligible. However, upon stimulation with the Gram-negative bacterial endotoxin lipopolysaccharide (LPS), these cells release substantial amounts of the free radical NO (Zielasek et al., 1992; Chao et al., 1992). Reports that psychoactive cannabinoids such as THC inhibit NO production by murine macrophages and macrophage-like cells suggest that cannabinoids may exert similar effects on microglial cells. An *in vitro* model of macrophage multi-factor activation in response to interferon-gamma (IFNγ) plus LPS (i.e., IFNγ/LPS) was employed to assess the effect of cannabinoids on inducible NO production by neonatal rat microglial cells.

2. DIFFERENTIAL INHIBITION OF NITRIC OXIDE PRODUCTION BY CP55940 VERSUS ITS PAIRED ENANTIOMER CP56667

Enriched cultures (Fig. 1) of rat microglial cells ($5X10^5$ cells/ ml), isolated as previously described (Waksman et al., 1999), were preincubated for 8 h with either vehicle, the synthetic cannabinoid CP55940, or its paired enantiomer CP56667. The analog CP55940 exhibits high affinity binding for the CB_1 cannabinoid receptor (K_i = 0.9 nM) (Compton et al., 1993). In contrast, its paired enantiomer CP56667 exhibits an approximate 60-fold less binding affinity (K_i = 62 nM) for the CB_1 receptor. Thus, these paired enantiomers are stereoselective in terms of their specific interaction with the CB_1 receptor. Furthermore, the paired enantiomers have been shown to exert activities which are correlative to their structural relationships as relate to receptor binding (Compton et al., 1993). Following incubation with cannabinoid or vehicle, cultures were subjected to IFNγ/LPS activation. NO release (at 24 h post-activation) was represented as percent inhibition compared with the IFNγ (10U/ml)/ LPS (20μg/ml)-activated vehicle control (Fig. 2A). The cannabinoid analog CP55940 exerted a dose-dependent inhibition of NO release when compared with the activated vehicle control. Maximal inhibition of NO production, approximately 50% when compared with activated vehicle-treated cells, was

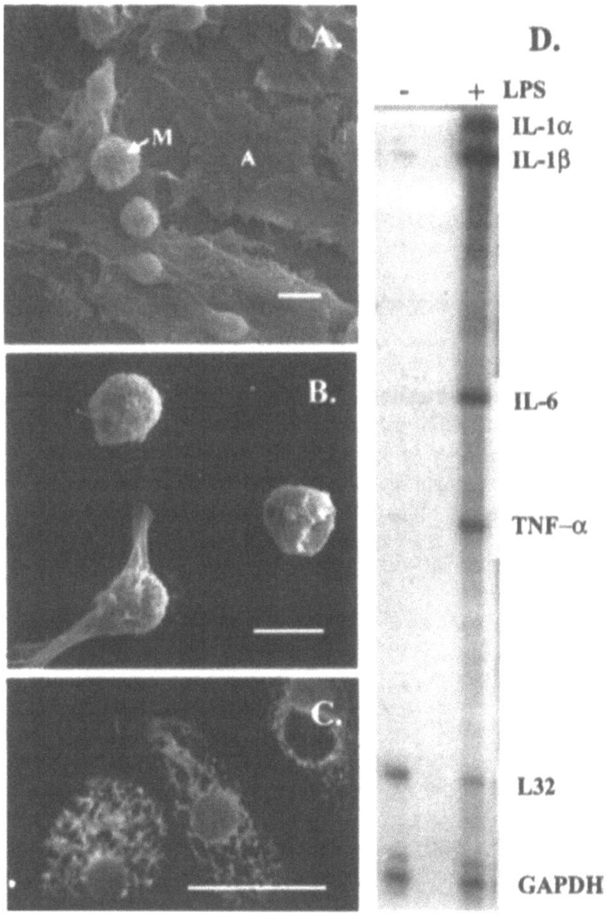

Figure 1. *In vitro* model of neonatal rat microglial cell activation. (A). Scanning Electron Micrograph (SEM) of mixed astrocyte and microglial cells. Mixed neonatal rat glial cells were isolated and allowed to adhere to coverslips overnight. The cells were fixed with 4% glutaraldehyde followed by 1% osmium tetroxide, dehydrated, immersed in hexamethyldisilazane, air-dried, and coated with gold. Cultures were viewed using a Jeol JSM 820 Scanning Electron Microscope at an accelerating voltage of 15 kV. (A=astrocyte; M=microglial cell). (B). SEM of isolated rat microglial cells. Rat microglia cells were isolated by agitation (2h, 37°C, 180 rpm) from mixed glial cultures 10 - 14 days post-seeding, allowed to adhere to coverslips overnight and examined by SEM. (C). Immunofluorescence of microglia purified from mixed glial cultures. Microglial cell cultures were greater than 96% pure as determined by immunofluorescent staining using FITC-conjugated isolectin-B4, a marker for identifying microglia. The bars in panels A, B, and C represent 10μm. (D). Multiprobe RNase Protection Assay (RPA) of cytokine expression in rat microglial cells. Isolated microglia were untreated (-) or treated (+; 1 μg/ml LPS) for 6 h. Total RNA was isolated from microglia using Trizol (Life Tech., Grand Island, NY) reagent. Cytokine mRNA species were detected using RPA with the RiboQuant rCK-1 probe set (Pharmingen, San Diego, CA), resolved on a 6% polyacrylamide gel containing 6 M urea, and imaged on a 445SI Phosphorimager (Molecular Dynamics, Sunnyvale, CA). The bands indicative of L32 and GAPDH represent constitutively expressed transcripts used for standardization of inducible cytokine message levels. Treatment of purified microglial cultures with LPS resulted in induction of message for IL-1α, IL-1β, IL-6, and TNF-α consistent with a state of microglial activation. A similar inducible cytokine mRNA profile was obtained when microglia were treated with IFNγ (10 U/ml) plus LPS (20 μg/ml).

measured for cells treated with 8X10^{-6} M CP55940. The drug dose-dependent inhibition exerted by CP55940 on NO release was found to be significantly ($P < 0.05$, Student's t-test) greater than that exerted by the less active paired enantiomer CP56667 at each comparable concentration tested. Thus, CP55940 exerted an enantiomeric stereoselective inhibition of NO release by rat microglial cells. In addition, cells pretreated with 5X10^{-6} M CP55940 and activated with LPS/IFNγ exhibited approximately a two-fold less level of NO release at 48 and 72 h post-activation when compared with cells pretreated with CP56667 (data not shown). Collectively, these results indicate that CP55940 selectively inhibits the production of NO rather than causing a delay in its release.

3. REVERSAL OF CP55940-MEDIATED INHIBITION OF NO PRODUCTION BY THE CB$_1$-SELECTIVE ANTAGONIST SR141716A

The data indicating a drug dose-dependent differential effect of CP55940 versus that of CP56667 on NO production were consistent with a cellular action mediated through a cannabinoid receptor. In order to provide additional evidence for a role of a cannabinoid receptor in the mediation of NO production, the effect of pretreatment of microglial cells with the CB$_1$-selective antagonist SR141716A (Rinaldi-Carmona et al., 1994) antecedent to

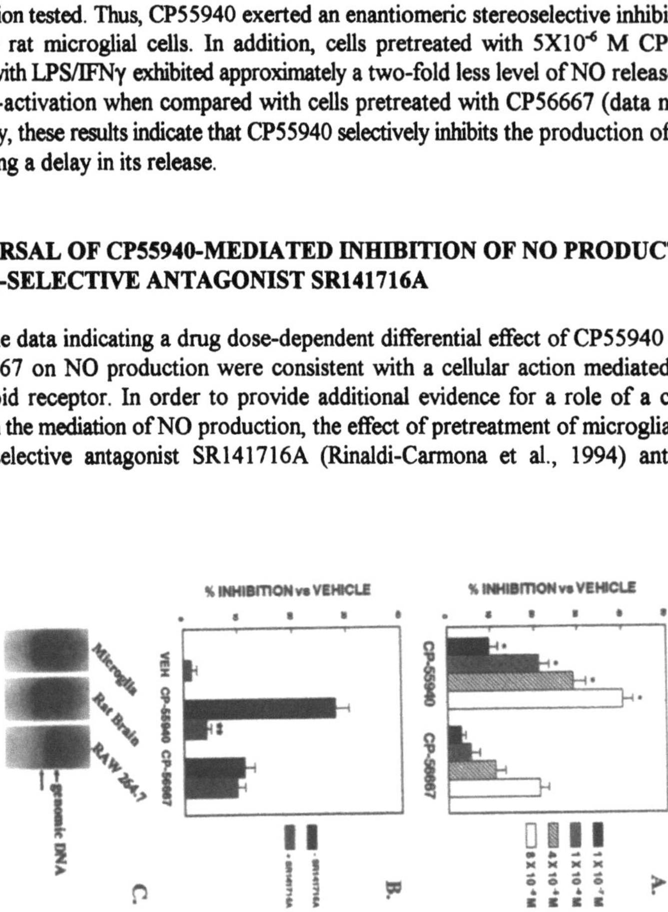

Figure 2. Cannabinoid-mediated inhibition of nitric oxide production. (A). Differential inhibition of NO release by the cannabinoid agonist CP55940 versus its paired enantiomer CP56667. Microglial cells were treated with drug or vehicle for 8 h and then were exposed to 20 µg/ml LPS plus 10 U/ml IFNγ. Culture supernatants were assayed for nitrite 24 h later using the Griess reagent and measurement of absorbance at 550 nm using a Spectramax 250 enzyme-linked immunosorbent assay reader (Molecular Devices Corp., Sunnyvale, CA). Nitrite release from vehicle-treated cultures was 25.4 ± 3.3 [µM/10^6 cells/ml]. Results are expressed as percent inhibition versus vehicle and are the mean ± S.E.M of triplicate wells. Similar results were obtained in three identical experiments. The high affinity cannabinoid CP55940 exerted a dose-dependent inhibition of NO release from rat microglial cells. The drug dose-dependent inhibition was significantly greater ($P < 0.05$, Student's t-test) than that exerted by its paired enantiomer CP56667 at each comparable concentration. (B). Reversal of CP55940-mediated inhibition of NO release by SR141716A. Microglial cells were pretreated with 5X10^{-7}M SR141716A prior to exposure to 5X10^{-6} M CP55940 or CP56667 and LPS/IFNγ activation. Results (mean ± S.E.M of triplicate wells) are expressed as percent inhibition versus vehicle control (**P < 0.01 versus -SR141716A). Nitrite accumulation in vehicle-treated cultures was 29.3 ± 3.5 [µM/ 10^6 cells/ml].

exposure to CP55940 and cell activation with IFNγ/LPS was investigated (Fig. 2B). Pretreatment (2h) of microglial cells with SR141716A ($5X10^{-7}$ M) resulted in a reversal of the inhibitory effects exerted by CP55940 ($5X10^{-6}$ M). SR141716A ($5X10^{-7}$ M) administered alone had no effect on the release of NO by microglial cells.

4. MOLECULAR IDENTIFICATION OF CB_1 EXPRESSION IN RAT MICROGLIAL CELLS

In order to confirm the presence of CB_1 in rat microglial cells, mutagenic reverse transcription-polymerase chain reaction (MRT-PCR) was performed. PCR-amplified cDNA from CB_1 mRNA possessing a unique *MspI* restriction site was digested yielding a cleavage product of 623 bp if cells contained CB_1 message. Total RNA from whole mouse brain or from purified rat astrocytes (not shown) served as a positive control. Total RNA from murine RAW 264.7 macrophage-like cells, which express only CB_2 receptors, served as a negative control. MRT-PCR, using total RNA extracted from highly purified rat microglial cells, revealed the presence of CB_1 mRNA (Fig. 3). As expected, no amplified gene product indicative of the expression of CB_1 mRNA in RAW264.7 cells was detected.

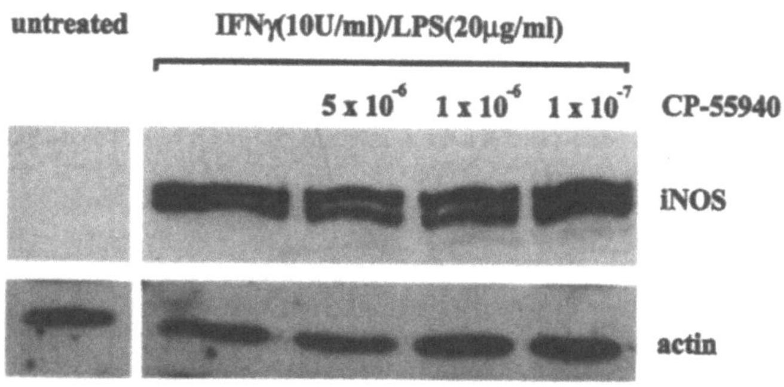

Figure 3. Identification of CB_1 mRNA in rat microglial cells. Southern analysis of MRT-PCR products amplified from total RNA from neonatal rat microglial cells, whole rat brain, and RAW264.7 murine macrophage-like cells. Total RNA was subjected to reverse transcription using an oligonucleotide primer containing a single base mismatch generating a unique *MspI* restriction site. The reverse transcription products then were amplified by PCR using a pair of highly conserved oligonucleotide primers specific for the rat sequence. The primers were designed as described previously (Pettit et al., 1996). The PCR amplification products were digested with *MspI* and subjected to electrophoretic separation on a 1.5% agarose gel. The DNA was transferred to a nylon membrane and hybridized to a ^{32}P random-primed cDNA fragment specific for rat CB_1 derived from pCD-SKR6 (Matsuda et al., 1990). CB_1 mRNA was detected in rat microglial cells and rat brain as demonstrated by the presence of two products following digestion with *MspI*. No amplicon indicative of the presence of CB_1 mRNA was detected from total RNA of RAW264.7 cells.

5. TREATMENT OF MICROGLIAL CELLS WITH CP55940 DOES NOT AFFECT LEVELS OF iNOS PROTEIN

In order to determine whether the CP55940-mediated inhibition of inducible NO production was effected at the level of inducible nitric oxide synthase (iNOS) protein expression, Western immunoblotting studies were performed. Isolated microglial cells were

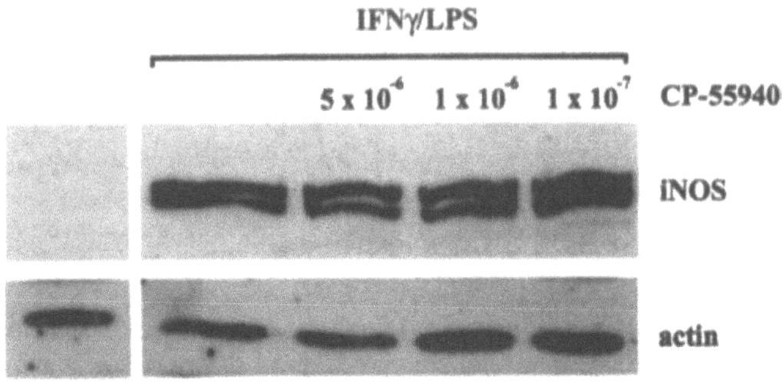

Figure 4. Western immunoblot of purified microglial cells indicating that the potent cannabinoid analog CP55940 does not alter the inducible expression of iNOS protein. Purified microglia were treated for 8 h with CP55940 at the indicated concentrations and exposed to 20 μg/ml LPS and 10 U/ml rat IFNγ for 24 h. Cells were solubilized at 4°C in 1% SDS, 1 mM sodium vanadate, and 10 mM Tris-HCl, pH 7.4. After SDS-PAGE separation and transfer to PVDF-plus membrane (Micron Separations, Inc., Westborough, MA) and 2 h blocking by casein in TBS (10 mM Tris, 0.9% NaCl, pH 7.4; Pierce, St. Louis, MO), membranes were incubated with an antibody to mouse iNOS (Transduction Lab., Lexington, KY) diluted 1:5000 in blocker. After washing with TBS, membranes were incubated with goat anti-rabbit IgG-horseradish peroxidase (HRPO) as the secondary antibody, developed by enhanced chemiluminescence using the SuperSignal CL-HRP substrate (Pierce), and blots were exposed on Kodak XAR imaging film (Eastman Kodak Co., Rochester, NY). The blots were stripped and reprobed using an antibody to actin (1:100; Sigma, St. Louis, MO) to allow for standardization of protein expression. Treatment with CP55940 had no major effect on levels of the inducible expression of iNOS.

treated for 8 h with CP55940, subjected to IFNγ/LPS activation, and whole cell homogenates were assessed by SDS-polyacrylamide gel electrophoresis and Western immunoblotting using an anti-murine iNOS antibody which also recognizes rat iNOS protein. No iNOS protein was detected in homogenates from cells not subjected to IFNγ/LPS activation. In contrast, high level of iNOS was detected in homogenates from IFNγ/LPS-treated microglial cells. Pretreatment with $1X10^{-7}$ M - $5X10^{-6}$ M THC (data not shown) or CP55940 (Fig. 4) had no major effect on levels of iNOS in cells subjected to IFNγ/LPS treatment. These results suggest that the cannabinoid-induced inhibition of NO production is due either to effects on co-factors requisite for iNOS activity or to the functional activity of the iNOS protein itself.

6. DISCUSSION

Exogenous and endogenous cannabinoids have been shown to alter the functional capabilities of immune cells *in vivo* and *in vitro*, including the elicitation of NO (Jeon et al., 1996; Coffey et al., 1996). Burnette-Curley et al. (1993) demonstrated that psychotropic cannabinoids inhibited inducible NO production by the murine macrophage-like cell line RAW264.7. However, to date it has not been established whether cannabinoids exert a similar effect on microglial cells. In the present studies, a multi-factor macrophage activation protocol, which entailed exposure to IFNγ in concert with LPS, was used to assess the effect of cannabinoids on inducible NO production by rat neonatal microglial cells.

Experiments employing cannabinoid paired enantiomers indicated that the synthetic cannabinoid CP55940 exerted a differential dose-dependent inhibition of inducible NO production when compared with that exerted by the less bioactive enantiomer CP56667. These results implicate a cannabinoid receptor in the mediation of inducible NO by microglial cells since enantiomeric stereoselectivity is a requisite for establishing receptor functional linkage to a specified functional activity. To date, two cannabinoid receptor subtypes have been identified (Matsuda et al., 1990; Munro et al., 1993). CB_1 has been localized primarily in neural tissues while CB_2 has been identified in cells of the immune system (Galiègue et al., 1995). Thus, in order to confirm the paired enantiomer data implicating a functional linkage between a cannabinoid receptor and inhibition of inducible NO production, and to obtain insight regarding the cannabinoid receptor subtype involved, experiments using the CB_1 receptor-selective antagonist SR141716A were performed. Pretreatment of microglial cells with SR141716A resulted in a reversal of the CP55940-mediated inhibition of NO production by LPS/IFNγ-activated cells, consistent with a functional role for the CB_1 receptor in this inhibitory process.

Evidence for CB_1 receptor gene expression in microglial cells was corroborated by MRT-PCR which allows for unequivocal discrimination of cDNA amplified from mRNA versus that amplified from residual contaminating genomic DNA (Taniguchi et al., 1994). This technique has been used previously for identification of CB_1 and CB_2 mRNA in rat brain and immune cells, respectively (Pettit et al., 1996). MRT-PCR confirmed the presence of CB_1 receptor message in total RNA obtained from highly purified neonatal rat microglial cells.

Finally, Western immunoblotting experiments were performed using an antibody to iNOS, which catalyzes the synthesis of NO, in order to obtain insight as to the site of action at which cannabinoids exert their inhibitory effect. Neither THC nor CP55940 exerted a major effect on levels of iNOS elicited in response to IFNγ/LPS. These observations are consistent with the proposition that cannabinoids do not alter levels of iNOS message and cognate protein but, instead, either affect co-factors requisite for iNOS activity or alter the functional activity of the iNOS itself. In this context, iNOS has multiple functional domains which consist of sites for binding calmodulin, NADPH, and the flavoproteins FMN and FAD. Cannabinoid effects on the structural stability of iNOS could alter the functionality of any of these sites. Indeed, effects on iNOS protein stability mediated through TGF-β have been reported (Vodovotz et al., 1993). Alternatively, cannabinoids could have an indirect effect on iNOS. That is, cannabinoids could alter intracellular compartmentation of iNOS and affect expression of intracellular factors which impact iNOS activities.

The collective data support a linkage between CP55940 and the inhibition of NO production, at least in part, through the CB_1 receptor. These results expand on our current knowledge concerning the role of cannabinoid receptors in the modulation of immune cell function, particularly in relationship to the brain. However, the step in NO production at which the receptor-mediated action is effected remains to be defined. The data suggest, in addition, that select cannabinoid compounds may be effective for ablating the elicitation of proinflammatory mediators especially under conditions of chronic neuropathological disease. In this context, application of such homogeneous and pure compounds would obviate the potential risks of the use of marijuana as a therapeutic substance since these compounds would be free of contaminants with carcinogenic and other properties deleterious to human health. Nevertheless, *in situ* hybridization and radioligand binding studies indicate that the CB_1 receptor is expressed on a variety of cells in the brain in addition to microglial cells. Thus, consideration of the use of cannabinoids as therapeutic agents should be tempered by the recognition that a variety of untoward neurological effects could be exerted in addition to ablation of proinflammatory mediator production.

ACKNOWLEDGMENTS

The authors wish to thank Ms. C. Boothe for excellent technical assistance. This research was supported by NIH awards: DA05832, DA05247, and [1]T32DA07027.

REFERENCES

Burnette-Curley D, Marciano-Cabral F, Fischer-Stenger K and Cabral GA (1993) Δ-9-tetrahydrocannabinol inhibits cell contact-dependent cytotoxicity of bacillus Calmétte-Guérin-activated macrophages. *Int J Immunopharmacol* **15**: 371-382.

Cabral GA, Toney DM, Fischer SK, Harrison MP and Marciano-Cabral F (1995) Anandamide inhibits macrophage-mediated killing of tumor necrosis factor-sensitive cells. *Life Sci* **56**: 2065-2072.

Chao CC, Hu S, Molitor TW, Shaskan EG and Peterson PK (1992) Activated microglia mediate neuronal cell injury via a nitric oxide mechanism. *J Immunol* **149**: 2736-2741.

Coffey RG, Yamamoto Y, Snella E and Pross S (1996) Tetrahydrocannabinol inhibition of macrophage nitric oxide production. *Biochem Pharmacol* **52**: 743-751.

Compton DR, Rice K, De Costa BR, Razdan RK, Melvin LS, Johnson MR and Martin B R (1993) Cannabinoid structure-activity relationships: Correlation of receptor binding and *in vivo* activities. *J Pharmacol Exp Ther* **265**: 218-226.

Devane WA, Hanus L, Breuer A, Pertwee RG, Stevenson LA, Griffin G, Gibson D, Mandelbaum A, Etinger A and Mechoulam R (1992) Isolation and structure of a brain constituent that binds to the cannabinoid receptor. *Science* **258**: 1946-1949.

Felder CC, Veluz JS, Williams HL, Briley EM and Matsuda LA (1992) Cannabinoid agonists stimulate both receptor- and non-receptor-mediated signal transduction pathways in cells transfected with and expressing cannabinoid receptor clones. *Mol Pharmacol* **42**: 838-845.

Fischer-Stenger K, Dove Pettit DA and Cabral GA. (1993) Δ^9-Tetrahydrocannabinol inhibition of tumor necrosis factor-α: Suppression of post-translation events. *J Pharmacol Exp Ther* **267**: 1558-1565.

Galiègue S, Mary S, Marchand J, Dussossoy D, Carrière D, Carayon P, Bouaboula M, Shire D, Le Fur G and Casellas P (1995) Expression of central and peripheral cannabinoid receptors in human immune tissues and leukocyte subpopulations. *Eur J Biochem* **232**: 54-61.

Gehrmann J and Kleihues P (1994) Neuropathology of CNS HIV infection. *Curr Diagn Pathol* **1**:121-130.

Howlett AC, Qualy JM and Khachatrian LL (1986) Involvement of Gi in the inhibition of adenylate cyclase by cannabimimetic drugs. *Mol Pharmacol* **29**: 307-313.

Jeon YJ, Yang KH, Pulaski JT and Kaminski NE (1996) Attenuation of inducible nitric oxide synthase gene expression by Δ^9-tetrahydrocannabinol is mediated through the inhibition of nuclear factor-kB/Rel activation. *Mol Pharmacol* **50**: 334-341.

Kaminski NE, Koh WS, Yang KH, Lee M and Kessler FK (1994) Suppression of the humoral immune response by cannabinoids is partially mediated through inhibition of adenylate cyclase by a pertussis toxin-sensitive G-protein coupled mechanism. *Biochem Pharmacol* **48**: 1899-1908.

Matsuda LA, Lolait SJ, Brownstein MJ, Young AC and Bonner TI (1990) Structure of a cannabinoid receptor and functional expression of the cloned cDNA. *Nature* **346**: 561-564.

Merril JE and Martinez-Maza O (1990) Cytokines in AIDS-associated neurons and immune system dysfunction, in *Neurobiology of Cytokines Part B: Methods in Neuroscience* (DeSouza EB ed) pp 243-266, Academic Press, Inc., San Diego, CA.

Munro S, Thomas KL and Abu-Shaar M (1993) Molecular characterization of a peripheral receptor for cannabinoids. *Nature* **365**: 61-65.

Pettit DA, Anders DL, Harrison MP and Cabral GA (1996) Cannabinoid receptor expression in immune cells. *Adv Exp Med Biol* **402**: 119-129.

Rinaldi-Carmona M, Barth F, Heaulme M, Shire D, Calandra B, Congy C, Martinez S, Maruani J, Nelait G, Caput D, Ferrara P, Soubrie P, Breliere JC and Le Fur G (1994) SR141716A, a potent and selective antagonist of the brain cannabinoid receptor. *FEBS Lett* **350**: 240-244.

Tang JL, Lancz G, Specter S and Bullock H (1992) Marijuana and immunity: tetrahydrocannabinol-mediated inhibition of growth and phagocytic activity of the murine cell line, P388D1. *Int J Immunopharmacol* **14**: 253-262.

Taniguchi S, Watanabe T, Nakao A, Seki G, Uwatoko S and Kurokawa A (1994) Detection and quantitation of EP_3 prostaglandin E2 receptor mRNA along mouse nephron segments by RT-PCR. *Amer J Physiol* **266**: c1453-c1458.

Vodovotz Y, Bogdan C, Paik UJ, Xie QW and Nathan C (1993) Mechanisms of suppression of macrophage nitric oxide release by transforming growth factor beta. *J Exp Med* **178**: 605-613.

Waksman Y, Olson JM, Carlisle SJ and Cabral GA (1999) The central cannabinoid receptor (CB1) mediates inhibition of nitric oxide production by rat microglial cells. *J Pharmacol Exp Ther* **288**: 1357-1366.

Zielasek J, Tausch M, Toyka KV and Hartung H. (1992) Production of nitrite by neonatal rat microglial cells/brain macrophages. *Cell Immunol* **141**: 111-120.

MODULATION OF CB1 mRNA UPON ACTIVATION OF MURINE SPLENOCYTES

Sasha N. Noe , Catherine Newton, Raymond Widen,
Herman Friedman, and Thomas W. Klein

Department of Medical Microbiology & Immunology
University of South Florida, College of Medicine
Tampa, FL 33612.

ABSTRACT

There is significant evidence that cannabinoids have the ability to exert immunomodulatory effects. The identification of cannabinoid receptors in immune tissues has therefore led to questions about whether these immunomodulatory effects occur via these cannabinoid receptors. The cannabinoid receptor 1 (CB1), although expressed primarily in the brain, is also expressed in lower amounts in peripheral tissues. Of interest to us is the fact that CB1 is expressed in immune tissues such as spleen, albeit at lower levels than the peripheral cannabinoid receptor, CB2. To examine the function of CB1 in immune cells, activation experiments were performed using different stimuli e.g., anti-CD3, phorbol 12-myristate 13-acetate (PMA)/Ionomycin (Io), and PMA/Io + IL-2. Whole spleen cells were cultured in the presence of different stimuli for 0, 2, 4, and 24 hours, harvested at each time point, RNA isolated, and RT-PCR performed. FACS analysis was also performed using CD69 (an early activation marker) to determine whether cells were actually being activated. Results from anti-CD3 stimulation indicated a decrease in CB1 mRNA expression following activation. CB1 mRNA expression in murine splenocytes that were stimulated with PMA/Io in the presence or absence of IL-2 was also modulated. Expression of the message was enhanced upon stimulation with PMA/Io and PMA/Io + IL-2, however, stimulation with PMA/Io + IL-2 led to a stronger increase within 2 to 4 hours with CB1 returning to at or below baseline levels by 24 hours. Expression of CD69 was detected in all stimulated samples thereby indicating that the splenocytes were becoming activated. In summary, anti-CD3 stimulation appeared to decrease CB1 mRNA expression while PMA/Io + IL-2 stimulation significantly increased CB1 mRNA expression. These results demonstrate that the expression of CB1 mRNA is modulated upon cellular activation and that this modulation is dependent on the stimulus that is used.

Neuroimmune Circuits, Drugs of Abuse, and Infectious Diseases
Edited by Herman Friedman *et al.*, Kluwer Academic/Plenum Publishers, 2001

INTRODUCTION

Cannabis is the generic name for various preparations that are derived from the hemp plant, Cannabis sativa. The primary psychoactive constituent in cannabis, delta-9-tetrahydrocannabinol or THC, is a polycyclic, aromatic hydrocarbon, and is one of over 60 cannabinoids present in marijuana smoke. The understanding of the mechanism of action of THC has become more clear with the identification of cannabinoid receptors as well as endogenous cannabinoid ligands. A cDNA from rat brain cortex, termed CB1, was isolated and characterized, and was found to encode a cannabinoid receptor (1). CB1 was later demonstrated in human tissues (2). The cDNA for rat and human CB1 encodes for a guanine-nucleotide-binding regulatory protein (G protein)-coupled receptor of 473 and 472 amino-acid residues, respectively, with 97.3% identity between these species. The mouse CB1 cDNA was later cloned and sequenced (3). There have been several lines of evidence that THC has immunomodulatory effects. These include the inhibition of mitogen-induced T-lymphocyte proliferation, (4), (5), the inhibition of γ-interferon production (6), and the suppression of induction of cytolytic function of cytotoxic T cells (7). Although CB1 and its mRNA are predominantly found in the brain (8), (9) the mRNA has been detected in several peripheral tissues, including spleen, adrenal glands, heart, lung, prostate, uterus, ovary, testis, bone marrow, thymus, colon, stomach, placenta, kidney, and tonsils. (10), (11), (12), and (13). The expression of CB1 in immune cells suggests that it may play a functional role in immune homeostasis. Although evidence exists that demonstrates the presence of CB1 mRNA in the periphery our understanding of the exact distribution and function of CB1 in immune cells remains unclear. It was therefore of interest to examine the expression and distribution of CB1 mRNA in murine immune cells and to determine whether is expression is modulated upon cell activation.

MATERIALS AND METHODS

RNA ISOLATION

BALB/c female mice (Jackson Labs, Bar Harbor, ME) were purchased at 5-8 weeks of age, housed under specific pathogen free conditions, fed, and cared for at our accredited animal facility. 10-12 week old mice were killed and spleens aseptically removed. Red blood cells were lysed in a hypotonic NH_4Cl solution and subsequently washed in HBSS. Total cellular RNA was isolated from BALB/c splenocytes using TRI REAGENT (Molecular Research Center, Inc., Cincinnati, OH). All samples were treated with DNase I (2 U/μg of RNA in Tris-$MgCl_2$ buffer; Boehringer Mannheim, Indianapolis, IN) for 30 min at 37°C before RT-PCR.

RT-PCR

RT-PCR was performed as previously described with some alterations (14). Briefly, messenger RNA was reversed transcribed using oligo dT, RNasin, and AMV RT (Promega, Madison, WI) followed by PCR amplification. Primers for PCR were selected using the DNASIS program and are based on the submitted cDNA sequence of the mouse brain CB1 cDNA in Genbank. The sequences for the CB1 primers were as follows: sense 5'-TGG TGT ATG ATG TCT TTG GG-3' and antisense 5'-ATG CTG GCT GTG TTA TTG GC-3' and amplify a region of 324bp of CB1, which is found in the NH_2-terminal end of the molecule. Primers for β_2 - microglobulin (15) were amplified as an internal control to control for sample-to-sample variation in the amount of starting mRNA. Amplified products were resolved on a 2% agarose-ethidium bromide gel, and autoradiographed.

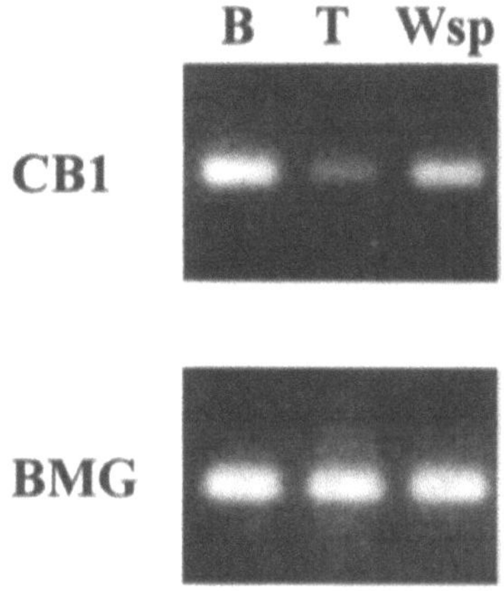

Figure 1. Ethidium bromide stained gel of RT-PCR results indicating CB1 mRNA expression in enriched B and T cells, and whole splenocytes. RNA was isolated from each cell subpopulation and treated with DNase. RT-PCR was performed using primers for CB1 and BMG. The percent enrichment for each cell subpopulation is as follows: B cells (B): 93% CD19, 0.6% CD3, 0.1% NK, T cells (T): 0.7% CD19, 87% CD3, 8% NK, and Whole Spleen cells (Wsp): 38% CD19, 40% CD3, 10% NK.

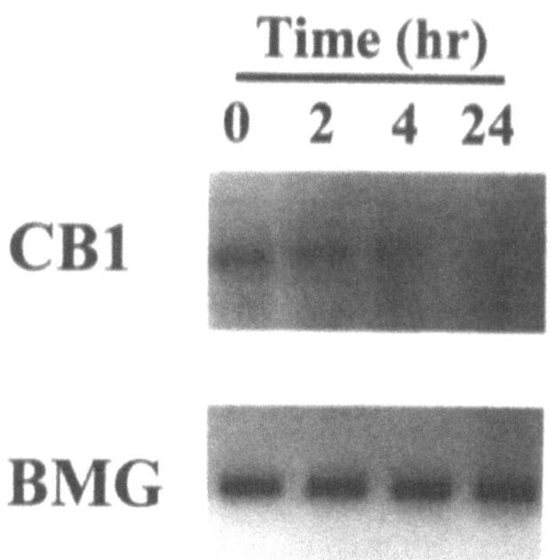

Figure 2. Ethidium bromide stained gel of RT-PCR results demonstrating CB1 mRNA expression in PMA/Io stimulated murine splenocytes. RNA was isolated from cultured cells at each time point (0, 2, 4, and 24 hrs), DNase treated, and RT-PCR performed using primers for CB1 and BMG. CB1 mRNA decreases following activation of murine splenocytes with PMA/Io.

T CELL ENRICHMENT

Spleens were removed and processed as mentioned above. T cells were then purified by negative selection using a mouse T cell enrichment column (R&D System, Minneapolis, MN). The mouse T cell enrichment column binds B cells via F(ab)-surface immunoglobulin (Ig) that are interacting with glass beads coated with anti-Ig, and monocytes bind via Fc interactions to beads coated with Ig. Three spleens were processed at a time for each experiment and the pooled cells were suspended in RPMI 1640. The resulting population was analyzed by flow cytometry (CD19+ for B cells, CD3+ for T cells, and NK1.1 for NK cells).

ENRICHMENT OF B CELLS

Negative selection using magnetic bead separation was used to isolate and purify B cells from whole spleen as described above (Stem Sep, Vancouver, BC). Spleens were removed and processed as mentioned above. Flow cytometry was used to determine the percent enrichment of the sub-populations.

FLOW CYTOMETRY

Cells were harvested at each time point, labeled with either anti-CD3 (FITC), anti-CD19 (PE), or anti-CD69 (TRICOLOR) and analyzed to determine sub-population purity and whether cell activation occurred. CD69 is an early activation marker present on B cells, T cells, NK cells and macrophages.

CELL ACTIVATION

Spleen cells were isolated and cultured in the presence of either anti-CD3 (1μg/ml), PMA (10 ng/ml)/Io (10^{-7} M), or PMA/Io + IL-2 (20 ng/ml) for 0, 2, 4, 24 hrs. Cells were harvested at each time point, analyzed for CD69 expression, RNA extracted and RT-PCR and densitometry performed.

RESULTS AND CONCLUSIONS

The results of RT-PCR on enriched immune cell sub-populations demonstrated that although CB1 mRNA is expressed in whole spleen cells, T cells and B cells, the expression of message is higher in B cells than T cells (Fig. 1). The percent enrichment of B cells as determined by flow cytometry was 93% CD19, 0.6% CD3, and 0.1 % NK cells while the percent enrichment for T cells was 0.7% CD19, 87% CD3, and 8 % NK cells. The composition of B and T cells prior to enrichment was 38% CD19, 40% CD3, and 10% NK cells. These results concur with previous studies examining CB1 expression in human peripheral blood mononuclear cells (PBMC's) where human B cells were found to express more CB1 mRNA than human T cells [10]. It was then of interest to determine whether the expression of CB1 mRNA was modulated upon cellular activation. Therefore, whole spleen cells were cultured in the presence of phorbol 12-myristate 13-acetate (PMA)/Ionomycin (Io) in the presence or absence of IL-2. This stimulus was chosen to potently activate both T cells and B cells in culture. RNA was isolated and RT-PCR was performed. Stimulation of whole spleen cells with PMA and Io led to a slight decrease in CB1 mRNA expression (Fig. 2), however, in the presence of IL-2, PMA and Io markedly enhanced CB1 mRNA expression at 2 hours (Fig. 3). To ensure that the cells were actually becoming activated using this activation protocol, flow cytometry was performed at the respective time points and the expression of CD69 on T cells and B cells (an early activation marker) was examined. CD69 expression was upregulated at 2, 4, and 24 hrs both in the presence and absence of IL-2 (Table 1) indicating that indeed the cells were becoming activated when these stimulation protocols were used. Although mouse T cells appear to express low levels of CB1 mRNA, a protocol for stimulation of T cells was used

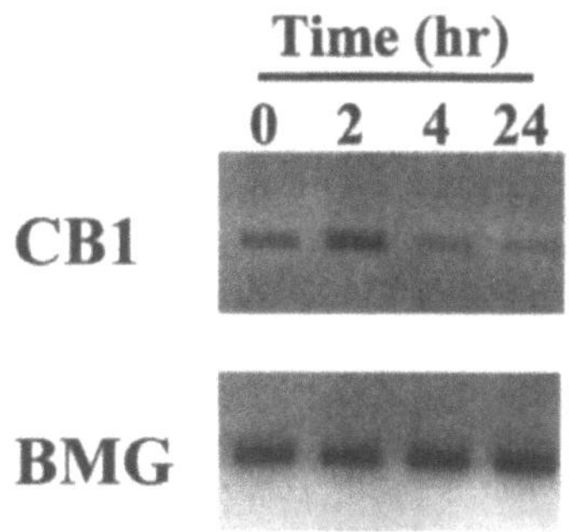

Figure 3. Ethidium bromide stained gel of RT-PCR results demonstrating CB1 mRNA expression in PMA/Io + IL-2 stimulated murine splenocytes. RNA was isolated from cultured cells at each time point (0, 2, 4, and 24 hrs), DNase treated, and RT-PCR performed using primers for CB1 and BMG. CB1 mRNA markedly increases at 2 hours following activation of murine splenocytes with PMA/Io in the presence of IL-2.

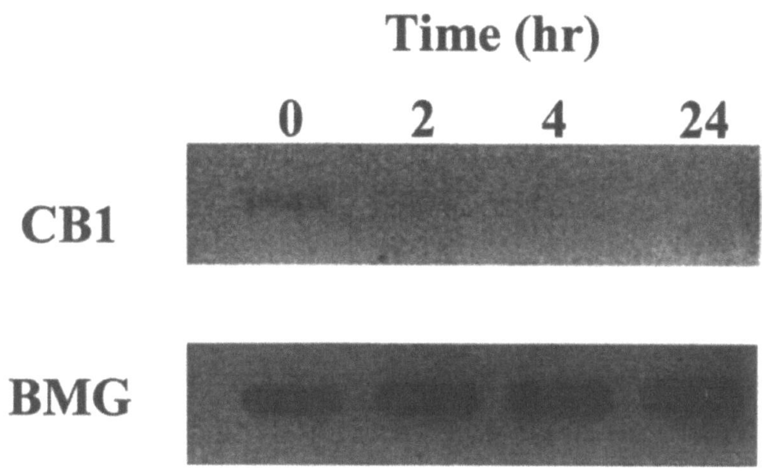

Figure 4. Ethidium bromide stained gel of RT-PCR results demonstrating CB1 mRNA expression in anti-CD3 stimulated murine splenocytes. RNA was isolated from cultured cells at each time point (0, 2, 4, and 24 hrs), DNase treated, and RT-PCR performed using primers for CB1 and BMG. Results indicate that CB1 mRNA decreases following activation of murine splenocytes with anti-CD3.

in the following experiments. This protocol was chosen based on previous findings that have demonstrated that CB1 mRNA is increased in Jurkat cells, a human T cell line, following activation (16). Similar to murine T cells, Jurkat cells express only low levels of CB1 mRNA at the basal level, however, after stimulation, CB1 mRNA expression increased. To determine whether modulation of CB1 message was occurring as a result of activation of murine T cells in this mixed culture, whole spleen cells were cultured in the presence of anti-CD3. Preliminary results demonstrate that stimulation of murine whole spleen cells with anti-CD3 leads to a decrease in the expression of CB1 mRNA (Fig. 4). The effect of anti-CD3 stimulation on CB1 mRNA expression was similar to what was observed with PMA/Io stimulation in the absence of IL-2. It has been demonstrated that PMA and Io stimulation can mimic anti-CD3 stimulation, therefore, the possibility exists that in these experiments stimulation with PMA and ionomycin alone may be primarily activating T cells, while the addition of IL-2 may be activating other cells. Altogether these results suggest that CB1 mRNA is modulated upon cellular activation, and this modulation is dependent on the stimulus that is used. There is also a possibility that IL-2 may be playing a specific role in CB1 mRNA expression. IL-2 has the ability to activate many cellular transcription factors that in turn can activate other genes. Further experiments will need to be performed to determine whether IL-2 is playing a role in CB1 gene regulation. Currently experiments are being performed to determine whether CB1 mRNA may be modulated in a different manner if stimulated with anti-CD3 in the presence of IL-2, as well as using anti-CD40, a potent B cell stimulus, in the presence or absence of IL-2. These studies will help us to identify what happens with the expression of CB1 mRNA when cells are stimulated in a T or B cell dependant manner, as well as further elucidating the role of cannabinoid receptors in the immune system.

ACKNOWLEDGEMENTS

Supported by NIDA Grant DA10683 and DA07245

REFERENCES

1. Matsuda, L.A., et al., Structure of cannabinoid receptor and functional expression of the cloned cDNA. *Nature*, 1990. **346**: p. 561-564.
2. Gerard, C.M., et al., Molecular cloning of a human cannabinoid receptor which is also expressed in testis. *Biochem. J.*, 1991. **279**: p. 129-134.
3. Chakrabarti, A., E. Onaivi, and G. Chaudhuri, Cloning and sequencing of a cDNA encoding the mouse brain-type cannabinoid receptor protein. *DNA Sequence*, 1995. **5**: p. 385-388.
4. Nahas, G.G., et al., Inhibition of cellular mediated immunity in marihuana smokers. *Science*, 1974. **183**: p. 419-420.
5. Luo, Y.D., et al., Effects of cannabinoids and cocaine on the mitogen-induced transformations of lymphocytes of human and mouse origins. *Int J Immunopharmac*, 1992. **14**: p. 49-56.
6. Blanchard, D.K., et al., In vitro and in vivo suppressive effects of delta-9-tetrahydrocannabinol on interferon production by murine spleen cells. *Int. J. Immunopharm.*, 1986. **8**: p. 819-824.
7. Klein, T.W., et al., Marijuana components suppress induction and cytolytic function of murine cytotoxic T cells in vitro and in vivo. *J Toxicol Environ Health*, 1991. **32**: p. 465-477.
8. Herkenham, M., et al., Cannabinoid receptor localization in brain. *Proc. Natl. Acad. Sci. USA.*, 1990. **87**: p. 1932-1936.
9. Matsuda, L.A., T.I. Bonner, and S.J. Lolait, Localization of cannabinoid receptor mRNA in rat brain. *J Comp Neurol*, 1993. **327**(4): p. 535-50.

10. Galiegue, S., et al., Expression of central and peripheral cannabinoid receptors in human immune tissues and leukocyte subpopulations. *Eur J Biochem*, 1995. **232**(1): p. 54-61.
11. Gerard, C.M., et al., Molecular cloning of a human cannabinoid receptor which is also expressed in testis. *Biochem J*, 1991. **279**(Pt 1): p. 129-34.
12. Kaminski, N.E., et al., Identification of a functionally relevant cannabinoid receptor on mouse spleen cells that is involved in cannabinoid-mediated immune modulation. *Mol Pharmacol*, 1992. **42**(5): p. 736-42.
13. Shire, D., et al., An amino-terminal variant of the central cannabinoid receptor resulting from alternative splicing [published erratum appears in J Biol Chem 1996 Dec 27;271(52):33706]. *J Biol Chem*, 1995. **270**(8): p. 3726-31.
14. Noe, S.N., et al., Cannabinoid receptor agonists enhance syncytia formation in MT-2 cells infected with cell free HIV-1MN. *Adv Exp Med Biol*, 1998. **437**: p. 223-9.
15. Ehlers, S., et al., Kinetic analysis of cytokine gene expression in the livers of naive and immune mice infected with Listeria monocytogenes. The immediate early phase in innate resistance and acquired immunity. *J Immunol*, 1992. **149**(9): p. 3016-22.
16. Daaka, Y., H. Friedman, and T.W. Klein, Cannabinoid receptor proteins are increased in Jurkat, human T-cell line after mitogen activation. *J. Pharmacol. Exp. Ther.*, 1996. **276**: p. 776-783.

DOWNREGULATION OF CANNABINOID RECEPTOR 2 (CB2) MESSENGER RNA EXPRESSION DURING *IN VITRO* STIMULATION OF MURINE SPLENOCYTES WITH LIPOPOLYSACCHARIDE

Sumi Fong Lee, Catherine Newton, Raymond Widen, Herman Friedman, and Thomas W. Klein

Department of Medical Microbiology & Immunology
University of South Florida, College of Medicine
Tampa, FL 33612

ABSTRACT

Cannabinoid receptor 2 (CB2) has been identified as the most abundant cannabinoid receptor subtype in the immune system. Bacterial lipopolysaccharide (LPS) is a potent stimulant of B cells, inducing proliferation and differentiation into antibody secreting cells. It has been reported that CB2 receptor expression is upregulated during human, tonsillar B cell activation through CD40. It was of interest to investigate the expression of CB2 mRNA using another B cell activator, LPS. Using northern blot analysis, we measured CB2 mRNA levels in murine splenocytes and enriched B cells. Results indicated that the 4.0 kb CB2 transcript was 2 fold higher in abundance in murine B cells than in whole splenocyte preparations. This observation confirmed data from others and from our previous RT-PCR studies that the expression of CB2 mRNA is more abundant in B cells. Upon LPS stimulation, CB2 transcripts were decreased 46% and 42% at 4 hours and 24 hours, respectively, when compared to unstimulated populations. An examination by flow cytometry of the CD69, early activation marker, on splenocytes, showed that the majority of the B cells were activated at 24 hrs. Thus, these results suggested that LPS stimulation of murine B cells caused a decrease in CB2 mRNA expression in contrast to the increase observed following human B cell stimulation through CD40.

INTRODUCTION

The principal psychoactive constituent of cannabis, Δ^9-tetrahydrocannabinol (THC), exerts its psychoactivity through an interaction with a G protein-coupled, seven-transmembrane receptor, denoted CB1, which is found mainly in the brain (Matsuda et al., 1990). A second receptor was later cloned from a human macrophage cell line (Munro et al., 1993) and this CB2 receptor shared 44% overall identity with CB1 and was located mainly in peripheral organs. A murine CB2 receptor has also been cloned from a mouse

Neuroimmune Circuits, Drugs of Abuse, and Infectious Diseases
Edited by Herman Friedman *et al.*, Kluwer Academic/Plenum Publishers, 2001

splenocyte cDNA library displaying a 3.7 kb sequence and containing a coding region sharing 82% overall identity with human CB2 (Shire et al., 1996). In addition to these receptors, endogenous ligands, anandamide (Devane et al., 1992; Smith et al., 1994) and 2-arachidonyl glycerol, have been discovered and shown to modulate immune functions (Mechoulam et al., 1995). Human CB2 mRNA has been found in varying amounts in human lymphocyte subpopulations with the order being the following: B cells > NK cells > monocytes > polymorphonuclear neutrophil cells > CD8 cells > CD4 cells. In addition, the CB2 protein has been detected in tonsils by immunohistology, and it appeared to be restricted to B cell enriched areas (Galieque et al., 1995). Furthermore, cannabinoids have been found to enhance human, tonsillar B cell proliferation apparently through CB2 receptor (Carayon et al., 1998; Derocq et al., 1995). Our laboratory has reported that CB1 message is increased following LPS stimulation of macrophages and stimulation of T cell lines (Klein et al., 1995; Daaka et al., 1996). Here we report on results with CB2 mRNA expression and show that splenic B cells had higher levels of CB2 message and that CB2 expression in splenocytes was decreased upon LPS stimulation.

MATERIALS AND METHODS

Cell Culture

Spleens were aseptically collected from 10-12 week-old female BALB/c mice (National Cancer Institute-Harlan, Fredricksburg, MD). Splenocytes (2 x 10^6 /ml) were suspended in RPMI 1640 with 10% fetal calf serum, and then cultured in the presence or absence of LPS (*Escherichia coli*, Sigma Chemical, St. Louis, MO; 5 µg/ml) for 2, 4, and 24 hrs. In some experiments, the B cell population was purified from splenocytes with magnetic cell depletion techniques using a negative selection kit (Stem Cell technologies, Vancouver, Canada). The purity of B cells was 92% as determined by flow cytometry.

Flow Cytometry

Cultured splenocytes were stained with anti-CD3(FITC), anti-CD19(PE), anti-CD69(biotin) (PharMingen, San Diego, CA), and streptavidin Tricolor (Caltag Laboratories, Burlingame, CA). The percentage of CD19+/CD69+ and CD3+/CD69+ cells were analyzed with a FACStar (Becton Dickinson, Mountain View, CA) equipped with an Argon ion laser tuned to 488nm at 15mW. To test the purity of B cells, the cells were stained with anti-CD3(FITC) and anti-CD19(PE) and then analyzed.

Probe Preparation

To generate a ^{32}P labeled CB2 antisense RNA probe for Northern blotting, the CB2 gene was sub-cloned into plasmid pGEM7Zf(-) (Promega, Madison, WI) which contained T7 and SP6 phage RNA polymerase promoters. The pGEM-CB2 clone was digested with HindIII and then used in an *in vitro* transcription reaction. A MAXIscript kit (Ambion, Austin, TX) was used, T7 phage RNA polymerase was added, ^{32}P UTP was used in the reaction as the labeling agent. The β actin antisense RNA probe was produced in the same way using pTRI-β–actin-Mouse control template DNA as the template provided by the manufacture. CB2 sense strand RNA was synthesized by using SP6 phage RNA polymerase and then used as a hybridization control.

RNA Isolation and Northern Blotting

Total RNA was isolated from whole splenocytes and enriched B cells using TRI Reagent (Sigma) following the protocols provided by the manufacture. The RNA was then analyzed using the NorthernMax kit (Ambion). In brief, 10 μg of total RNA from each sample was electrophoresed on 1% formaldehyde gel and downward transferred onto a BrightStar-Plus positively charged nylon membrane (Ambion). The membrane was hybridized with CB2 antisense RNA probe and β actin antisense RNA probe simultaneously at 65°C overnight in Prehyb/Hyb solution, or 2 hours in ZIP-Hyb solution. The membrane was then washed and analyzed by a phosphoimager (Molecular Dynamics, Sunnyvale, CA) with ImageQuaNT (Molecular Dynamics) software. The amount of CB2 expression was normalized as the ratio of count per minute (CPM) of CB2 divided by CPM of β actin. Commercially prepared mouse spleen poly (A) RNA was purchased (Ambion) and used as a CB2 mRNA expression positive control. Chinese hamster ovary cells (Pro5)(ATCC CRL-1781; Manassas, VA) were used as a negative control in Northern blot analysis.

RESULTS AND CONCLUSIONS

In this study, RNA was extracted from either murine splenocytes, enriched B-cells, or LPS-stimulated splenocytes. The RNA samples were analyzed by Northern blotting for CB2 and β-actin mRNA expression. As illustrated in Figure 1, a 4 kb CB2 and a 2.1 kb β-actin band were observed in splenic RNA. The finding of this sized transcript in murine

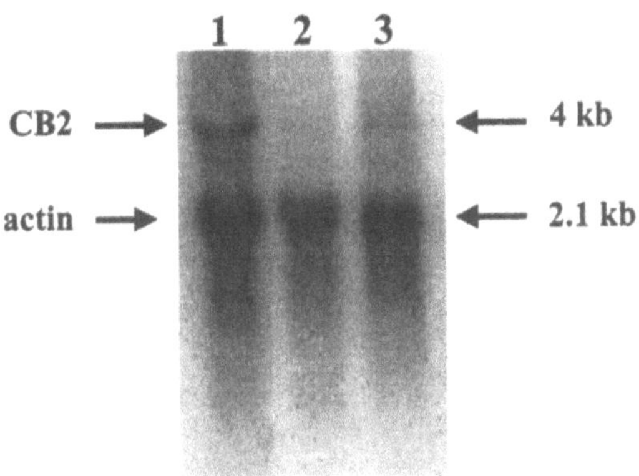

Fig. 1. Northern blot showing CB2 and β-actin mRNA transcripts. Ten μg of total RNA from whole splenocytes (lane 1) and Pro5 (lane 2) and 2.5 μg of poly (A) RNA from spleen (lane 3) were probed with ^{32}P UTP labeled CB2 antisense RNA probe and β-actin antisense RNA probe.

splenocytes corresponds to previous reports in mice (Condie et al., 1996; Schatz et al., 1997). The density of the CB2 band in poly (A) RNA preparation was weaker than expected. This may reflect that the poly (A) RNA was extracted from whole spleen tissue rather than the purified splenic leukocytes used in the other samples (Fig. 1). Quantitation of the CB2 blot densities was calculated based on densitometry readings and ratio comparisons to the β-actin bands. These results showed that the amount of transcript from enriched B cells was almost double the amount found in total splenocyte preparations (0.52 ratio versus 0.29 ratio). This supported previous findings using RT-PCR showing B cells expressed more

CB2 than other lymphocyte subsets (Galieque et al., 1995). Human B cells have also been shown by flow cytometry to have higher CB2 surface protein (Carayon et al., 1998).

Very few studies have addressed the issue of CB2 gene modulation following immune cell activation. One report showed that CB2 mRNA and protein expression were increased within 24 hrs following stimulation with anti-CD40 antibody suggesting that cell activation led to an increase in gene expression (Carayon et al., 1998). Interestingly, however, in that study CB2 mRNA expression sharply decreased to baseline 24 hrs later (i.e., 48 hrs after stimulation). Bacterial LPS is also a polyclonal B cell activator but works through a different signaling cascade than CD40 and so we examined the effect of LPS on CB2 expression. Murine splenocytes were cultured for 0, 2, 4, and 24 hrs with or without LPS. As indicated in Table 1, without LPS, there was a 46% increase in transcript level in splenocytes after 2 hrs of incubation. This increase could be due to the stimulatory signals encountered by the splenocytes when placed in culture as it declined as culture times were extended. Upon LPS stimulation, however, there was only a slight increase (14%) at 2 hrs followed by a steady decline through the 24 cultivation (Table 1). This CB2 reduction was more prominent at 4 hr and 24 hrs, with a 46% and 42% decrease, respectively, compared to controls. Therefore, the effect of LPS appeared to decrease the CB2 expression. Similar effects were detected in other experiments with a higher dose of LPS (10μg/ml) and employing a different Northern blotting procedure (data not shown).

It was also of interest to examine these cell populations for the surface expression of CD69, an early activation marker on lymphocytes (Ziegler et al., 1994). Therefore, duplicate populations of cells were set up and at appropriate times were stained for CD3 (T cells), CD19 (B cells) and CD69. As indicated in Table 1, the results showed a continuing increase in $CD69^+$/ $CD19^+$ cells throughout the 24 hr incubation period ending with 63% of

Table 1. Effect of LPS stimulation of murine splenocytes on CB2 mRNA and on CD69 surface expression.

Hour of Culture[1]	Northern Blot Analysis CB2/β-actin[2] Media	LPS (5ug/ml)	Flow cytometry[3] Percent of CD69+CD19+
0	0.293	—	1.59
2	0.428	0.333	12.45
4	0.377	0.204	29.61
24	0.345	0.201	63.35

[1] Splenocytes cultured with or without LPS.
[2] RNA data are expressed in ratios of the band densities.
[3] Flow cytometry data are expressed as the percentage of $CD19^+/CD69^+$ cells.

$CD69^+$B cells, which represented 93% of the total B-cells (data not shown). This compared to only 6% for T cells (data not shown). Thus, together these data suggest that LPS-induced activation of splenic B cells is accompanied by a corresponding decrease in CB2 message. Down-regulation of human CB2 receptors has been observed in human B cells during differentiation and the CB2 message level in response to CD40 activation was observed to downregulate after an initial upregulation (Carayon et al., 1998). Since LPS can stimulate B cells to not only proliferate but also differentiate (Deenick et al., 1999), it is possible that the reduced CB2 receptor expression is preferentially linked to differentiation rather than proliferation. Interestingly, LPS has been observed to downregulate other seven-transmembrane receptors, including the chemokine receptors, CCR2 in mice (Zhou et al., 1999) and CXCR1 and CXCR2 in humans (Khandaker et al., 1998). Additional studies are needed to determine the precise regulatory effect of LPS on CB2 expression and the role of CB2 expression in B cell proliferation and differentiation.

ACKNOWLEDGMENTS

Supported by NIDA Grant DA10683 and DA07245

REFERENCES

Carayon, P., Marchand, J., Dussossoy, D., Derocq, J. M., Jbilo, O., Bord, A., Bouaboula, M., Galiegue, S., Mondiere, P., Penarier, G., LeFur, G., Defrance, T. & Casellas, P. 1998. Modulation and functional involvement of CB2 peripheral cannabinoid receptors during B-cell differentiation. *Blood*, 92, 3605.

Condie, R., Herring, A., Koh, W. S., Lee, M. & Kaminski, N. E. 1996. Cannabinoid induction of adenylate cyclase-mediated signal transduction and interleukin 2 (IL-2) expression in the murine T-cell line, EL4.IL-2. *J. Biol. Chem.*, 271, 13175.

Daaka, Y., Friedman, H. & Klein, T. W. 1996. Cannabinoid receptor proteins are increased in Jurkat, human T-cell line after mitogen activation. *J. Pharmacol. Exp. Ther.*, 276, 776.

Deenick, E. K., Hasbold, J. & Hodgkin, P. D. 1999. Switching to IgG3, IgG2b, and IgA is division linked and independent, revealing a stochastic framework for describing differentiation. *J Immunol*, 163, 4707.

Derocq, J., Segui, M., Marchand, J., LeFur, G. & Casellas, P. 1995. Cannabinoids enhance human B-cell growth at low nanomolar concentrations. *FEBS Letters*, 369, 177.

Devane, W. A., Hanus, L., Breuer, A., Pertwee, R. G., Stevenson, L. A., Griffin, G., Gibson, D., Mandelbaum, A., Etinger, A. & Mechoulam, R. 1992. Isolation and structure of a brain constituent that binds to the cannabinoid receptor. *Science*, 258, 1946.

Galieque, S., Mary, S., Marchand, J., Dussossoy, D., Carriere, D., Carayon, P., Bouaboula, M., Shire, D., Le Fur, G. & Casellas, P. 1995. Expression of central and peripheral cannabinoid receptors in human immune tissues and leukocyte subpopulations. *Eur. J. Biochem.*, 232, 541.

Khandaker, M. H., Xu, L., Rahimpour, R., Mitchell, G., DeVries, M. E., Pickering, J. G., Singhal, S. K., Feldman, R. D. & Kelvin, D. J. 1998. CXCR1 and CXCR2 are rapidly down-modulated by bacterial endotoxin through a unique agonist-independent, tyrosine kinase-dependent mechanism. *J. Immunol.*, 161, 1930.

Klein, T. W., Newton, C., Zhu, W., Daaka, Y. & Friedman, H. 1995. Minireview: D^9-tetrahydrocannabinol, cytokines and immunity to *Legionella pneumophila*. *Proc. Soc. Exp. Biol. Med.*, 209, 205.

Matsuda, L. A., Lolait, S. J., Brownstein, M. J., Young, A. C. & Bonner, T. I. 1990. Structure of cannabinoid receptor and functional expression of the cloned cDNA. *Nature*, 346, 561.

Mechoulam, R., Ben-Shabat, S., Hanus, L., Ligumsky, M., Kaminski, N. E., Schatz, A. R., Gopher, A., Almog, S., Martin, B. R., Compton, D. R., Pertwee, R. G., Griffin, G., Bayewitch, M., Barg, J. & Vogel, Z. 1995. Identification of an endogenous 2-monoglyceride, present in canine gut, that binds to cannabinoid receptors. *Biochem. Pharm.*, 50, 83.

Munro, S., Thomas, K. L. & Abu-Shaar, M. 1993. Molecular characterization of a peripheral receptor for cannabinoids. *Nature*, 365, 61.

Munro, S., Thomas, K. L. & Abu-Shaar, M. 1993. Molecular characterization of a peripheral receptor for cannabinoids. *Nature*, 365, 61.

Schatz, A. R., Lee, M., Condie, R. B., Pulaski, J. T. & Kaminski, N. E. 1997. Cannabinoid receptors CB1 and CB2: A characterization of expression and adenylate cyclase modulation within the immune system. *Toxicol. appl. Pharm.*, 142, 278.

Shire, D., Calandra, B., Rinaldi-Carmona, M., Oustric, D., Pessegue, B., Bonnin-Cabanne, O., Le Fur, G., Caput, D. & Ferrara, P. 1996. Molecular cloning, expression and function of the murine CB2 peripheral cannabinoid receptor. *Biochim. Biophys. Acta*, 1307, 132.

Smith, P. B., Compton, D. R., Welch, S. P., Razdan, R. K., Mechoulam, R. & B.R., M. 1994. The pharmacological activity of anandamide, a putative endogenous cannabinoid, in mice. *J. Pharmacol. Exp. Ther.*, 270, 219.

Zhou, Y., Yang, Y., Warr, G. & Bravo, R. 1999. LPS down-regulates the expression of chemokine receptor CCR2 in mice and abolishes macrophage infiltration in acute inflammation. *J. Leuk. Biol.*, 65, 265.

Ziegler, S. F., Ramsdell, F. & Alderson, M. R. 1994. The activation antigen CD69. *Stem Cells*, 12, 456.

CB1 AND CB2 RECEPTOR mRNA EXPRESSION IN HUMAN PERIPHERAL BLOOD MONONUCLEAR CELLS (PBMC) FROM VARIOUS DONOR TYPES.

Liang Nong, Catherine Newton, Herman Friedman, and Thomas W. Klein

Department of Medical Microbiology and Immunology, University of South Florida, Tampa, FL 33612

ABSTRACT

Marijuana cannabinoid receptors (CBR), CB1 and CB2, are G protein-coupled receptors reported to be expressed in brain as well as cells of the periphery. Human peripheral blood mononuclear cells (PBMCs) are reported to express CBR mRNA with CB2 expression higher than CB1 and expression in B cells higher than other cells. However, it is not known if the mRNA expression is constant among individuals of differing ages, gender, or ethnic origins. In the present study, PBMCs were obtained from a limited number of normal donors of both genders, of ages ranging from 21 to 55, and from Caucasian, and Asian ethnic origin. Using semi-quantitative RT-PCR, we confirmed previous reports that CB2 mRNA expression was higher than CB1 in PBMCs and in addition demonstrated that this basic profile was observed when stratified by age, gender, or ethnic origin. The latter results suggest that CBR expression is relatively constant across the human population.

INTRODUCTION

Marijuana cannabinoids cause biological changes through both receptor mediated and non-receptor mechanisms (Felder et al., 1992). Within the past decade, two subtypes of cannabinoid receptors (CBRs) have been identified. The CB1 receptor (Matsuda et al., 1990) was found predominantly in the brain and testis with some evidence of expression in cells of the immune system (Kaminski et al., 1992). The CB2 subtype was cloned from an immune cell line and appears to be the primary type in peripheral tissues including the immune system (Munro et al., 1993). Animal studies have suggested that cannabinoids are immunomodulators; however, the role of cannabinoid receptors in this modulation is unclear as well as the relative expression of CBRs among different immune cell subtypes (Klein et al., 1998). Subpopulations of PBMCs have been reported to express CB1 and CB2 mRNA as well as CB2 protein, with CB2 mRNA expression higher in these cells than CB1 (Galieque et al., 1995). However, little is known about the variation among donors from different demographic groups, therefore, the possibility exists that CBR expression may vary considerably within the human population. In the present study, by using semi-quantitative RT-PCR, we have obtained preliminary results suggesting that the CBR mRNA expression in PBMCs is relatively constant when donors are stratified according to age, gender, and ethnic origin.

MATERIALS AND METHODS

Isolation of Peripheral Blood Mononuclear Cells

Normal, healthy donors were recruited from University of South Florida community and were all given informed consent. Collected blood (15 mls) was diluted in PBS (1:1) and then layered over HISTOPAQUE-1077 (Sigma, St. Louis, Mo). The PBMCs were collected from the top layer following centrifugation (400g, 20 min) and washed 2 times in PBS.

RNA Preparation and RT-PCR Analysis

Total RNA was extracted from PBMCs with TRIzol Reagent (Life Technologies, Rockville, MD) according to standard procedures. RNA quality was assessed by analyzing the integrity of 18S and 28S rRNA in a 1% agarose formaldehyde gel with ethidium bromide staining. Total RNA was quantified using a RiboGreen RNA Quantitation Kit (Molecular Probe, Eugene, OR). RNA was treated with RNase-free DNase (Life Technologies) to eliminate any residual genomic DNA and re-quantified.

For RT-PCR, RNA (1 μg in 20 μl) was reverse-transcribed with Omniscript Reverse Transcriptase Kit (QIAGEN, Valencia, CA) according to manufacturer's recommendations. PCR was performed using Taq PCR Core Kit (QIAGEN). RT product (1μl) in a 25 μl volume was amplified in a Mastercycler Gradient thermal cycler (Eppendorf Scientific, New York) under optimum conditions. The annealing temperature was 63°C and the cycle number was 39 for both CB1 and CB2. The β-actin conditions were 58°C for 30 cycles. The primers used were the following: CB1 sense primer, 5'-caccttccgcaccatcaccac-3'; CB1 antisense primer, 5'-gtctcccgcagtcatcttctcttg-3'; CB2 sense primer, 5'-catggaggaatgctgggtgac-3'; CB2 antisense primer, 5'-gaggaaggcgatgaacaggag-3'; ß-actin sense primer, 5'-tgatggtgggcatgggtcag-3' and ß-actin antisense primer, 5'-gtgttggcgtacaggtcttt-3'. The expected sizes of the amplicons were 166 bp and 604 bp for CB1 and CB2, respectively, and 765 bp for ß-actin. PCR reactions were electrophoresed in 2% agarose gels containing ethidium bromide. The bands were measured using a densitometer (BioRad, Hercules, CA) and normalized to the corresponding β-actin amplicon and presented as the ratio of CB1 or CB2 to β-actin. PCR reactions were controlled to run in the exponential phase of the reaction, wherein the amount of PCR product is proportional to the input amount of template (Murphy et al., 1990).

RESULTS AND CONCLUSIONS

A previous report has shown that human PBMCs and peripheral blood neutrophils express mRNA for both CB1 and CB2 (Galieque et al., 1995). Interestingly, when the PBMCs were separated into lymphocyte subpopulations, a rank order of CBR expression (measured by RT-PCR) was observed such that B cells expressed the greatest amount of CBR message and T cells expressed the least. In addition, CB2 mRNA was reported to be expressed to a greater extent in the immune cells than CB1 (Galieque et al., 1995). However, there was no information presented on the demographics of the donors and it was not clear from the study whether or not the CBR mRNA level varied significantly from donor to donor. Questions concerning message variability from donor to donor become more important when attempting to determine the level of CBR expression in various immune cells and in various disease states including marijuana use. We therefore designed our experiments to test for possible differences in CBR mRNA levels in healthy, non-drug abusing individuals of both sexes, two different age groups, and of Caucasian and Asian ethnic origin.

Blood was collected from local laboratory workers, faculty, and students and PBMCs were isolated by density gradient centrifugation. Total RNA was isolated from cell pellets and rigorously treated with DNase to eliminate contamination with genomic DNA. PCR reactions for all three primer pairs were optimized for maximum amplicon production and the amount of input cDNA controlled to result in product levels in the linear range of the amplification curve in order to be able to detect either increases or decreases in cDNA concentration. To control for variations among the samples in input cDNA due to technical errors, each sample was normalized based on the housekeeping gene, ß-actin, and reported as the ratio of CBR gene product to β-actin gene product. To analyze the data, we divided the donor results into multiple groups based on age (two groups), gender (female and male) or ethnicity (Caucasian and Asian). The number of donors per group was as follows: age, 11 between 21–39 years old and 17 between 40-55 years old; gender, 8 females and 20 males; and ethnicity, 20 Caucasians and 8 Asians. As demonstrated in Figure 1, there were no differences in the mean ratio of CB1 to β–actin mRNAs when the donors were stratified based on age (A), gender (B) or ethnicity (C). Similar results were obtained when CB2 message was analyzed with no differences observed in mRNA levels (Figure 2). However,

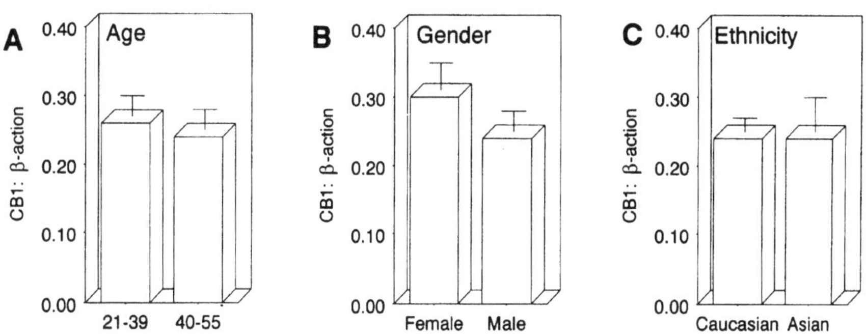

Figure 1. Analysis of CB1 mRNA expression in human PBMCs in different normal donor groups. The data are expressed as the ratio ± SEM of CB1 to B-actin amplicon densities. Donors grouped according to age (A), gender (B), and ethnicity (C).

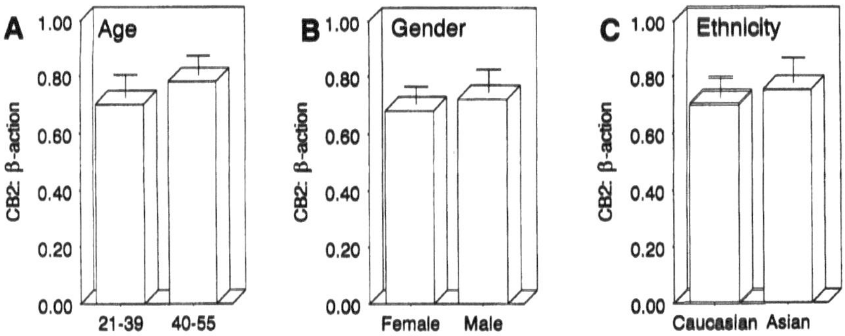

Figure 2. Analysis of CB2 mRNA expression in human PBMCs in different normal donor groups. The data are expressed as the ratio ± SEM of CB2 to B-actin amplicon densities. Donors grouped according to age (A), gender (B), and ethnicity (C).

the level of CB2 mRNA expression was consistently higher than CB1 (0.71±0.06 versus 0.25±0.02) among the donors confirming the report by Galieque that the CB2 subtype is more prominently expressed in human peripheral white blood cells (Galieque et al., 1995).

Previous studies on drug-abusers and animals injected with cannabinoids, as well as *in vitro* models employing immune cells, have demonstrated that marijuana and cannabinoids are immunomodulator. Our current study, using semi-quantitative RT-PCR, demonstrates that both CBR genes are expressed at a relatively constant level in PBMCs from a range of normal human volunteers. Furthermore, as previously reported (Galieque et al., 1995), CB2 expression was higher than CB1 in all donor groups. These conclusions are, at best, preliminary because of the limited number of donors studied. Also, several important groups such as the aged and blacks were not included in this study and inclusion of these donors could possibly modify our conclusions. Additional study groups are also of interest to us. For example, marijuana smoking has been suggested to modulate immune function (Baldwin et al., 1997). It is possible that marijuana use might cause a change in the CBR expression on immune cells. Furthermore, CBR expression has been linked to leukemogenesis (Valk et al., 1997) and therefore CBR analysis of bone marrow cells from these individuals might provide some interesting clues as to the biology of these receptors. Continued analysis of the expression and function of CB1 and CB2 in immune cells will provide a better understanding of their function in immune homeostasis and drug abuse.

ACKNOWLEDGMENTS

This work supported by grants DA03646, DA10683, and DA07245 from the National Institute on Drug Abuse.

REFERENCES

Baldwin, G. C., Tashkin, D. P., Buckley, D. M., Park, A. N., Dubinett, S. M. & Roth, M. D. 1997. Marijuana and cocaine impair alveolar macrophage function and cytokine production. *Am. J. Respir. Crit. Care Med.*, 156, 1606.

Felder, C. C., Veluz, J. S., Williams, H. L., Briley, E. M. & Matsuda, L. A. 1992. Cannabinoid agonists stimulate both receptor- and non-receptor-mediated signal transduction pathways in cells transfected with and expressing cannabinoid receptor clones. *Mol. Pharm.*, 42, 838.

Galieque, S., Mary, S., Marchand, J., Dussossoy, D., Carriere, D., Carayon, P., Bouaboula, M., Shire, D., Le Fur, G. & Casellas, P. 1995. Expression of central and peripheral cannabinoid receptors in human immune tissues and leukocyte subpopulations. *Eur. J. Biochem.*, 232, 54.

Kaminski, N. E., Abood, M. E., Kessler, F. K., Martin, B. R. & Schatz, A. R. 1992. Identification of a functionally relevant cannabinoid receptor on mouse spleen cells that is involved in cannabinoid- mediated immune modulation. *Mol. Pharm.*, 42, 736.

Klein, T., Newton, C. & Friedman, H. 1998. Cannabinoid receptors and immunity. *Immunol. Today*, 19, 373.

Matsuda, L. A., Lolait, S. J., Brownstein, M. J., Young, A. C. & Bonner, T. I. 1990. Structure of cannabinoid receptor and functional expression of the cloned cDNA. *Nature*, 346, 561.

Munro, S., Thomas, K. L. & Abu-Shaar, M. 1993. Molecular characterization of a peripheral receptor for cannabinoids. *Nature*, 365, 61.

Murphy, L. D., Herzog, C. E., Rudick, J. B., Fojo, A. T. & Bates, S. E. 1990. Use of the polymerase chain reaction in the quantitation of mdr-1 gene expression. *Biochemistry*, 29, 10351.

Valk, P., Verbakel, S., Vankan, Y., Hol, S., Mancham, S., Ploemacher, R., Mayen, A., Lowenberg, B. & Delwel, R. 1997. Anandamide, a natural ligand for the peripheral cannabinoid receptor is a novel synergistic growth factor for hematopoietic cells. *Blood*, 90, 1448.

EFFECT OF COCAINE ON CHEMOKINE AND CCR-5 GENE EXPRESSION BY MONONUCLEAR CELLS FROM NORMAL DONORS AND HIV-1 INFECTED PATIENTS

Madhavan P. N. Nair[1], Supriya Mahajan[1], Kailash C. Chadha[2], Narayanan M. Nair[1] Ross G. Hewitt[3], Santosh K.Pillai[1], Priya Chadha[1], Prathibha C. Sukumaran[1] and Stanley A. Schwartz[1]

[1]Buffalo General Hospital
[2]Roswell Park Cancer Institute
[3]Erie County Medical Center
State University of New York at Buffalo
at the Buffalo General Hospital
100 High Street
Buffalo, NY 14203

INTRODUCTION

Although the CD4 molecule is the primary receptor for HIV-1, several studies have shown that in addition to CD4, other co-receptors such as CCR3, CCR5 may be required for efficient viral entry. Recent studies show that HIV suppressing β-chemokines, such as, RANTES, MIP-1α and MIP-1β can bind to HIV-1 entry co-receptors, and, therefore, are important host factors that can influence susceptibility to M-tropic HIV-1 (Broder[1] and Cocchi[2]). Cocaine is one of the most widely abused drugs in the U.S. The last decade has witnessed a great, entangled epidemic of cocaine abuse and HIV-1 infections. Several studies have described the association of cocaine use with susceptibility to and progression of HIV-1 infections (Donahoe[3], Webber[4] and Bagasra[5]). We hypothesize that cocaine can mediate these pathologic effects through modulation of HIV suppressing chemokine and their receptors. The present study examines the effect of cocaine on MIP-1β synthesis by lymphocytes from normal and HIV infected subjects and MIP-1β and CCR 5 gene expression by normal PBMC.

Materials and Methods

The HIV infected subjects were recruited from the Immunodeficiency Services Clinic of the Erie County Medical Center after obtaining appropriate consents. HIV-1 infected subjects were not using cocaine at the time of the study, and their age varied from 25 to 40 years. The peripheral blood mononuclear cell (PBMC) were separated over a cushion of Ficoll-Hypaque. Triplicate cultures containing $3x10^6$ cells/ml of complete medium received either different concentrations of cocaine, or LPS or LPS plus cocaine. Cocaine hydrochloride was provided by the National Institute on Drug Abuse (Rockville, MD) and used at 10^{-6}, 10^{-9} and 10^{-12} M final concentrations.

Neuroimmune Circuits, Drugs of Abuse, and Infectious Diseases
Edited by Herman Friedman *et al.*, Kluwer Academic/Plenum Publishers, 2001

RNA Isolation and Reverse Transcriptase-Polymerase Chain Reaction (RT-PCR)

Cell cultures treated with various concentrations of cocaine were harvested at 8 hr and cytoplasmic RNA was extracted by an acid guanidinium thiocynate-phenol-chloroform (AGPC) method as described by Chomczynski[6]. After extraction, RNA was reverse transcribed to make a DNA copy for use in PCR using a Perkin-Elmer kit (cat. #N808-0143) according to the directions of the manufacturer. The following primer sequences were used in the experiments: β-actin:5'-GTGGGGCGCCCCAGGCACCA-3'(upstream), 5'CTCCTTAATGTCACGCACGATTTC-3'(downstream) 548bp MIP-1β:5'-CCAAACCAAAAGAAGCAAGC3'(upstream), 5'AGAAACAGTGACAGTGGACC-3'(downstream) 320bp CCR5:5'-CTCGGATCCGGTGGAACAAGATGGATTAT-3'(upstream) 5'- CTCGTCGACATGTGCACAACTCTGACTG-3'(downstream) 1117bp.

RESULTS

Cocaine Selectively Suppresses LPS-Induced β Chemokine Production by Lymphocytes from HIV Infected Patients

Data presented in Table1 show the effect of cocaine on LPS-induced MIP-1β production by lymphocytes from normal donors and HIV infected patients. Lymphocytes from normals and HIV-1 infected patients cultured in media alone produced similar levels of MIP-1β (155 and 142 pg/ml of respectively, p<.16). Addition of LPS (10ug/ml) to cultures of lymphocytes from normal donors significantly enhanced MIP-1β production to 1529 pg/ml (p<. 0001) compared to untreated culture (155pg/ml). Lymphocytes from HIV-1 infected patients cultured with similar concentrations of LPS also produced enhanced level of MIP-1β (1490 pg/ml, p<.0001) compared to untreated control culture (142pg/ml). Cocaine did not affect LPS-induced MIP-1β production by normal lymphocytes. Normal PBMC cultured with LPS plus 10^{-6}, 10^{-9} and 10^{-12} M concentrations of cocaine produced 1479, 1500, and 1471 pg/ml respectively compared to 1529 pg/ml produced by culture treated with LPS alone. In contrast, cocaine significantly suppressed LPS-induced MIP-1β production by PBMC from HIV-1 infected patients in a dose-

Table 1. Effect of cocaine on LPS-induced MIP-1β production by PBMC from Normal Donors and HIV infected Patients

LPS (10ug/ml)	Compound added Cocaine M	Normal MIP-1β titer (pg/ml)	HIV Infected
0	0	155.0 ± 15.2	142.2±10.8
LPS	0	1529.9 ± 80.4	1490.0 ± 78.0
LPS	10^{-12}	1471.0 ± 33.0 (NS)	1184.6 ± 28.1 (p< 0.04)
LPS	10^{-9}	1500.4 ± 70.1 (NS)	1153.5 ± 76.8 (p< 0.04)
LPS	10^{-6}	1479.5 ± 66.2 (NS)	930.6 ± 64.4 (p< 0.007)

PBMC ($3x10^6$/ml) from normal and HIV infected patients were cultured alone, or with LPS + different concentrations of cocaine for 24 hr and culture supernatants were quantitated for MIP-1β by ELISA (R&D System, Mineapolis, MN). The data are the mean ± SD values of 5 independent experiments performed using PBMC from 5 different normals and 5 different HIV infected patients. The control levels of MIP-1β producted by FBMC from normals and HIV-1 infected subjects were similar (p = 0.1). Statistical significance of difference was calculated by "t"test.

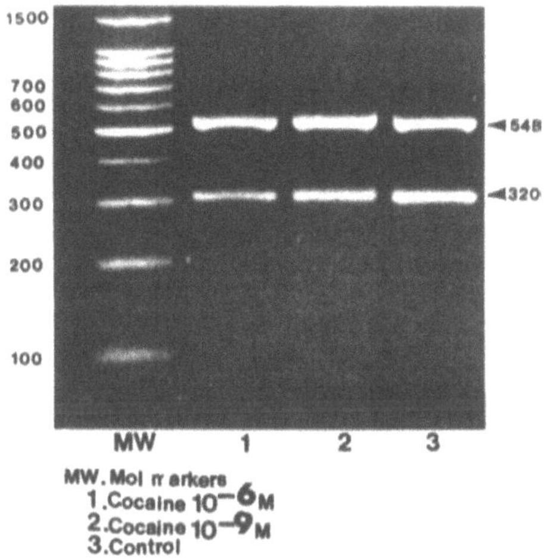

Figure 1A. Cocaine suppresses MIP1-β gene expression by lymphocytes as measured by RT-PCR. PBMC ($3x10^6$/ml) from normal donors were treated with 10^{-6} and 10^{-9} M cocaine for 8 hr, mRNA extracted, reverse transcribed, and amplified with MIP1-β and the house-keeping gene β-actin primers in the same tube and electrophoresed. cDNA from amplified PCR products of MIP-1β and β-actin banded at 320bp and 548bp respectively. Lane MW; molecular weight markers, Lane 1; cocaine 10^{-6}M, Lane 2; cocaine 10^{-9} M, Lane 3; untreated control. This data represents results from a single experiment. The experiment was repeated independently 4 times using PBML from 4 different subjects with similar results.

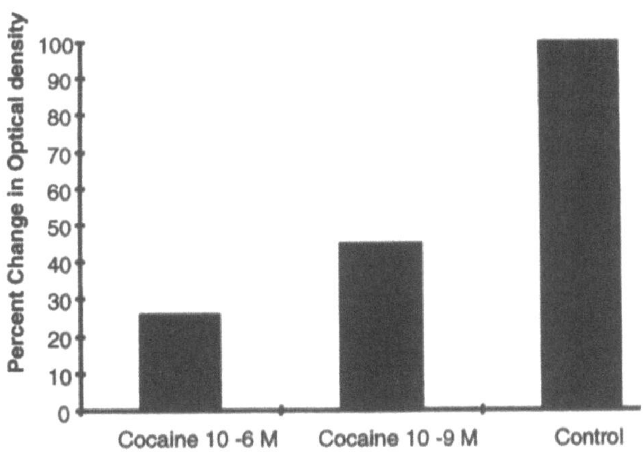

Figure 1B. Quantitation of changes in MIP-1β gene expression. Percent change in laser densitometry reading of the photographic negatives of experimental values after normalization with respective β-actin values and then compared with control values from Figure 1A Lane 1; cocaine 10^{-6}M, Lane 2;. cocaine 10^{-9} M, Lane 3; untreated control.

dependent manner; the levels of MIP-1β produced were 930, (P<0.007), 1153 (P<.04) and 1184pg/ml (P<.04) respectively for 10^{-6}, 10^{-9} and 10^{-12} M concentrations of cocaine compared to 1490 pg/ml produced by PBMC from HIV infected patients treated only with LPS. These results demonstrate that cocaine selectively suppresses LPS-induced MIP-1β production by PBMC from HIV infected patients while leaving cells from normal donors unaffected.

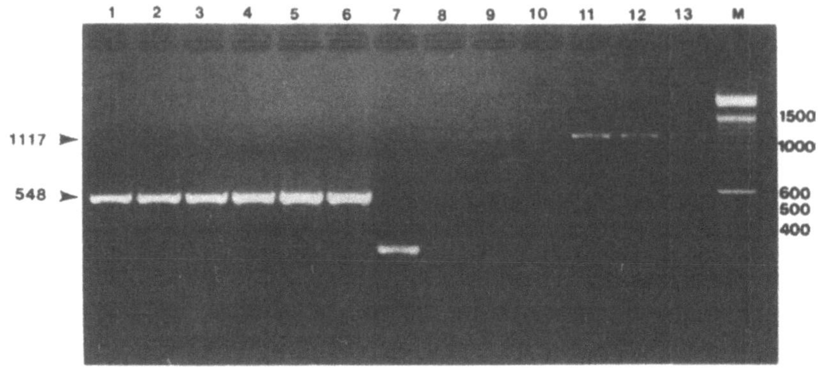

Figure 2A. Cocaine upregulates CCR5 gene expression by lymphocytes as measured by semiquantitative RT-PCR. RNA from PBML ($3x10^6$/ml) untreated and cocaine treated cultures for 8 hr, were reverse transcribed and amplified with CCR5 primers and electrophoresed. 1-6. β-actin, 7. Reagent Control, 8. Cocaine 10^{-9} M, 9. Cocaine 10^{-6} M, 10. Control, 11. Cocaine 10^{-9} M, 12. Cocaine 10^{-6} M, 13. Control M.MW. Results are expressed for two experiments (8-10 and 11-13).

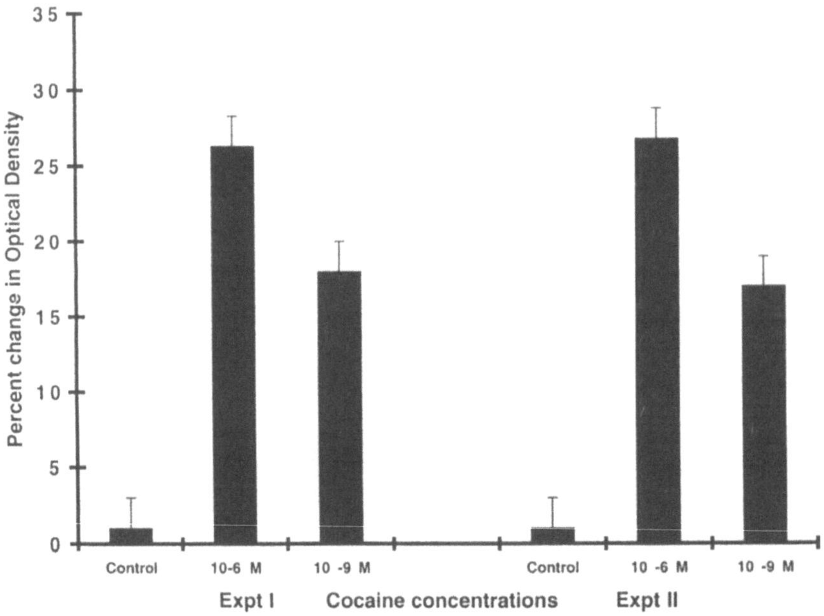

Figure 2B. Quantitation of changes in CCR5 gene expression. Percent change in the densitometry reading of the photographic negatives after normalization with respective β-actin values and then compared with control values from Figure 2A. Cocaine significantly upregulated CCR5 gene expression in dose-dependent manner for both experiments.

Cocaine Modulates the Expression of MIP-1β and CCR5 Gene Expression by PBMC

Figure 1A demonstrates that treatment of PBML with cocaine suppressed transcription of the MIP-1β gene. Lane MW reflects molecular weight markers. Cocaine at 10^{-6} M

(Lane1, OD=38) and 10^{-9} M (Lane 2, OD=66) significantly suppressed MIP-1β gene expression compared to control culture (Lane 3, OD=146). Figure1B shows the percent change of OD values from Figure 1A. Cocaine at 10^{-6}, and 10^{-9} M concentrations significantly suppressed MIP-1β gene expression in a dose dependent manner; the % suppression was 73.9% and 54.7% respectively for 10^{-6} M (Lane1) and 10^{-9}M (Lane 2) compared to untreated control culture (Lane 3).

Figure 2A shows the effects of cocaine on gene expression of the CCR5 chemokine receptor/HIV co-receptor. Results are expressed in Figure 2A for two experiments using separate PBML. Lanes 1-6 demonstrate β-actin gene expression of control and cocaine treated cultures. Treatment of cultures with cocaine did not affect the constitutively expressed β-actin gene (Lanes 1-2 and 4-5) and was comparable to the control (Lanes 3 & 6). Control cells (for expt #1) incubated in medium alone for 8 hr demonstrated detectable CCR5 gene expression (Figure, 2A Lane 10, OD=38). Cultures treated with cocaine at 10-6M (Lane 9, OD=48; 27% upregulation) and 10-9M (Lane 8, OD=45; 17% upregulation) concentrations significantly upregulated CCR5 gene expression compared to control culture (Lane10, OD=38). In the experiment #2, control PBML incubated in medium alone for 8 hr demonstrated detectable CCR5 gene expression (Lane13, OD=41). Cultures treated with cocaine at 10-6 (Lane 12, OD=52; 26% upregulation) and 10-9M (Lane 11, OD=48; 18% upregulation) concentrations also upregulated CCR5 gene expression compared to control culture (Lane 13, OD=41). Data presented in Figure 2B show the % change of the densitometry reading after normalized with β-actin from control and treated cultures of these two experiments. Cocaine significantly upregulated CCR5 gene expression in a dose-dependent manner compared to control culture.

DISCUSSION

Our studies demonstrate that cocaine selectively suppresses LPS-induced MIP-1β production by PBMC from HIV infected patients while leaving cells from normal donors unaffected. These results suggest that although lymphocytes from HIV infected patients who are not using cocaine can produce normal levels of MIP-1β in response to external stimuli as presented, lymphocytes from HIV infected patients who use cocaine may produce lower levels of MIP-1β in response to external stimuli such as infections thereby increasing the risk of acquisition or progression of HIV-1 infection. Our results suggest that one of the mechanisms of cocaine associated immunosuppression in HIV infected subjects may be cocaine-mediated inhibition of HIV protective chemokine(s) by PBMC. In patients with active HIV infections, the virus uses an expanded range of co-receptors, including the chemokine receptors, CCR3, CCR2β, CCR5 and CXCR4; this adaptation was associated with progression to AIDS. In our investigations, we demonstrates that cocaine can upregulate CCR5 gene expression by PBMC in a dose-dependent manner. These studies support our hypothesis that cocaine is a co-factor in susceptibility to HIV infection and disease progression and suggest that cocaine may mediate these effects by two previously unrecognized, interrelated and self reinforcing mechanisms: upregulation of HIV entry coreceptors and inhibition of host protective chemokines. Our studies on the modulation of β-chemokine, MIP-1β and HIV-1 entry co-receptor, CCR5 by cocaine in HIV-1 infected subjects may yield new information on the role of drugs of abuse in the natural history of HIV-1 infections.

Acknowledgements

Supported by NIH grants RO3 NIDA 1119-01, R01 MH 47225, R01 DA 10632, RO1 DA 12366 and the Margaret Duffy and Robert Cameron Troup Memorial Fund of the Buffalo General Hospital, Kaleida Health.

REFERENCES

1. C.C. Broder, and R.G. Collman, Chemokine receptors and HIV, *J. Leukoc. Biol.* 62:20 (1997).
2. F. Cocchi, , A.L. DeVico, A. Garzino-Demo, S.K. Arya, R.C. Gallo, and P. Lusso, Identification of RANTES, MIP-1α MIP-1β as the major HIV suppressive factors produced by CD8+ T cells, *Science* 270:1811 (1995).
3. R.M. Donahoe, Drug abuse and AIDS: Causes for the connection, *NIDA Res. Monogr.* 96:181 (1990).
4. M.P. Webber, E.E. Schoenbaum, M.N. Gourevitch, D. Buono, and R.S. Klein, A prospective study of HIV disease progression in female and male drug users, *AIDS* 4:257 (1999).
5. O. Bagasra, and R.J. Pomerantz, Human immunodeficiency virus type 1 replication in peripheral blood mononuclear cells in the presence of cocaine, *J. Infec. Dis.* 168:1157 (1993).
6. P. Chomczynski, and N. Saachi, Single step method of RNA isolation by acid guanidinium thiocyanate-phenol-chloroform extraction, *Anal.Biochem.* 162:156 (1987).

INTRAVENOUS COCAINE ABUSE: A RODENT MODEL FOR POTENTIAL INTERACTIONS WITH HIV PROTEINS

Departments of Anatomy/Neurobiology
Neurology and Graduate Center for Toxicology
University of Kentucky Medical Center
Lexington, Kentucky, 40506

INTRODUCTION

Drug abuse is a major cause of morbidity in young adults and has vast socioeconomic consequences for society. Combined with the fact that this population is also highly susceptible to contracting HIV infection, drug abuse has major health consequences. It is estimated that nearly 1.7 million Americans are cocaine users, (half of which use crack), 2.4 million report heroin use and 4.9 million have used methamphetamine at sometime of their lives. Drug abusing populations initially either ingest, smoke or snort these drugs; however, with continued use abusers switch to intravenous drug use either due to development of tolerance or development of nasal ulcers. Injecting drug users are at higher risk for contracting HIV infection and at developing neurological and systemic complications. According to latest report by UNAIDS and WHO (1999) in nearly one third Americans infected with HIV, injection drug use is a risk factor making drug abuse the fastest growing vector for spread of HIV infection.

Establishing a well-characterized model for intravenous drug abuse and HIV neurotoxicity will allow studies to delineate the mechanisms of drug/HIV interactions and common pathways of toxicity. One critical consideration with regard to establishing an in vivo rodent model of gp120/Tat neurotoxicity in cocaine abuse is the route of cocaine administration. Commonly used routes of cocaine administration to rodents have been via intraperitoneal (IP) injection, subcutaneous (SC) injection or oral/intragastric (PO) administration. These exposure routes exhibit moderate and sustained plasma levels of cocaine compared to administration of cocaine via intravenous (IV) bolus that results in a characteristic rapid peak concentration followed by precipitous clearance (Booze et al., 1997). The route of administration of the drugs of abuse in animal models should mimic the pharmacokinetics of human drug abuse, i.e. IV bolus (Evans et al., 1996). This is particularly important for cocaine and morphine since they have short half-lives. Recently, much attention has been paid to development of animal models that closely mimic the human drug abuse (Booze et al., 1997; Hutchings et al., 1991).

The general applicability of models for IV drug self-administration has been constrained by the use of chronic indwelling catheters that are exteriorized and tethered. A recent technical innovation is a SC implantable catheter as a port for the routine and repeated IV administration of drugs to group-housed rats (Mactutus et al., 1994). Using this implantable access port, cocaine-induced locomotor activity can be evaluated in freely moving rats following IV injections (Wallace et al., 1996; Booze et al., 1999). In our studies of adult male rats, IV administration of 3.0 mg/kg cocaine elevated arterial levels of cocaine by 30 seconds to approximately 2,500 ng/ml (Booze et al., 1997). This is comparable to the peak arterial levels associated with euphoric and cardiovascular effect in male human volunteers (Evans et al., 1996). Given that the risk factor for HIV infection in drug users is via needle sharing and injection drug use, the IV route of cocaine administration in rats is clearly preferable due to the similar route of administration and kinetic profile found in studies of at-risk drug abusers

Thus, in the present studies we have developed an in vivo rodent model for delivering cocaine via the intravenous route. In addition, we have begun to explore the neurotoxicity of HIV neurotoxic proteins in the rat striatum. Ultimately, we will use this model to study the modulatory effects of cocaine on gp120/Tat induced neurotoxicity.

METHODS

Cocaine Injections: Adult (3-month-old, N=25) male Sprague-Dawley rats were maintained according to NIH guidelines. All animals were surgically implanted with vascular catheters as described by Mactutus et al. (1994). Following at least 24hr recovery from surgery, the animals were assigned to 3 groups that received either 0.5, 1.0 or 3.0 mg/kg cocaine IV. All injections were delivered over 15 sec in a volume of 0.2 ml and were followed by flushing for 15 sec with 0.2ml saline. Arterial withdrawals commenced immediately after injection for 20 sec and 400 ul samples were collected at 8 time points (0.5, 1.0, 1.5, 2.0, 5.0, 10, 20, and 30 min) from each animal. Methods for sample extraction were according to Booze et al., 1997. Plasma samples were analyzed by single ion monitoring following injection onto a Hewlett-Packard 5890 gas chromatograph-5970 mass spectrometer equipped with a 7673 autoinjector.

HIV Proteins: We used recombinant gp120 from HIV-SF2 provided as a gift by Chiron corporation. Recombinant gp120 was made in a mammalian cell line (CHO cells) and purified to 100% gp120 products (95% has a molecular wt. of 120kD and the remaining are break down products of gp120 as determined by western blot analysis). Tat 1-72 (first exon) was synthesized in our laboratory. Tat produced in our laboratory is >98% pure, contains mainly the monomeric form of Tat and can transactivate HIV-LTR (Ma et al., 1997).

Striatal Injections and Tissue Processing: Adult male SD rats (N=15) were stereotaxically injected in the striatum with either vehicle (saline), Tat (1,5,50 μg) or gp120 (100,250,400 ng). Bregma: M-L 2.5 mm; A-P 0.3 mm; D-V 4.8 mm; 1 μl volume/60 sec injection. Animals were perfused 7 days post-injection. Vibratome cut free floating sections were stained with cresyl violet and monoclonal glial fibrillary acidic protein (GFAP). Assessments of lesion size (cresyl violet staining) and GFAP response were conducted. Lesion size was assessed via computerized morphometry on consecutive striatal tissue sections. GFAP response was defined as the number of GFAP immunopositive cells in the striatum. GFAP immunoreactivity is expressed relative to the non-injected (control) side of the striatum. Details of cell counting methods are as given in Bansal et al., in prep.

RESULTS

Surgical implantation of a subcutaneous IV injection port was performed according to Mactutus et al., 1994 and Booze et al., 1996; 1999 (Figure 1). The IV administration of cocaine with carotid artery sampling produces a pronounced, but transient, peak in arterial kinetics (Figure 2). The IV administration of cocaine with carotid artery sampling initiated immediately after termination of drug injection, produced a pronounced, but transient, dose-response peak in arterial concentrations with dose-independent elimination kinetics. In the rat (Figure 2, solid lines) peak plasma concentrations of cocaine were noted at 30 seconds as a linear ($r=0.60$, $p \leq 0.004$) dose-dependent function (370 ± 14, 755 ± 119, 2553 ± 898 ng/ml for the 0.5, 1.0 and 3.0 mg/kg groups, respectively). For comparison, data from Evans et al., 1996 shows comparable kinetics for humans (Figure 2, dashed lines). By 20 min post-injection (data not shown) the arterial plasma concentrations had decreased by an order of magnitude for all groups.

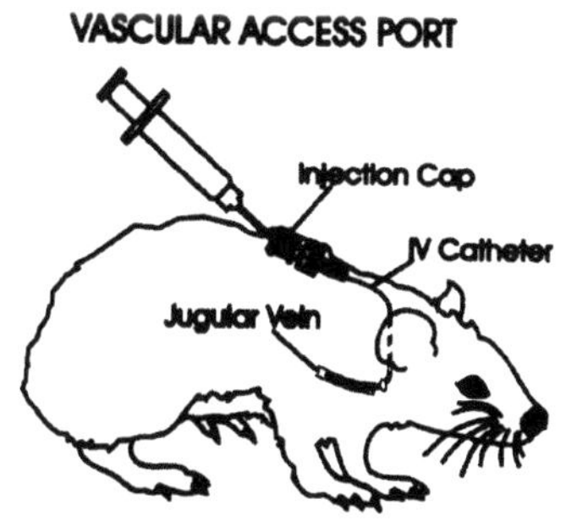

Figure 1. Schematic drawing of the sub-cutaneous IV access port for administering cocaine IV in rats. Using this port system, animals may be group-housed and maintain a flexible dosing regime.

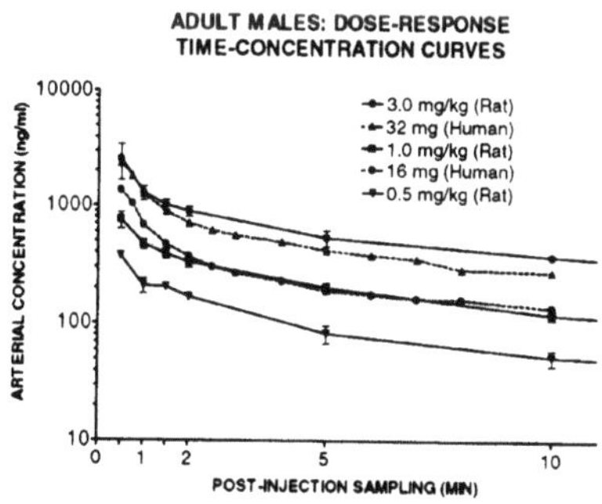

Figure 2. Dose curves following IV cocaine administration in rats (solid lines) and humans (dashed lines). Bolus IV cocaine produces a similar pharmaco-kinetic profiles in both rats and humans.

Injection of either gp120 or Tat into the rat striatum produced lesions and increased GFAP immunoreactivity, relative to contralateral and vehicle injected striatum. The ipsilateral side injected with gp120 or Tat showed increased staining. The astroglial cells were increased in size and the number of processes were also increased (Figure 3), whereas the changes in the astrocytes on noninjected or vehicle injected side were not significant. Injection of gp120, 250 ng and higher produced a striatal lesion (area of tissue loss) and a graded increase in the striatal GFAP positive cells. A dose of 100 ng did not produce a lesion but we saw marked astroglial changes. Injection of Tat, 5μg and higher produced a striatal lesion and astroglial changes, whereas Tat 1μg did not show any significant change relative to control side.

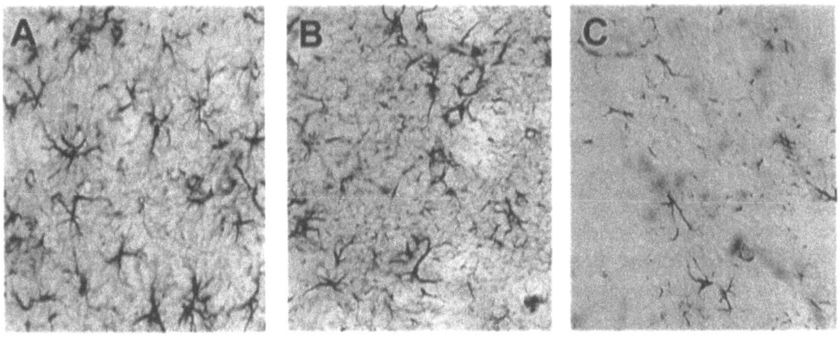

Figure 3. Micrograph (x200) of striatum of rats injected with 250 ng gp120 (A), 5 μ g tat (B) and saline (C). The processes of astrocytes and astrocytes are increased in number on the gp120 or tat injected side relative to saline injected side.

CONCLUSIONS

In summary, we have developed in vivo approaches for determining the mechanistic interactions between cocaine and HIV neurotoxic proteins. Further studies are needed to delineate the mechanisms of their interaction and to find the common pathways of their toxicity, which can be exploited to prevent the progression of this deadly disease in HIV infected population so as to allay the sufferings of these people. For this, we will use this well characterized model of intravenous cocaine use to study the modulatory effects of cocaine on the gp120 and tat induced neurotoxicity.

ACKNOWLEDGEMENTS

Supported by NIDA DA11337, DA 13137, DA09160 (RMB) and NINDS NS39253, NS38428, NS39184 (AN).

REFERENCES

Bansal, A., Nath, A., Maragos, W., Mactutus, C. F. Hauser, K. F. and Booze, R.M. (2000) Neurotoxicity of HIV-1 proteins gp120 and Tet in the rat striatum. *Brain Res.*, in press

Booze, R.M., Lehner, A.F., Wallace, D.R., Welch, M.A. and Mactutus, C.F. (1997) Dose-response cocaine pharmacokinetics and metabolic profile following intravenous administration and arterial sampling in unanesthetized freely moving male rats. *Neurotoxicology Teratology*, 19:7-15.

Booze, R.M., Wood, M.L., Welch, M.A., Berry, S. and Mactutus, C.F. (1999). Estrous cyclicity and behavioral sensitization in female rats following repeated intravenous cocaine administration. *Pharmacol. Biochem. Behav.* 64: 605-610.

Evans, S.M., Cone, E.J. and Henningfield, J.E. (1996) Arterial and venous cocaine plasma concentration in humans: Relationship to route of administration, cardiovascular effects and subjective effects. *J. Pharmacol. Exp. Ther.*, 279:1345-1356.

Hutchings, D.E., Dow-Edwards, D. (1991) Animal models of opiate, cocaine, and cannabis use. *Clin Perinatol* 18: 1-22.

Ma, M. and Nath, A. (1997) Molecular determinants for cellular uptake of Tat protein of human immunodeficiency virus type 1 in brain cells. *J. Virology, 71:2495-2499.*

Mactutus, C.F., Herman, A.S. and Booze, R.M. (1994) Chronic intravenous model for studies of drug (ab)use in the pregnant and/or group-housed rat: An initial study with cocaine. *Neurotoxicology Teratology, 16:183-91.*

UNAIDS and WHO, *AIDS epidemic update*: December 1999.

Wallace, D.R., Mactutus C.F. and Booze, R.M. (1996) Repeated intravenous cocaine administration: Locomotor activity and dopamine D2/D3 receptors. *Synapse*, 23:152-163.

SUBSTANCE P RECEPTOR MEDIATED MACROPHAGE RESPONSES

Ian Marriott and Kenneth L. Bost

Department of Biology
University of North Carolina at Charlotte
Charlotte, NC 28223

INTRODUCTION

Macrophages play an important role in the initiation of immune responses. They phagocytose immunogens and pathogens, present antigens to T lymphocytes, and provide an important source of soluble and cell surface co-stimulatory molecules required for optimal immune activation. Over the years our laboratory has focused on the importance of substance P receptor expression by macrophages and how this receptor signals macrophage activation. These studies have indicated that substance P/substance P receptor interactions serve as a positive signal for macrophage activation and function. Signaling through the substance P receptor can induce NF-kB translocation to the nucleus, and serve as a co-stimulus for macrophage-derived IL-12p70. Furthermore, mice challenged with an intracellular pathogen of macrophages, *Salmonella*, were more susceptible to infection when substance P/substance P interactions were antagonized. Taken together, these studies demonstrate that substance P receptor expression contributes in a positive manner to normal activation and function of macrophages in a protective host response.

SUBSTANCE P-INDUCED INTRACELLULAR SIGNALLING

The substance P receptor is a G-protein coupled seven transmembrane spanning molecule which has been extensively investigated with regard to intracellular signaling in a variety of cell types[1]. Agonists interacting with the substance P receptor are thought to initiate an elevation of intracellular calcium in some cell types. Therefore we began a series of studies to address the importance of intracellular calcium mobilization in macrophage function, and to determine whether substance P receptors might use such a signaling pathway. It was clear from early work that increased mobilization of intracellular calcium in murine macrophages was a potent and rapid signal for macrophage activation, as indicated by induction of IL-6[2]. These results suggested an important, calcium-mediated signaling pathway was present and could augment macrophage function. What was not clear from these results, or from the literature, was a clear delineation of endogenous factors which could induce significant mobilization of intracellular calcium in these cells. We demonstrated that two known macrophage activating factors, interferon gamma and lipopolysaccharide, were incapable of inducing mobilization of intracellular calcium in isolated macrophages[3]. Furthermore, the kinetics of IL-6 mRNA induction by calcium-mobilizing agonists were significantly more rapid than that seen with lipopolysaccharide or interferon gamma. Together these results suggested calcium-dependent and independent signaling pathways for activation of macrophages.

To address whether substance P receptors could induce rapid mobilization of intracellular calcium in isolated macrophages, changes in intracellular calcium following treatment with substance P were monitored fluorimetrically. Despite substance P receptor expression by these cells, no substance P-induced calcium mobilization could be detected[4]. Co-stimulation with lipopolysaccharide and substance P induced no mobilization, whereas other agonists (thapsigargin or platelet activating factor) were quite effective in increasing intracellular calcium concentrations. We concluded from these studies that substance P did not induce any detectable mobilization of intracellular calcium within minutes of interaction with macrophage receptors. Conversely, substance P induced rapid activation of NF-kB and initiated its translocation to the nucleus in these cells[4]. Thus the predominant signaling pathway for substance P in macrophages is not calcium-dependent, but does involve activation of the transcription factor, NF-kB.

SUBSTANCE P RECEPTOR-MODULATED MONOKINE SECRETION

Since NF-kB is an important transcription factor in the regulation of cytokine production, the demonstration that NF-kB is activated following interaction with substance P suggested that macrophages might respond to this neuropeptide with increased monokine secretion. Monokines such as IL-6 and IL-12, but not TGF-beta, are known to be upregulated following translocation of NF-kB to the nucleus. Therefore we began a series of studies to investigate substance P-induced monokine secretion.

IL-12 is a heterodimer, composed of IL-12p35 and IL-12p40 subunits[5]. The IL-12p40 subunit can be secreted in excess over the biological agonistic heterodimer following stimulation[6], therefore it is important to determine what form of IL-12 is being produced. By itself, substance P can induce significant IL-12p40 production in otherwise unstimulated murine macrophages[7]. However, it was also clear that merely signaling via the substance P receptor alone was not sufficient to induce substantial secretion of IL-12p70. Co-stimulation of substance P with lipopolysaccharide synergistically induced secretion of IL-12p70, demonstrating that agonists interacting with the substance P receptor could be an important co-stimulus for production of this monokine.

While IL-12 functions as a pro-inflammatory mediator and augments the development of cell mediated immune responses[8], significant levels of monokines, like transforming growth factor-beta (TGF-beta), can severely limit such immune responses[9]. If signaling through the substance P receptor on macrophages is pro-inflammatory, we questioned what effect this neuropeptide might have on secretion of a monokine with anti-inflammatory properties. While substance P by itself elicited modest increases in TGF-beta expression, this neuropeptide could dramatically diminish lipopolysaccharide or interferon-gamma induced TGF-beta production[10]. This effect could be blocked with specific, competitive antagonists of substance P, demonstrating the receptor-mediated nature of the response. Therefore these studies suggested a previously unrecognized mechanism where substance P may act as a proinflammatory mediator by limiting the production of excessive levels of TGF-beta.

Taken together, these in vitro studies suggested that substance P could augment the initiation of macrophage-mediated immune responses in several ways. The finding of substance P-induced IL-12 co-stimulation was especially intriguing since this monokine is critical for optimal cell mediated immune responses. We next questioned whether substance P/substance P receptor interactions might contribute to the protective cell mediated immune response using an in vivo murine model of intracellular pathogenesis.

EXPRESSION OF PREPROTACHYKININ AND SUBSTANCE P RECEPTOR mRNA IN VIVO FOLLOWING ORAL CHALLENGE WITH *SALMONELLA*

If substance P and its receptor are to play an important role in mucosal immune responses against an intracellular pathogen, such as *Salmonella*, expression of these molecules at sites of infection might be expected. We questioned whether the mRNA encoding substance P (i.e. preprotachykinin mRNA) and its receptor might be upregulated at mucosal sites following oral inoculation with *Salmonella*. Mice were challenged with *Salmonella* and at varying times post infection, RNA from the Peyer's patches, mesenteric lymph nodes, and spleen were subjected to reverse transcription-polymerase chain reaction. Beta and gamma preprotachykinin mRNA expression was significantly elevated in lymphoid organs within hours following oral inoculation with *Salmonella*[11]. Increases in preprotachykinin mRNA expression from leukocytes within the mesenteric lymph nodes was localized within the adherent cell populations. This result suggested that adherent macrophages may represent one cell population responding to this bacterial invasion by increased preprotachykinin mRNA expression. This finding supported previous in vitro work suggesting that macrophages could be an extra-neuronal source of substance P[12].

A similar experimental design was used to demonstrate increases in substance P receptor mRNA expression following infection with *Salmonella*. This enteric pathogen induced rapid and dramatic increases in substance P receptor mRNA expression in mucosal lymphoid organs[13]. The kinetics of increased substance P receptor mRNA expression followed the appearance of viable bacteria into that particular organ. This correlation suggested that the presence of *Salmonella* was responsible for these increases.

Taken together, these studies demonstrated an unexpected relationship. Following invasion of *Salmonella* at mucosal sites, there was a rapid and dramatic upregulation of mRNA that encoded substance P and its receptor. These results suggested that this receptor/ligand interaction might play a role in the host response against this intracellular pathogen. Increased expression of the mRNAs for substance P and its receptor suggested additional in vivo studies would be required to demonstrate the functional importance of such expression.

INCREASED SUSCEPTIBILITY TO SALMONELLOSIS FOLLOWING ANTAGONISM OF SUBSTANCE P/SUBSTANCE P RECEPTOR INTERACTIONS IN VIVO

Our initial studies utilized mice which had received neonatal treatment with capsaicin. Such treatment has been shown to severely limit the presence of afferent, sensory neurons in the peripheral organs of rodents[14]. Such neurons contain high concentrations of neuropeptides such as substance P, vasoactive intestinal peptide, and somatostatin, which are released in response to neurogenic stimulation. We challenged groups of capsaicin treated mice with *Salmonella* given orally and compared their rates of survival with groups of control neonates treated with vehicle alone. As shown in Figure 1, mice receiving treatment with capsaicin were significantly more susceptible ($p < 0.01$) to death via salmonellosis than mice treated with vehicle. These results were the first to suggest that afferent sensory neurons could contribute in a positive manner to a protective host response against an intracellular pathogen of macrophages.

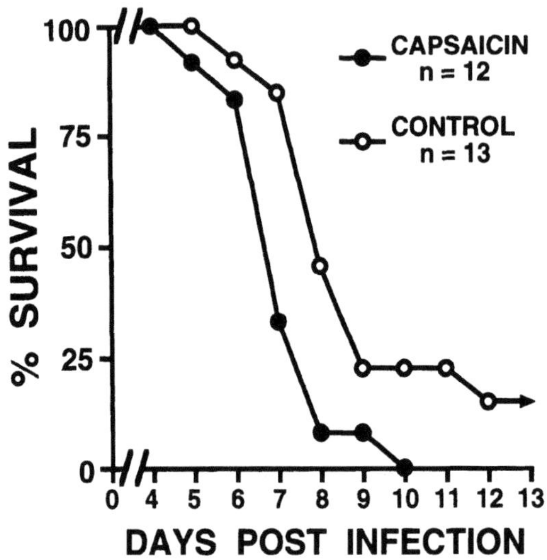

Figure 1. Neonatal capsaicin treatment decreases survival of mice challenged with oral inoculations of *Salmonella*. Two day old neonatal mice were treated with capsaicin or with a vehicle control. When treated mice reached approximately 12 weeks of age, they were challenged with oral *Salmonella dublin* (5 X 10^5 organisms) and monitored for the ability to survive this infection.

The capsaicin-depletion study shown in Figure 1 could not identify the nature of the contribution made by the peripheral nervous system to this immune response. Therefore we focused our attention on substance P as a possible explanation for this observation. Mice were implanted with osmotic pumps which released a constant amount of a substance P antagonist, spantide II. Following oral challenge with *Salmonella*, groups of mice were monitored for the development of salmonellosis and for immune responsiveness. Mice receiving the substance P antagonist had increased salmonellosis and decreased survival when compared to mice receiving a control peptide[13]. Furthermore, spantide II treated mice had reduced IL-12 and interferon gamma responses to *Salmonella* than that seen in control groups. These results were not due to differences in the initial dissemination of *Salmonella* from the gut or due to global immune dysfunction induced by the substance P antagonist. Rather, these results demonstrated a marked suppression of the protective cell mediated immune response against *Salmonella*. Since this bacteria is an intracellular pathogen of macrophages, these results are consistent with substance P receptor-mediated activation of this cell population during protective responses.

MODULATION OF SUBSTANCE P RECEPTOR EXPRESSION ON MACROPHAGES

We have demonstrated that signaling through substance P receptors can augment protective cell mediated immune responses, presumably through IL-12 induced interferon gamma production. However, studies by us and others[15] suggest that substance P can also augment humoral immune responses. The molecular basis for such a finding is not clear, though we have suggested that B lymphocytes at certain stages of differentiation might express functional substance P receptors.

To identify potential soluble factors which might affect substance P receptor expression on macrophages, we have begun in vitro studies. For these studies, peritoneal macrophages were isolated and cultured in the presence of IL-4 (30 pg/ml), interferon gamma (30 pg/ml), lipopolysaccharide (500 ng/ml) or substance P (10 nM). After 8 hours, RNA was extracted and reverse transcription-polymerase chain reaction (RT-PCR) performed to quantify substance P receptor mRNA expression. Surprisingly, both IL-4 or interferon gamma could augment substance P receptor mRNA expression by murine macrophages over the constitutive expression seen in non-treated controls (Figure 2).

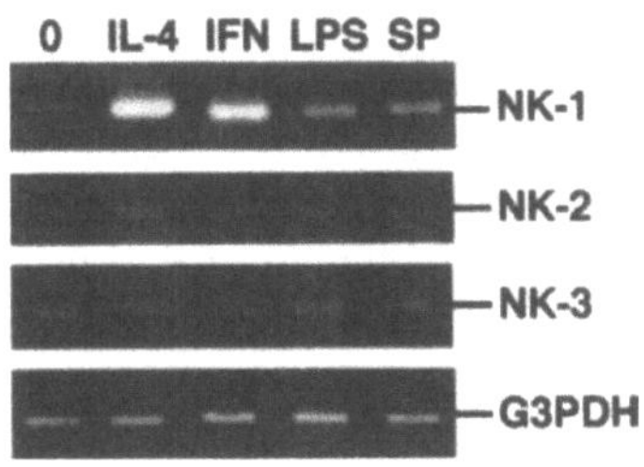

Figure 2. Quantification of neurokinin receptor mRNA expression following treatment of cultured, peritoneal macrophages with soluble factors. RNA was isolated from macrophages treated with the indicated factors (untreated, 30 pg/ml IL-4, 30 pg/ml interferon gamma, 500 ng/ml lipopolysaccharide, or 10 nM substance P) for 8 hours. RT-PCR was used to quantify expression of NK-1, NK-2, NK-3, or glyceraldehyde 3-phosphate dehydrogenase mRNAs. Results are shown as amplified products electrophoresed on ethidium bromide stained agarose gels.

These results have important implications for the interaction of macrophages with T lymphocytes. T helper type 1 cells can secrete interferon gamma and support optimal cell mediated immune responses, whereas T helper type 2 cells can secrete IL-4 and support optimal humoral immune responses. The fact that these two very different cytokines can act on macrophages to induce substance P receptor expression suggests that amplification of cell mediated and humoral immune responses may both be sensitive to signaling through this receptor on macrophages. If these results can be extrapolated to in vivo models, pathogens or immunogens which induce IL-4 or interferon gamma might augment substance P-receptor mediated proinflammatory mechanisms by increasing receptor expression in macrophages.

The mammalian tachykinins, substance P, neurokinin A (substance K), and neurokinin B, have preferential binding affinities for the neurokinin receptor subtypes, NK-1, NK-2 and NK-3, respectively. Since some tissues can express all three receptor subtypes, it was also of interest to determine if NK2 and NK3 receptor mRNAs could be modulated on macrophages. Interestingly, mRNA for NK2 and NK3 receptors could be detected in these cells. However, while NK1 receptor mRNA expression was significantly upregulated by T helper type 1 and T helper type 2 cytokines, the same modulation was not observed for NK2 or NK3 mRNA expression. Thus, based on these results, NK1 receptor expression seems the most sensitive to modulation by macrophage differentiation factors.

SUBSTANCE P RECEPTOR EXPRESSION AND AMPLIFICATION OF IMMUNE RESPONSES

A hallmark of the developing immune response is the synergistic interaction between cells and their products in an effort to amplify the response. Such amplification and positive feedback loops serve to recruit cells to the site of infection while providing activation signals which continue to expand the response. Based on our knowledge to date, it appears that substance P receptor expression plays an important role in this amplification in the initial stages of a developing immune response. As shown in Figure 3, ligation of substance P receptors results in an intracellular signal which includes translocation of NF-kB to the nucleus. This signal serves as a co-stimulus to increase cytokine production, like IL-12, which could augment T helper type 1 responses. The resulting interferon gamma production from T helper type 1 cells (or CD8 or Natural Killer cells) would upregulate substance P receptor expression on macrophages, further augmenting the response. Conversely, IL-4 production by T helper type 2 cells during the development of a humoral immune response would also upregulate substance P receptor expression on macrophages. Speculatively, this could create a positive feedback loop for the augmentation of antibody responses even though the nature of the substance P-induced macrophage signal to T helper type 2 cells has not been defined.

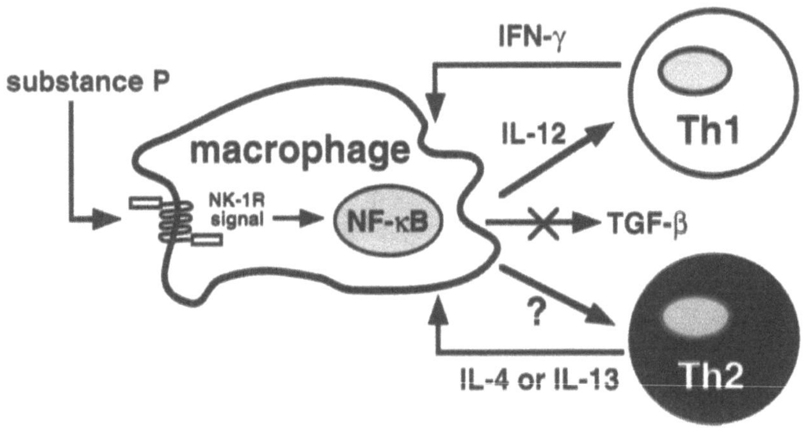

Figure 3. Schematic representation of substance P-mediated macrophage activation and the possible effect on T lymphocyte responses. Neuronally derived substance P interacting with substance P (NK1) receptors on macrophages induce NF-kB translocation to the nucleus. This receptor signaling could limit excess production of TGF-beta, and augment IL-12 secretion. Substance P-induced IL-12 production could then augment interferon gamma production by T helper type 1 (TH1) lymphocytes. Interferon gamma secretion by TH1 or IL-4 or IL-13 secretion by T helper type 2 (TH2) lymphocytes could also serve as a positive signal to augment substance P receptor expression by macrophages.

Inherent in the schematic shown in Figure 3 is the concept that substance P receptor expression and macrophage activation must be viewed in the context of the immunogenic stimulus. For example, an endogenous pathogen, such as *Salmonella*, would induce predominantly a T helper type 1 response that could be augmented by ligation of the substance P receptor. Alternatively, an exogenous immunogen would stimulate an antibody mediated immune response which could also be augmented by a substance P/substance P receptor interaction. Thus, we do not view substance P/substance P receptor interactions as dictating the direction of the developing immune response so much as we suggest that this interaction will facilitate the response regardless of the direction.

It is clear that peptidergic neurons are present in the gut and in mucosal lymphoid organs in particular. Such neurons contain substantial amounts of neuropeptides, including substance P, and would therefore be positioned as a source of ligand to interact with substance P receptors. It is also possible that stimulated macrophages might also be a source of smaller quantities of substance P, which might act in an autocrine or paracrine fashion. What is not clear from present work in the field is a clear delineation of the stimuli which can induce neurons or macrophages to secrete substance P during the initial events following interaction with a pathogen or immunogen.

SUMMARY

Taken together, these studies demonstrate an important role for substance P receptor expression by macrophages. The results to date suggest proinflammatory signals mediated by this receptor, and it is clear that substance P can act synergistically with other factors to stimulate macrophage activity. Antagonism of substance P/substance P receptor interactions in vivo profoundly affect immunity against *Salmonella*. This model provides evidence that an optimal host response against this intracellular pathogen of macrophages requires signaling through the substance P receptor. The ability of interferon gamma or IL-4 to upregulate substance P receptor mRNA expression on macrophages suggests that substance P-mediated amplification loops might involve either T helper type 1 or T helper type 2 responses. Thus, depending upon the immunologic stimulus, substance P could contribute to cell mediated as well as humoral immune responses.

Several important questions remain. Since the antigen processing and presenting function is an important macrophage activity, the effect of signaling through the substance P receptor on these events has not been defined. Furthermore, since macrophages are only one type of antigen presenting cell, it will be important to determine the role of substance P receptor expression in the activity of dendritic cells. We anticipate that these ongoing investigations will further define the positive contributions that substance P/substance P receptor interactions have in the initiation of immune responses.

ACKNOWLEDGMENTS

This work was supported by the National Institutes of Health grant AI32976.

REFERENCES

1. C.A. Maggi. The mammalian tachykinin receptors. *Gen. Pharmacol.* 26:911 (1995).
2. K.L. Bost, and M.J. Mason. Thapsigargin and cyclopiazonic acid initiate rapid and dramatic increases of IL-6 mRNA expression and IL-6 secretion in murine peritoneal macrophages. *J. Immunol.*, 155:285 (1995).
3. I. Marriott, K.L. Bost, and M. J. Mason. Differential kinetics for induction of IL-6 mRNA expression in murine peritoneal macrophages: evidence for calcium dependent and independent signaling pathways. *J. Cell. Physiol.* 177:232 (1998).
4. I. Marriott, M.J. Mason, A. Elhofy, and K.L. Bost. Substance P activates NF-kB independent of elevations in intracellular calcium in murine macrophages and dendritic cells. *J. Neuroimmunol.* in press, 2000.
5. U. Gubler, A.O. Chua, D.S. Schoenhaut, C.M. Dwyer, W. McComas, R. Motyka, N. Nabavi, A.G. Wolitzky, P.M. Wuinn, P.C. Familletti, and M.K. Gately. 1991. Coexpression of two distinct genes is required to generate secreted bioactive cytotoxic lymphocyte maturation factor. *Proc. Natl. Acad. Sci. (USA)* 88:4143 (1991).
6. K.L. Bost, and J.D. Clements. Intracellular *Salmonella dublin* induces substantial secretion of the 40-kilodalton subunit of Interleukin-12 (IL-12) but minimal secretion of IL-12 as a 70-kilodalton protein in murine macrophages. *Infect. Immun.* 65:3186 (1997).
7. T. Kincy-Cain, and K.L. Bost. Substance P-induced IL-12 production by murine macrophages. *J. Immunol.* 158:2334 (1997).
8. T. Kincy-Cain, J.D. Clements, and K.L. Bost. Endogenous and exogenous interleukin-12 augment the protective immune response in mice orally challenged with *Salmonella dublin*. *Infect. Immun.* 64:1437 (1996).
9. S.M. Wahl. Transforming growth factor beta: the good, the bad, and the ugly. *J. Exp. Med.* 180:1587 (1994).
10. I. Marriott, and K.L. Bost. Substance P dramatically diminishes lipopolysaccharide and interferon-gamma induced TGF beta-1 production by cultured murine macrophages. *Cell. Immunol.* 183:113 (1998).
11. K.L. Bost. Expression of preprotachykinin mRNAs in lymphoid organs of normal and orally immunized mice: involvement of macrophages in preprotachykinin mRNA expression in vivo. *J. Neuroimmunol.* 62:59 (1995).
12. D.W. Pascual, and K.L. Bost. Substance P production by P388D1 macrophages: a possible autocrine mechanism of interaction. *Immunology* 71:52 (1990).
13. T. Kincy-Cain, and K.L. Bost. Increased susceptibility of mice to *Salmonella* infection following in vivo treatment with the substance P antagonist, spantide II. *J. Immunol.* 157:255 (1996).
14. M. Fitzgerald. Capsaicin and sensory neurons-a review. Pain 15:109 (1983).
15. K.L. Bost, and D.W. Pascual. Substance P: a late acting B lymphocyte differentiation co-factor. *Am. J. Physiol.* 262:C537 (1992).

LYMPHOCYTE MODULATION BY SEVEN TRANSMEMBRANE RECEPTORS: A BRIEF REVIEW OF SESSION 1

Burt M. Sharp

Department of Pharmacology
University of Tennessee, Memphis
Memphis, TN 38163

Although the specific function of many of the seven transmembrane G-protein coupled receptors expressed by immune cells has not been clarified, the picture is coming into focus, especially regarding the chemokine receptors. The opening session of this conference, with its central theme on the expression, function and interaction between 7 transmembrane receptors for cannabinoids, opioids and chemokines, offered new insights into the immunomodulatory effects of these ligands. The presentations provided experimental data which advanced our knowledge of the following topics: the lymphoid organ-specific and mitogen-dependent expression of delta opioid (DOR), kappa opioid (KOR) and cannabinoid receptors; the expression and function of cannabinoid receptors on microglial cells; the interaction between opioid and chemokine receptors; and the effects of mu opioid receptors (MOR) on T cell apoptosis. This report will summarize these topics by briefly considering the contributions from individual laboratories.

In his Symposium Lecture, Guy Cabral, from the Medical College of Virginia, summarized converging lines of evidence showing that a variety of seven transmembrane G-protein-coupled receptors expressed in the brain are identical to, or have similar subtypes, as those found on immune cells. Included are receptors for cannabinoids, chemokines, opiate alkyloids, and opioid peptides. Two cannabinoid receptor subtypes with the following tissue expression patterns have been identified: the first, designated as CB_1, has been localized primarily to neural tissue, but has also been identified in testis and cells of the immune system; the second, termed CB_2, has only been detected in cells of the immune system, including B and T lymphocytes, macrophages, and natural killer cells. Recently, cannabinoid receptors have been identified on microglial cells in the brain, where they constitute a resident population of macrophages capable of migration, differentiation, and proliferation.

Upon activation, microglia produce cytokines and nitric oxide (NO). Both pharmacological and molecular studies indicate that cannabinoids inhibit the production of inducible NO by microglia, an action that may depend on the CB_1 receptor subtype. Cannabinoid-mediated inhibition of inducible NO production is stereoselective. For example, the high affinity CB_1 enantiomer, CP55940, dose-dependently inhibited endotoxin-induced NO release from neonatal rat cortical microglial cells. In contrast, the inhibitory effect of CP56667, a lower-affinity paired enantiomer, was minimal. In addition, the inhibitory action of CP55940 was reversed by pretreating microglia with the CB_1-selective antagonist, SR141716A. Multiple molecular and biochemical approaches that include the mutagenic reverse transcription-polymerase chain reaction (MRT-PCR), Western immunoblot analysis using a CB_1-specific antibody, and the cellular co-localization of CB_1 receptor protein with a microglial marker reactive to *Griffonia simplicifolia* isolectin B_4, have all confirmed the presence of the CB_1 receptor on rat microglial cells. Although there is much to be learned about the physiological and pathological conditions in which endogenous cannabinoids affect immune function, the presence of cannabinoid receptors on immunocytes suggests that cannabinoids are immunomodulators.

Recent studies on the expression and function of DORs by murine T-cells were presented by Burt Sharp. *In vitro* experiments with splenocyte cultures showed that crosslinking the T-cell receptor (TCR) with anti-CD_3 induced the expression of DOR mRNA by T-cells, through transcriptional mechanisms. In the absence of mitogenic stimulation, *in vitro* incubation also enhanced DOR mRNA expression. This depended directly on cell density and was unrelated to the release of a soluble factor(s).

A TCR-dependent *in vivo* model for the induction of DOR mRNA and protein by splenocytes was described. Staphylococcal enterotoxin B (SEB), a superantigen recognized by T-cells expressing Vβ7, 8.1, 8.2 and 8.3 alleles that is involved in the toxic shock syndrome, was administered to Balb/c mice; 8 and 24 hours thereafter, splenic DOR mRNA levels were significantly elevated. Immunofluorescence analysis showed a significant increase in the fraction of splenocytes that expressed DOR. This increase was found predominantly in T-cells, and approximately two-thirds of the T-cell subpopulation expressed DOR by 15-24 hours after stimulation by SEB. Additional studies showed that DOR agonists inhibited the anti-CD_3–induced phosphorylation of the MAPKs, ERKs 1 and 2. Thus, T-cells appear to express the same DOR that has been reported in the brain. However, DORs are present, for the most part, on T-cells which have been activated by stimulation through the TCR. DORs appear to suppress T-cell responses to stimulation through the TCR.

Using a FITC-labeled κ agonist, FITC-AA, followed by an amplification procedure and flow cytometry, Jean Bidlack has shown that κ opioid receptors were detected on peripheral immune cells and on human microglia. These studies showed that κ opioid receptors are expressed on greater than 60% of thymocytes and peritoneal macrophages from 6-8 week old male C57Bl/6ByJ mice. In contrast, κ opioid receptors were detected on less than 30% of CD4+ or CD-8+ splenic T-cells and on 15% of B cells (CD45R+).

Recent investigations showed that mitogenic stimulation of $CD4^+$ and $CD8^+$ splenocytes from 6-8 male week old C57Bl/6ByJ mice increased the amount of κ opioid receptor expressed per cell. Splenocytes were cultured with increasing concentrations of the mitogens, Con A and PHA, for varying time intervals. The detection of CD69, a marker for early lymphocyte activation, was used to characterize the activated population of cells. Both Con A and PHA increased the expression of the κ opioid receptor by CD8+ cells. Con A increased receptor expression by $CD4^+$ cells, and had a modest effect on $CD8^+$ cells. In contrast to these effects, the fraction of cells expressing the κ opioid receptor was not affected. Thus, in contrast to the effects of SEB on the T-cell expression of DORs, this study demonstrates that mitogens increased the amount of the κ opioid receptor expressed by $CD4^+$ and $CD8^+$ cells without altering the percentage of cells expressing the receptor.

Several laboratories have reported that μ, δ, and κ opioids inhibit phagocytosis, whereas morphine and the endogenous opioids induce chemotaxis by human monocytes and neutrophils. At the same time, pre-treatment with opioids inhibits human monocyte and neutrophil responses to complement-derived chemotactic factors, and to the chemokines MIP-1α, RANTES, MCP-1 and IL-8. Recent studies have shown that this opioid-induced inhibition is due to the process of heterologous desensitization of chemokine receptors following the stimulation of MORs or DORs. Thomas Rogers presented results from his laboratory showing that the chemotactic responses to μ and δ opioids of murine thymocytes, human peripheral blood monocytes, and human keratinocytes were inhibited by pre-treatment with the chemokines MIP-3β, RANTES, and MCP-1, but not IL-8. The inhibition of opioid receptor-induced chemotaxis following chemokine receptor activation appeared to coincide with phosphorylation of the opioid receptor. These findings also suggest that the function of neuronal opioid receptors located within the central nervous system and peripherally may be affected by chemokines produced within the brain and by the peripheral immune system.

In addition to heterologous desensitization, other interactions take place between opioid and chemokine networks. Thomas Rogers showed that the expression of the CC chemokine receptor (CCR) 2 was significantly elevated within the thymus following treatment with the MOR agonist, DAMGO. Moreover, DAMGO increased the expression of IL-10, MCP-1 and RANTES in human peripheral blood leukocytes.

Finally, Pravin Singhal presented novel findings showing that opiates promote T cell apoptosis. Morphine, as well as the MOR specific peptide agonist DAGO, not only enhanced T cell apoptosis, but also activated caspase-8 and downstream effector caspases, and cleaved PARP into 116 and 85 kDa proteins. In addition, opiates activated of c-Jun NH_2-terminal kinase (JNK) and inhibited ERK activation. Although these studies suggest a role for MORs in opiate-induced T cell apoptosis, the presence of

MORs on T cells is controversial. Other laboratories have been unable to detect MOR mRNA using RT/PCR assays of human peripheral T cells, whereas saturable morphine binding to sonicated, activated lymphocytes was reported. Thus, this study suggests that opiate-induced apoptosis may be mediated through caspase-8 and its downstream effector pathway. In addition, the JNK pathway may be involved in this opioid-induced apoptosis, which may be independent of MORs.

In summary, DORs, KORs and cannabinoid receptors are expressed on T-cells and other peripheral and central cells (e.g. microglia) involved in host defense. Expression of opioid receptors depends on the state of cellular activation within a specific lymphoid compartment. Opioid receptors appear to suppress the activation of TCR-dependent intracellular signaling molecules (e.g. ERKs 1 and 2) and the activation of other 7-transmembrane G-protein coupled receptors, such as those for certain chemokines. Conversely, certain chemokines appear to inhibit the activation of opioid receptors and the ensuing chemotactic responses. MOR agonists can elicit these effects on chemokine receptors and on chemotaxis, and MOR agonists also induce T-cell apoptosis. However, the cellular target mediating the activity of MOR agonists on immunocytes remains to be defined.

These presentations have advanced our knowledge of the cellular targets and intracellular effects of opioids and cannabinoids on the immune system. The specific physiological and pathological conditions in which endogenous opioids and cannabinoids interact with these molecular targets and suppress immune function remains to be determined.

NEUROAIDS: RETROVIRAL PATHOLOGY AND DRUGS OF ABUSE
Session Summary

Phillip K. Peterson and Howard S. Fox

At the dawn of the twenty-first century, the intersecting pandemics of acquired immunodeficiency syndrome (AIDS) and drug abuse represent two critical threats to global health. The etiologic agents of these diseases, i.e., human immunodeficiency virus (HIV) and substances of abuse, have a lot in common. Both are commonly acquired by the intravenous route, they often afflict the same risk groups, prevention is inadequate and cure of the diseases they cause is at present either not feasible (HIV infection) or frequently unsuccessful (drug abuse). The fact that these agents also share the central nervous system (CNS) as a target organ formed the basis for this session in which new insights into the neuropathogenesis of HIV and the impact of injection drug use (IDU) on HIV-related encephalopathy were reviewed.

In recent years, it has become clear that the neuropathogenesis of HIV is complex and involves not only the viral proteins gp120, gp41, and tat but products of activated glial cells, as well. Eugene Major's laboratory has highlighted the involvement of astroglia to HIV encephalopathy. His group has demonstrated that astrocytes, which are the predominant cell type in the brain, can be productively infected by HIV, albeit at a lower level than that seen with microglia. However, the nature of the receptors on astrocytes which mediate viral entry is unclear. Unlike in CD4 lymphocytes and microglia, the CD4 receptor is not involved, since astrocytes are CD4-negative. Productive infection of astrocytes is non-cytopathic, relatively short-lived, and is followed by a latent stage of infection in which proviral DNA is not expressed. Reactivation of HIV can be provoked however, by tumor necrosis factor -α and interleukin -1β. Thus, astrocytes may serve as an important reservoir for HIV in the brain. Also, latent infection of astrocytes is associated with a functional defect, i.e., decreased glutamate uptake.

In the past several years, many research laboratories have focused on the involvement of β-chemokines in HIV neuropathogenesis. Major and his colleagues recently have found that tat protein selectively upregulates, via a nuclear factor-κB regulatory pathway, production of monocyte chemotactic protein (MCP)-1 by astrocytes. When compared to relevant control groups, patients with AIDS dementia were found to have markedly increased levels of MCP-1 in cerebrospinal fluid (CSF) and brain tissue. The potential importance of this β-chemokine in HIV-related encephalopathy was further supported in studies of simian immunodeficiency virus (SIV) reported later in the session by Janice Clements.

The role of cells of the monocyte lineage in HIV neuropathogenesis has been clearly demonstrated by Howard Gendelman and his group. Working with a SCID mouse model, in which HIV-1-infected human monocytes or microglia are sterotactically

Neuroimmune Circuits, Drugs of Abuse, and Infectious Diseases
Edited by Herman Friedman *et al.*, Kluwer Academic/Plenum Publishers, 2001

inoculated into the putamen or candate nucleus, they have found histopathological changes (astriogliosis, microglial nodules, and neuronal degeneration) and behavioral abnormalities (cognitive and motor deficits) that closely mimic AIDS dementia. Brain macrophages or microglia appear to elicit these changes through the production of an array of neurotoxins, including platelet activating factor, quinolinate, and cytokines, and by suppressing neurotropic factors.

Gendelman's group has also been exploring the involvement of chemokines in the neuropathogenesis of HIV-1. They found that about 20-30% of neurons express the chemokine receptor, CXCR4, and that the natural ligand for this receptor, i.e., SDF-1, is produced by activated astrocytes. Because SDF-1 can induce Ca^{2+} influx in neurons, activation of CXCR4 receptors may play a role in HIV encephalopathy. Finally, Gendelman proposed that future therapeutic strategies for AIDS dementia should include not only combination anti-retroviral drugs, which have had a dramatic effect on this disease, but also anti-inflammatory and neuroprotective agents.

Walter Royal reviewed the findings from the ALIVE cohort of HIV-infected IDU. Diverse types of neurologic disease are observed in HIV-infected patients, but the potential effects of IDU on HIV-induced disease remain unclear. This is due to many factors, including a high rate of baseline abnormalities in IDU. One testing modality, the digit span test, did indeed reveal a significant abnormality in HIV-infected IDU. Although HIV targets the CNS early in the course of infection, clinical studies of HIV-infected and uninfected cohorts have revealed little evidence of severe neuropathologic consequences until CD4 counts are markedly reduced late in infection. Interestingly, in a study of patients diagnosed with HIV-associated dementia, a history of IDU correlated with a more rapid progression of the neurologic symptoms, and such patients showed increased gp41 and macrophage-marker immunostaining in the CNS at autopsy.

The role of particular viral variants in the induction of CNS disease has received much support from the work of Janice Clements. Her group has recently developed an exciting new model of predictable, rapid CNS disease induction in SIV-infected monkeys. In these experiments, co-infection of pigtailed macaques with the SIV/DeltaB670 strain and a molecular clone, denoted SIV/17E-Fr, leads to rapid CD4 T cell depletion and a high rate (80-90%) of encephalitis. Severe lesions and high levels of SIV RNA were often found in the basal ganglia, but also noted in other regions of the brain. Interestingly, plasma viral loads were not correlated with those found in the CSF or brain, indicating a compartmentalization of viral replication. Although high CSF viral loads were found early after infection in all animals, such high levels were only maintained in those animals that developed encephalitis.

Macrophage infiltration is a significant part of the neuropathological alterations occurring in HIV and SIV encephalitis. The MCP-1 chemokine, found by Eugene Major to be increased in the CSF and brains of those with AIDS dementia, was found by Dr. Clements to be similarly elevated in the CSF and brains of animals with encephalitis induced by the rapidly neurovirulent SIV infection.

The session concluded with Dr. Linda Chang describing elegant work on the use of new non-invasive techniques and their application to the CNS abnormalities arising in HIV-infection, drug abuse, and their combination. In HIV infection in general, a mild decrease in global brain volume, increase in CSF, and decrease in cerebral blood flow is found. Perfusion MRI revealed that in patients with early-stage cognitive deficits, cerebral blood flow is decreased. Interestingly, an increased blood flow was found in the subcortical white matter in these individuals.

Proton magnetic resonance spectroscopy studies of neurochemical metabolites in the brain were also studied. In 20 patients with minimal CNS disease, and 34 with dementia, the concentration of N-acetyl (NA) containing compounds was normal, suggesting no neuronal loss. However, increased concentrations of choline-containing compounds (CHO) and myoinositol (MI) were found in both groups, suggesting increased glial activity during early stages of the HIV-cognitive/motor complex. Such measures may indicate a surrogate marker for CNS disease in HIV infection. Interestingly, in a group of 14 patients treated with highly active antiretroviral therapy (HAART), the abnormalities in CHO and MI became normalized.

Effects of drugs of abuse and their effects on HIV-induced abnormalities were also studied by Dr. Chang. Using PET scanning, methamphetamine abusers have a large decrease in dopamine transport, as well as abnormal glucose metabolism. Proton magnetic resonance spectroscopy revealed a decrease in NA. In methamphetamine-using HIV-infected patients, higher concentrations of CHO and MI were found in the frontal lobe of drug users compared to HIV patients without a history of drug abuse. Lower NA concentrations in the basal ganglia were found in methamphetamine-using HIV-infected patients compared to HIV patients without a history of drug abuse. Methamphetamine thus may increase the harmful effects of HIV-1 on the brain. The effects of cocaine were also described. In HIV-infected individuals who abuse cocaine, decreased cerebral blood flow was found relative to infected individuals without a history of cocaine use. Interestingly, the combination of HIV infection and cocaine abuse results in lower levels of NA in the brain, but no increase in CHO or MI.

This session dealt with many of the current issues regarding the mechanisms of HIV pathogenesis in the CNS, and the challenge of identifying the processes altered by drug abuse. In the clinical situation, Dr. Royal summarized some of the difficulties in determining the effects of IDU on HIV neuropathogenesis, including the use of multiple drugs of abuse, inaccurate self-reports, other confounding cofactors such as co-infections, and the fact that the control group may in fact also use drugs. The availability of the pre-clinical models of neruoAIDS, such as the culture system of Dr. Major and the animal models of Drs. Gendelman and Clements, careful clinical studies as those described by Dr. Royal, and the development of sensitive non-invasive techniques illustrated by Dr. Chang will help elucidate the impact of drug abuse on HIV infection, and give us new insights into the etiology, pathogenesis, and treatment of these conditions.

THE EFFECTS OF DRUGS AND NEUROPEPTIDES ON IMMUNOMODULATION AND OPPORTUNISTIC INFECTION: SUMMARY OF SYMPOSIUM 2

Toby K. Eisenstein[1] and Herman Friedman[2]

[1] Temple University School of Medicine, Philadelphia, Pennsylvania
[2] University of South Florida College of Medicine, Tampa, Florida

The use of recreational drugs of abuse by a large number of individuals in this country and abroad has aroused serious concerns about the consequences of this activity. The relationship between drug abuse and increased incidence of infectious diseases has stimulated investigations of whether and how such drugs affect immune function, especially against infectious agents. This symposium session focused attention on newer information about immunological effects of drugs of abuse and neuropeptides on infectious diseases.

The overview speaker was Dr. Phillip K. Peterson of the University of Minnesota School of Medicine. He discussed "Microglia, Bugs and Drugs". When viewed from the perspective of the extraordinary number of pathogenic and opportunistic microorganisms that affect the central nervous system, the brain appears to be an "immunologically underprivileged site". His laboratory has been studying the effect of opioids and other psychoactive agents on human microglia. The results of studies in his laboratory indicate that the anti-inflammatory properties of certain psychoactive drugs, especially opioids, might be useful for treatment of CNS infection. Kappa opioid receptor agonist was shown to affect viral replication in microglial cell cultures of HIV-1, the cause of AIDS.

Dr. Toby K. Eisenstein of Temple University presented data showing that opioids, especially morphine, directly subverted resistance to opportunistic enteric organisms. She reported that mice implanted with a morphine slow-release pellet were markedly sensitized to oral infection with virulent *Salmonella typhimurium* as determined by survival response, mean survival time and Salmonella burdon in their Peyer's Patches, mesenteric lymph nodes and spleen. Studies using mini pumps to deliver opioids selective for kappa or delta opioid receptors showed that in addition to morphine, only delta-2 agonists had activity in enhancing Salmonella infection and it was much weaker than the activity of morphine. Both kappa agonists and delta-2 agonists delivered by pump were immunosuppressive in a plaque-forming assay measuring splenic capacity to mount an antibody response to sheep red blood cells. These studies define more closely the opiate pathway that regulates susceptibility to bacteria administered by the oral route.

Neuroimmune Circuits, Drugs of Abuse, and Infectious Diseases
Edited by Herman Friedman *et al.*, Kluwer Academic/Plenum Publishers, 2001

Dr. Kenneth L. Bost of the University of North Carolina summarized studies from his laboratory concerning the receptor to P-septins in the neurological system and initiation of the mucosal immune response. Antagonism of substance P binding to its receptor early in the mucosal immune response against intracellular pathogens such as Salmonella limited the host's capacity to resist infection. His laboratory defined the mechanisms for this finding by showing that substance P antagonists blocked induction of IL-12. Other aspects of his work focused on substance P receptor expression by macrophages and dendritic cells following infection *in vitro* and *in vivo*. He showed that P upregulated CD40 ligand and demonstrated a molecular mechanism by which the neuropeptide enhanced the mucosal immune response by direct effects on receptor expressing antigen presenting cells.

Dr. Thomas Klein of the University of South Florida in Tampa then discussed cannabinoid induced modulation of T helper cell immunity in a mouse model of infection with the opportunistic pathogen *Legionella pneumophila*. He and his laboratory found that tetrahydrocannabinol, the major psychoactive component of marijuana, suppressed development of T cells important in cell mediated immunity, especially against opportunistic infectious organisms by inhibiting production of the important cytokine IL-12 and the RNA for the receptor. In contrast, the level of activity of IL-4 was increased, thus shifting T-helper cell responses from a Th1 to Th2 profile. Both CB1 and CB2 receptors, as well as elevated corticosteroids, appeared involved in this effect.

Dr. Sulie Chang of Seton Hall University discussed work from her laboratory concerning opioids and endotoxins. She summarized studies showing that LPS plus IL-1 induces upregulation of the mu opioid receptor. In addition, morphine activates adhesion molecules on brain microvasculature endothelial cells and enhances their permeability. Thus under some conditions morphine may have a proinflammatory effect.

In summary, the results presented in this session showed that opioids, cannabinoids and neuropeptides such as substance P have marked effects on resistance to infection and to inflammatory responses that orchestrate the host response to infection. These effects are receptor mediated as evidenced by use of specific antagonists to demonstrate specificity. All of these substances were shown to alter cytokine profiles in important ways that impinge on innate host defenses and acquired immunity to pathogens. The talks suggested that subdivisions may exist between various classes of opioids with respect to their effects on immune responses as morphine was reported to potentiate inflammatory cell adhesion and kappa agonists were shown to inhibit responses to chemotactic agents.

INDEX

MIX
Papier aus verantwortungsvollen Quellen
Paper from responsible sources
FSC® C105338

If you have any concerns about our products,
you can contact us on
ProductSafety@springernature.com

In case Publisher is established outside the EU,
the EU authorized representative is:
Springer Nature Customer Service Center GmbH
Europaplatz 3, 69115 Heidelberg, Germany

Printed by Libri Plureos GmbH
in Hamburg, Germany